Lecture Notes in Computer Science

Commenced Publication in 1973
Founding and Former Series Editors:
Gerhard Goos, Juris Hartmanis, and Jan van Leeuwen

Peter Hertling   Christoph M. Hoffmann
Wolfram Luther   Nathalie Revol (Eds.)

Reliable Implementation of Real Number Algorithms: Theory and Practice

International Seminar
Dagstuhl Castle, Germany, January 8-13, 2006
Revised Papers

 Springer

Volume Editors

Peter Hertling
Universität der Bundeswehr München
Fakultät für Informatik
85577 Neubiberg, Germany
E-mail: peter.hertling@unibw.de

Christoph M. Hoffmann
Purdue University
Department of Computer Science
West Lafayette, IN 47907-1398, USA
E-mail: cmh@cs.purdue.edu

Wolfram Luther
Universität Duisburg–Essen
Fakultät für Ingenieurwissenschaften
Abteilung Informatik und Angewandte Kognitionswissenschaft
47048 Duisburg, Germany
E-mail: luther@inf.uni-due.de

Nathalie Revol
École Normale Supérieure de Lyon
Laboratoire de l'Informatique du Parallélisme
Projet Arenaire
46, allée d'Italie, 69364 Lyon CEDEX 07, France
E-mail: Nathalie.Revol@ens-lyon.fr

Library of Congress Control Number: 2008932924

CR Subject Classification (1998): G.1, B.2, B.8, F.2

LNCS Sublibrary: SL 1 – Theoretical Computer Science and General Issues

ISSN 0302-9743

ISBN 978-3-540-85520-0 Springer Berlin Heidelberg New York

Springer is a part of Springer Science+Business Media

springer.com

© Springer-Verlag Berlin Heidelberg 2008

Typesetting: Camera-ready by author, data conversion by Scientific Publishing Services, Chennai, India
Printed on acid-free paper   SPIN: 12513802   06/3180   5 4 3 2 1 0

Preface

A large amount of the capacity of today's computers is used for computations that can be described as computations involving real numbers. In this book, the focus is on a problem arising particularly in real number computations: the problem of verified or reliable computations. Since real numbers are objects containing an infinite amount of information, they cannot be represented precisely on a computer. This leads to the well-known problems caused by unverified implementations of real number algorithms using finite precision. While this is traditionally seen to be a problem in numerical mathematics, there are also several scientific communities in computer science that are dealing with this problem.

This book is a follow-up of the Dagstuhl Seminar 06021 on "Reliable Implementation of Real Number Algorithms: Theory and Practice," which took place January 8–13, 2006. It was intended to stimulate an exchange of ideas between the different communities that deal with the problem of reliable implementation of real number algorithms either from a theoretical or from a practical point of view. Forty-eight researchers from many different countries and many different disciplines gathered in the castle of Dagstuhl to exchange views and ideas, in a relaxed atmosphere. The program consisted of 35 talks of 30 minutes each, and of three evening sessions with additional presentations and discussions. There were also lively discussions about different theoretical models and practical approaches for reliable real number computations. For a short description of the full program of the seminar including a complete list of presentations at the seminar the reader is referred to the Web page of the seminar on the website http://www.dagstuhl.de/ of Schloss Dagstuhl. During the seminar, theories and results concerning computability over the real numbers or just over the real algebraic numbers were presented and discussed. Topics also included formal proofs to gain an extra level of confidence in the quality of computed results, be they computed with effective real numbers or with floating-point arithmetic. Other discussions concerned software libraries, systems, and platforms: exploiting the well-defined specifications of floating-point arithmetic to get accurate and provable results, using interval arithmetic to obtain reliable results, or assessing the quality of floating-point results. Another theme was computational geometry and solid modeling, mainly as far as robustness is concerned, along with proposed solutions. Note that in computational geometry the special problem of implementing real number algorithms reliably is complicated by the interplay of numerical predicates and hidden dependencies between them that arise from geometry theorems that may not be known. This creates opportunities for inconsistent decisions that lead to faulty data structures and, ultimately, to failure of the computation.

We will now introduce the topics and results presented in the 12 articles in this book. They are written by participants of the seminar, and most of them are based on presentations at the Dagstuhl Seminar. They represent a cross-section through the topics of the seminar. First, we describe two papers presenting computability notions specially suited for geometric computations and results concerning effective computations involving real algebraic numbers. These two papers lead to computational geometry and solid modelling, which is treated in five papers, one of them containing an overview and a critical discussion, and one of them considering scientific visualization. Then we discuss three papers dealing with software systems for reliable computations and with implementation issues. They involve floating-point arithmetic or interval arithmetic or both. Finally, two papers are concerned with applications. They present methods and algorithms for solving real-world problems.

Computability and Complexity over Real Numbers and Real Algebraic Numbers

One approach in computable analysis is to represent objects by bit streams and to perform computations by Turing machines on such bit streams. Thus, real numbers are represented in an approximating way. This leads to a notion of computation which is quite closely related to constructive mathematics and corresponds to reliable computation on a digital computer. An entirely different, algebraic theory of computability and complexity over the real numbers is based on the real random access machine model. The fundamental assumption in this model is that elementary arithmetic operations as well as comparisons over the real numbers can be performed in one step. In his contribution, C. Yap presents a variant of the first approach in which it is possible to decide whether a given number is equal to zero or not. This approach is based on the so-called Exact Geometric Computation mode, which is encoded in several well-known libraries for geometric computation such as LEDA, CGAL, and the Core library. C. Yap's paper addresses the problem to provide a theoretical foundation for the Exact Geometric Computation mode of computation.

In the implementation of the Exact Geometric Computation mode and in the implementation of algorithms in computational geometry in general, effective operations on real algebraic numbers play an important role. I.Z. Emiris, B. Mourrain, and E.P. Tsigaridas analyze the bit complexity, that is, the complexity in the Turing machine model, of several operations involving real algebraic numbers. In particular, they consider real root isolation of univariate integer polynomials, sign evaluation and comparison of real algebraic numbers, and the problem of simultaneous inequalities. They give an overview of existing approaches and unify, simplify and improve them. In addition, they present results of numerical experimentations using these algorithms.

Computational Geometry and Solid Modeling, Robustness Problems

D. Michelucci, J.M. Moreau, and S. Foufou give a critical survey of a number of approaches for dealing with the robustness and inaccuracy problem in computational geometry. Specifically, they compare the approach based on the idea of Exact Geometric Computation, as explained in the paper by C. Yap, the use of interval arithmetic, and either arithmetic or geometric probabilistic methods. Their conclusion is that geometric probabilistic approaches, such as the sampling of a surface or of a configuration space (for the motion planning problem) or Monte-Carlo and ray-tracing methods prove to be simple, tractable with the increasing power of computers, and robust.

Computational geometry is closely related to solid modeling. Geometric objects are often described by geometric data and by combinatorial data. Due to imprecise geometric embeddings these data may become inconsistent. V. Shapiro uses "inflated" representations, e.g., a segment is "thickened." He argues that usually it is assumed that the intended exact set is homotopy equivalent to a set corresponding to the given approximate geometric data. He shows how sufficient conditions for such a homotopy equivalence can be derived systematically from the Nerve Theorem.

When the problem deals with boundaries of patches of a trimmed surface, the inconsistency usually resides in the disjointness of the boundaries of the representations of theoretically adjacent patches. Then, one aims at changing the inconsistent boundaries to close and consistent boundaries. This is called transfinite interpolation. N.F. Stewart and M. Zidani show how in the case of sets defined by combined subdivision surfaces, one can use the Whitney extension theorem for transfinite interpolation. They also give a bound on the deviation of the normal vectors of the newly defined surface from the corresponding normal vectors of patches contained in the original description of the surface.

The last two papers dealing with geometric or solid modeling address reliability and accuracy issues. E. Dyllong uses a representation of objects by octrees, using intervals for reliability. She presents an algorithm for the determination of polyhedral convex enclosures of such objects, which is both fast and accurate. Accuracy is obtained via the use of the accurate dot-product for the tests of orientation and visibility.

The difficulties caused by imprecise computations due to finite precision floating-point arithmetic increase when one switches from geometric modeling to animations and scientific visualizations. L.E. Miller, E.L.F. Moore, T.J. Peters, and A. Russell consider the problem of computing the minimal distance between two curves. They propose an algorithm that is fast enough for interactive visualization purposes. Their algorithm consists in sampling couples of points on these two curves, in discarding quickly points that are not good candidates, and in performing Newton's method on the remaining pairs. For reliability reasons, they also provide an algorithm that determines a lower bound on this minimal distance, along with a guarantee on the quality of this lower bound.

Software Systems for Reliable Computations, Implementation Issues

The finite precision of floating-point arithmetic has been described repeatedly as one of the reasons for the difficulties in numerical and geometric computation. At least one wishes to have an implementation of floating-point arithmetic with correct rounding. In fact, in the IEEE 754-1985 standard for binary floating-point arithmetic it is required that all four arithmetic operations and the square root function are correctly rounded. In order to be able to round correctly $f(x)$ where f is an elementary function and x is a floating-point value, one needs to evaluate $f(x)$ with extra precision. V. Lefèvre, D. Stehlé, and P. Zimmermann looked for the worst cases for correct rounding of the exponential function in the IEEE 754r decimal64 format. In their contribution they describe this search and their findings.

The subtleties of implementations of floating-point arithmetic are also important in the contribution of B. Lambov. He is the author of the package RealLib for fast computations over the real numbers with arbitrary precision. In his contribution he presents an implementation of double precision interval arithmetic, which is part of this package. The implementation makes efficient use of the single-instruction-multiple-data SSE-2 instruction and register set extensions. He describes the ideas needed to fit interval arithmetic to this set of instructions, compares the performance of this implementation with the performance of other interval arithmetic packages, and discusses possible hardware extensions that might significantly increase the performance of interval arithmetic.

G.F. Corliss, R.B. Kearfott, N. Nedialkov, J.D. Pryce, and S. Smith have given themselves the mission to "gather, organize and make available" interval software and libraries for developers that are not experts in interval arithmetic. Their goal is to offer a library of interval tools that are seamlessly integrated and that can solve a large variety of problems. Their paper gives an overview of different aspects of this project. They discuss the planned overall structure of the library, the planned mathematical basis of the library (containment sets), how they plan to collect, organize, and integrate already existing work and to ask for new contributions, and the software engineering methodologies they want to use in order to ensure and improve the quality of the library. We wish them success!

Applications

The last two papers present methods and algorithms which have been developed in order to solve real-life problems.

E. Auer, A. Rauh, E. P. Hofer, and W. Luther present a new method for the integration of ODEs with initial value conditions, which yields a reliable enclosure of the trajectory. This method is implemented in the VALENCIA-IVP solver. This solver has been integrated into MOBILE, yielding the template-based tool SMARTMOBILE. The tool MOBILE enables the user to model mechanical systems by building them out of predefined components such as joints, balls, etc. directly as executable programs that can simulate the behavior of these systems. This behavior is governed by ODEs. The paper also discusses possible strategies for reducing the overestimation which is due to the wrapping effect in interval arithmetic.

The problem addressed in the paper by S. Kempken and W. Luther is the modeling of traffic in queuing and service networks. Usually a stochastic modeling is used, based on semi-Markov processes. The goal is to determine characteristic values for the considered network: probabilities and autocorrelation parameters. The authors express the autocorrelation parameters as a sum of exponential terms and then propose a method to obtain reliable and tight enclosures of the sought parameters, via linear programming techniques to obtain the lower and upper endpoints of intervals, when applicable. They use these enclosures to study the transient and steady states and to compute the time required to reach the steady state, i.e., the time where the transient state intersects the steady state.

The articles show that there are already many connections between the various disciplines concerned with the reliable implementation of real number algorithms. We believe that further cooperation and discussions between the different communities will be fruitful for the reliable solution of numerical and geometric problems.

Finally, we thank the participants of the Dagstuhl seminar for their talks and discussions, the authors of the papers in this book for their contributions and their cooperation in making this volume as accurate and readable as possible, and the referees for their careful work. We thank also Alfred Hofmann, the editor of the LNCS series of Springer, for making it possible to publish this post-seminar volume in the LNCS series, and the people at Schloss Dagstuhl for their hospitality and their great efficiency in organizing this seminar.

March 2008   Peter Hertling
                                          Christoph M. Hoffmann
                                                Wolfram Luther
                                                Nathalie Revol

Table of Contents

Validated Modeling of Mechanical Systems with SMARTMOBILE: Improvement of Performance by VALENCIA-IVP

Ekaterina Auer[1], Andreas Rauh[2], Eberhard P. Hofer[2], and Wolfram Luther[1]

[1] Faculty of Engineering, IIIS
University of Duisburg-Essen
D-47048 Duisburg, Germany
{Auer, Luther}@inf.uni-due.de

[2] Institute of Measurement, Control, and Microtechnology
University of Ulm
D-89069 Ulm, Germany
{Andreas.Rauh, EP.Hofer}@uni-ulm.de

Abstract. Computer simulations of real life processes can generate erroneous results, in many cases due to the use of finite precision arithmetic. To ensure correctness of the results obtained with the help of a computer, various kinds of validating arithmetic and algorithms were developed. Their purpose is to provide bounds in which the exact result is guaranteed to be contained. Verified modeling of kinematics and dynamics of multibody systems is a challenging application field for such methods, largely because of possible overestimation of the guaranteed bounds, leading to meaningless results.

In this paper, we discuss approaches to validated modeling of multibody systems and present a template-based tool SMARTMOBILE, which features the possibility to choose an appropriate kind of arithmetic according to the modeling task. We consider different strategies for obtaining tight state enclosures in SMARTMOBILE including improvements in the underlying data types (Taylor models), modeling elements (rotation error reduction), and focus on enhancement through the choice of initial value problem solvers (VALENCIA-IVP).

1 Introduction

Modeling and simulation of kinematics and dynamics of mechanical systems is unavoidable in many branches of modern industry and applied science, whether in a crane manufacturing firm, where the behavior of a new crane has to be simulated, or at a space exploration institute, where parameters of planets have to be obtained. During the last decades, computer assisted solutions gained in importance. Computers help to reduce the design and development time for new products and to substitute low cost "virtual" tests for expensive experiments on real life prototypes. Besides, various types of modeling and simulation software (MSS) have found a market in industry owing to their ability to generate models from systems' descriptions automatically.

P. Hertling et al. (Eds.): Real Number Algorithms, LNCS 5045, pp. 1–27, 2008.

In this article, we restrict the discussion to computer generated models that consist of systems of differential or algebraic equations (or, in more complicated cases, both). Modern modeling tools solve these equations with the help of floating point algorithms implementing explicit or implicit methods, such as Runge–Kutta methods. However, the computed results can be unreliable due to errors that are unavoidably generated by numerical solvers of differential and algebraic systems. Furthermore, inaccuracy is induced by idealization and order reduction of system models and not exactly known parameters. This forces developers of modeling software to look for more reliable options, one of which is the use of validated[1] algorithms. Examples are interval and Taylor model arithmetics. The main reasons for choosing a validated approach are, first, the guaranteed correctness of results and, second, the ability to allow for uncertainty in parameters, which helps to provide more realistic models or take into account measurement errors.

In [AKTT04, Aue07], we have already reported on an integrated tool for validated modeling and simulation of kinematics and dynamics of mechanical systems, namely SMARTMOBILE (*S*imulation and *m*odeling of dyn*a*mics in MOBILE: *r*eliable and *t*emplate-based) based on the open source non-validated MSS MOBILE. The former program is able to verify the properties of different kinds of systems including explicitly time-dependent and closed-loop ones by using interval arithmetic from PROFIL/BIAS and an interval IVP solver based on VNODE [Ned02]. Both tools will be described later. However, SMARTMOBILE is affected by certain disadvantages of validated arithmetic. First, validated computations are characterized by higher computational effort than their floating point counterparts. Second, the "naive" use can lead to considerable overestimation, making the results unusable. In this paper, we focus on possible strategies to cope with the latter difficulty.

One of the possible approaches to reducing overestimation might be the use of another initial value problem (IVP) solver, which performs better in the context of SMARTMOBILE, instead of VNODE. The recently developed solver VALENCIA-IVP[2] (*VAL*idation of state *ENC*losures using *I*nterval *A*rithmetic for *I*nitial *V*alue *P*roblems) [RHA06] provides such an opportunity. Therefore, the second focus of this paper is the description of this solver and its implementation in SMARTMOBILE.

The paper is structured as follows. First, the validated methods employed in SMARTMOBILE are reviewed in Section 2. The algorithm of the newly developed solver VALENCIA-IVP is presented in detail. In Section 3, we consider general approaches to validation of MSS and the example of validating MOBILE. For this purpose, the description of the main features of MOBILE and its verifying version SMARTMOBILE is necessary. In Section 4, we discuss strategies for overestimation reduction in the latter tool including rotation error elimination, employment of Taylor models, and the use of the above mentioned VALENCIA-IVP. Finally, the main results are summarized in Section 5.

---

[1] In this article, the commonly used terms *validated*, *guaranteed*, *verified*, and *reliable* are used interchangeably to denote the fact that the results are mathematically (and not empirically) proven to be correct.

[2] Further information about VALENCIA-IVP and free software are available at http://www.valencia-ivp.com

2 Validated Methods to Solve Initial Value Problems

In this Section, we describe several techniques for validated simulation of continuous-time systems used in SMARTMOBILE for modeling and analysis of multibody systems. First, the class of the considered problems is defined. Then, fundamentals such as validated arithmetics and algorithmic differentiation are referenced. Finally, we concentrate on interval IVP solvers, and especially on the newly developed VALENCIA-IVP.

2.1 Considered Initial Value Problems

Throughout this Section, IVPs given by sets of nonlinear ordinary differential equations (ODEs)

$$\dot{x}_s(t) = f_s(x_s(t), p(t), t) \tag{1}$$

are studied. These nonlinear ODEs (1) are assumed to be given in state-space representation, where $x_s \in \mathbb{R}^{n_s}$ is the state vector, and $p \in \mathbb{R}^{n_p}$ is a vector of uncertain and possibly time-varying system parameters.

Exact initial values of the state variables x_s are assumed to be unknown. However, guaranteed bounds for these initial states are given according to

$$x_s(t_0) \in \left[x_s^0\right] , \tag{2}$$

where $t_0 = 0$ without loss of generality. The uncertainties of the initial states are denoted by $\left[x_s^0\right] = \left[\underline{x}_s^0 ; \overline{x}_s^0\right]$; the parameter uncertainties are denoted by $[p(t)] = \left[\underline{p}(t) ; \overline{p}(t)\right]$, resp. In this notation, underlined variables denote lower interval bounds (infima) of all components of the corresponding vector, while overlined variables denote upper bounds (suprema).

In the case of time-varying parameters $p(t)$, their dynamics is assumed to be given in state-space representation

$$\dot{p}(t) = \Delta p(t) , \tag{3}$$

where both $p(t)$ and $\Delta p(t)$ might be bounded. If the variation rates $\Delta p(t)$ of the system parameters are unknown, $\inf([\Delta p(t)])$ and $\sup([\Delta p(t)])$ are infinite. For time-invariant parameters, i.e., if $p(t) = const$, the relation $\Delta p(t) = 0$ holds.

Since both dynamical models for $\dot{x}_s(t)$ and $\dot{p}(t)$ can be combined to a single set of ODEs

$$\dot{x}(t) = f(x(t), t) , \tag{4}$$

the discussion is restricted to the case of uncertain initial states after definition of an extended state vector $x(t) = \left[x_s^T(t) ; p^T(t)\right]^T$ with $x(t) \in \mathbb{R}^n$, $n = n_s + n_p$. Extension to interval uncertainties of the system parameters is straightforward and therefore is omitted here. In the case of time-invariant state equations, the notation (4) is abbreviated by $\dot{x}(t) = f(x(t))$.

Note that the structure of the considered dynamical systems may be explicitly time-varying. Such time-varying system models are of great importance in engineering. Typical applications are switchings between different control strategies for the transient behavior after setting a system into operation, for controlling the system near its steady state operating conditions, and modeling of mechanical systems with variable degrees of freedom. Since switching points are often state-dependent and therefore unknown a priori, these models might violate the assumptions for applicability of the validated solvers which are used to determine enclosures $[x(t)]$ containing all reachable states for $t \in [0 ; T]$, $T \geq 0$. Extensions of the simulation techniques which are applicable for systems with state-dependent switchings are described in [NvM02, Rih93, RKAH06] and the citations therein.

2.2 Interval and Taylor Model Arithmetics

Most validated techniques for solving IVPs rely on interval arithmetic [JKDW01]. Intervals describe real numbers, in general not exactly representable by floating point values, by defining the lower and upper bounds of the exact number in floating point arithmetic. Employing the concept of directed rounding [Soc85], one can obtain a verified enclosure of the exact result on a computer. A summary of the basic properties and definitions of interval arithmetic can be found in [JKDW01]. Numerous packages implement interval arithmetic, for example, PROFIL/BIAS [Knü94] or INTLAB [Rum99a, Rum99b]. These packages provide validated evaluation of algebraic expressions, basic functions such as trigonometric, exponential, or logarithmic ones, and matrix-vector-computations.

Another example of a validated arithmetic is a Taylor model-based one [RMB05]. A Taylor model of a function according to Berz is a high order polynomial approximation to its Taylor series expansion with floating point polynomial coefficients and an interval remainder term. This arithmetic is also called remainder-enhanced differential algebra (RDA). The idea is implemented in the package COSY INFINITY [BM02]. A detailed survey of Taylor forms and models can be found in [Neu02].

2.3 Algorithmic Differentiation

Since computation of derivatives is essential in many interval arithmetic algorithms such as those implemented in the validated IVP solvers VNODE or VALENCIA-IVP, automatic differentiation is extremely important. The task of algorithmic differentiation [Gri00], a variant of automatic differentiation, consists in finding derivatives of first and higher orders of a function f under the assumption that an analytic formula or a closed-form expression for f are unknown, but that the inner structure of f is made available through a computer program.

With the help of the operator overloading technique for algorithmic differentiation, which is for example implemented in FADBAD/TADIFF [BS96, BS97], derivatives of arithmetic expressions can be computed directly from their source code. In general, forward and backward approaches as well as the computation of Taylor coefficients can be distinguished [BS96].

2.4 Interval Solvers for Initial Value Problems: VNODE and VALENCIA-IVP

In the following, algorithmic details of the two selected validated IVP solvers VNODE and VALENCIA-IVP are summarized. In this description, the focus is on VALENCIA-IVP.

VNODE. The IVP solver VNODE uses PROFIL/BIAS to handle elementary routines for interval arithmetic computations. As most other solvers relying on interval arithmetic, VNODE is based on discretization of the time span, which is considered for simulation of the dynamical system.

Denote the solution with the initial condition $x(t_{k-1}) = x_{k-1}$ by $x(t;t_{k-1},x_{k-1})$ and the set of solutions $\{x(t;t_{k-1},x_{k-1}) \mid x_{k-1} \in [x_{k-1}]\}$ by $x(t;t_{k-1},[x_{k-1}])$. The goal of VNODE is to find interval vectors $[x_k]$ for which the relation $x(t_k;t_0,[x^0]) \subseteq [x_k]$, $k = 1,\ldots,L$, holds.

The kth time step consists of two stages [Ned99] (simplified):

1. *Proof of existence and uniqueness.* Compute a step size h_{k-1} and an a priori enclosure $[\tilde{x}_{k-1}]$ of the solution such that
   (i) $x(t;t_{k-1},x_{k-1})$ is guaranteed to exist for all $t \in [t_{k-1};t_k]$ and all $x_{k-1} \in [x_{k-1}]$,
   (ii) the set of solutions $x(t;t_{k-1},[x_{k-1}])$ is a subset of $[\tilde{x}_{k-1}]$ for all $t \in [t_{k-1};t_k]$.
   In the first stage of VNODE, Banach's fixed-point theorem is applied to the Picard iteration. (The Picard iteration is discussed in more detail in the description of VALENCIA-IVP in formula (7).)

2. *Computation of the solution.* Compute a tight enclosure $[x_k] \subseteq [\tilde{x}_{k-1}]$ of the solution of the IVP such that $x(t_k;t_0,[x^0]) \subseteq [x_k]$. The usual approach here is to compute the enclosure $[x_k]$ from the enclosure $[x_{k-1}]$ at the previous step, accurately taking into account the discretization error. The prevailing algorithm is as follows.

2.1. Choose a one-step method

$$x(t;t_k,x_k) = x(t;t_{k-1},x_{k-1}) + h_{k-1}\varphi(x(t;t_{k-1},x_{k-1})) + z_k\,,$$

where $\varphi(\cdot)$ is an appropriate method function, and z_k is the local error which takes into account discretization effects. The usual choice for $\varphi(\cdot)$ is a Taylor series expansion.

2.2. Find an enclosure for the local error z_k. For the Taylor series expansion of order $q-1$, this enclosure is obtained as $[z_k] = h_{k-1}^q f^{[q]}([\tilde{x}_{k-1}])$, where $f^{[q]}([\tilde{x}_{k-1}])$ is an enclosure of the q-th Taylor coefficient of the solution over the state enclosure $[\tilde{x}_{k-1}]$ determined by the Picard iteration in *Stage One*. Usually $q > 1$ is chosen.

2.3. Compute a tight enclosure of the solution. If mean-value evaluation for computing the enclosures of the ranges of $f^{[i]}([x_k])$, $i = 1,\ldots,q-1$, instead of the direct evaluation of $f^{[i]}([x_k])$ is used, tighter enclosures can be obtained.
   The so-called direct Taylor series method naively implements this idea, which overestimates the enclosure considerably in most cases. To reduce the overestimation, non-orthogonal (parallelepiped) or orthogonal (QR-factorization) coordinate transformations can be used [Loh01].

This approach to validated integration requires computation of Taylor coefficients and Jacobians. For this purpose, VNODE uses the algorithmic differentiation libraries FADBAD and TADIFF mentioned earlier. This solver validates the existence and uniqueness of the solution by the Taylor series method at *Stage One* and offers a choice between the interval Taylor series method with QR-factorization and the interval Hermite-Obreschkoff method [Ned99] at *Stage Two*. The user can select either a constant or a variable step size control strategy; only the constant order control strategy is supplied by VNODE so far. Note that further algorithms for *Stages One* and *Two* as well as further step size and order control strategies can be easily added to the core of this object oriented solver.

ValEncIA-IVP. In contrast to other validated techniques such as VNODE or COSY VI (cf. Subsection 2.5), VALENCIA-IVP [RHA06] aims at calculation of validated enclosures

$$[x_{encl}(t)] = x_{app}(t) + [R(t)] \tag{5}$$

of the solution $x(t)$ of IVPs (4). As already mentioned in Subsection 2.1, the considered system models can contain both uncertain parameters and uncertain initial conditions. The enclosures (5) are assumed to consist of a non-validated approximate solution $x_{app}(t)$ and guaranteed error bounds $[R(t)]$.

In general, two practically relevant possibilities for computation of the approximate solution $x_{app}(t)$ exist. First, $x_{app}(t)$ can be calculated analytically using an appropriate easy-to-solve reference system. Usually, such reference systems are obtained by linearization of the considered nonlinear state equations and can be further improved by suitable perturbation techniques [RHA06]. Second, arbitrary non-validated numerical techniques for the solution of IVPs can be applied to compute approximate solutions $x_{app}(t)$. Since the approximation quality and the effort for obtaining the approximate solutions strongly depend upon the dimension, complexity, and structure of the considered system, there are no general recommendations which of these two procedures is to be preferred. However, numerical reference solutions are advantageous if solvers for IVPs given by algorithmic representations — as they are considered in this paper, cf. Section 3 — are desired.

The term $[R(t)]$ in definition (5) is a vector of unknown interval enclosures of the approximation error for each component of the state vector. Tight bounds of this additive correction term are obtained by an iteration scheme based on applying Banach's fixed-point theorem to the ODE (4) after substituting (5) for $x(t)$ on both sides. Advanced interval techniques such as mean-value rule evaluation and monotonicity tests are used to reduce the influence of overestimation during calculation of the error bounds. The exact procedure and the C++ implementation of VALENCIA-IVP using PROFIL/BIAS and FADBAD are discussed below.

In contrast to the previously summarized techniques, existence and continuity of only the *first* derivative of the state equation f with respect to all states, parameters, and the time variable t is required, i.e., $f : D \mapsto \mathbb{R}^n$, $D \subset \mathbb{R}^n \times \mathbb{R}^1$ open, $f \in C^1(D, \mathbb{R}^n)$. Hence, application of VALENCIA-IVP is generally advantageous if the state equations are not highly differentiable.

Iteration Scheme and Proof for Conservativeness of the Resulting Enclosures. A constituent of most interval techniques to enclose the solution of IVPs is the integration of a set of ODEs $\dot{x}(t) = f(x(t), t)$ on a finite time interval $[0 ; T]$ according to

$$x(t) = x(0) + \int_0^t f(x(\tau), \tau) \, d\tau \subseteq x(0) + [0 ; t] \cdot f([B], [0 ; t]) \quad \text{with} \quad t \in [0 ; T] \ . \tag{6}$$

In (6), the interval $[B]$ is a bounding box enclosing all reachable states in the considered time interval $[0 ; t]$, which is usually computed by the Picard iteration

$$\left[B^{(\kappa+1)} \right] = [x^0] + [0 ; t] \cdot f \left(\left[B^{(\kappa)} \right], [0 ; t] \right) \ . \tag{7}$$

Here, the superscript (κ) denotes the number of the iteration step. The iteration for computation of the bounding box is initialized with $\left[B^{(0)}\right] = \left[x^0\right]$. If the complete time interval is considered as a special case, t has to be replaced by T in (7). The interval of the initial guess for $\left[B^{(0)}\right]$ is widened as long as $\left[B^{(1)}\right] \not\subseteq \left[B^{(0)}\right]$. If $\left[B^{(1)}\right] \subseteq \left[B^{(0)}\right]$, (7) is evaluated until $\left[B^{(\kappa+1)}\right] \approx \left[B^{(\kappa)}\right]$. If this iteration does not converge, or if the resulting bounding box is inacceptably large, the width of the considered time interval has to be reduced [DJvH02]. The maximum possible step size in the calculation of the bounding box is limited by the step size for which validated explicit Euler techniques converge. Almost all information about the dynamical behavior of the considered system is neglected by the assumption of constant interval bounds $[B]$.

In VALENCIA-IVP, the bounding box $[B]$ is no longer assumed to be constant as in the above-mentioned basic idea. The bounding box is replaced by a time-varying state enclosure (5) with unknown error terms $[R(t)]$.

Theorem 1. *Let $\dot{x}(t) = f(x(t),t)$ be an IVP with $x(0) \in \left[x^0\right]$ and $t \in [0\,;T]$, where $f : D \mapsto \mathbb{R}^n$, $D \subset \mathbb{R}^n \times \mathbb{R}^1$ open, $f \in C^1(D,\mathbb{R}^n)$. Then, all states which are reachable at the point of time t are enclosed by $[x_{encl}(t)] = x_{app}(t) + [R(t)]$ as defined in (5), where $[R(t)]$ is computed by the two-step procedure given below.*

First, an interval enclosure of all possible time derivatives $[\dot{R}(t)]$ of the error term is computed by the iteration formula

$$
\begin{aligned}
\left[\dot{R}^{(\kappa+1)}(t)\right] &= -\dot{x}_{app}(t) + f\left(\left[x_{encl}^{(\kappa)}(t)\right],t\right) \\
&= -\dot{x}_{app}(t) + f\left(x_{app}(t) + \left[R^{(\kappa)}(t)\right],t\right) \\
&=: r\left(\left[R^{(\kappa)}(t)\right],t\right) \ .
\end{aligned}
\tag{8}
$$

This iteration converges to a verified enclosure of $[\dot{R}(t)]$ if $\left[\dot{R}^{(\kappa+1)}(t)\right] \subseteq \left[\dot{R}^{(\kappa)}(t)\right]$. The iteration (8) is continued until $\left[\dot{R}^{(\kappa+1)}(t)\right] \approx \left[\dot{R}^{(\kappa)}(t)\right]$.

Second, verified integration of $\left[\dot{R}^{(\kappa+1)}(t)\right]$ with respect to time according to

$$
\begin{aligned}
\left[R^{(\kappa+1)}(t)\right] &\subseteq \left[R^{(\kappa+1)}(0)\right] + \int_0^t \left[\dot{R}^{(\kappa+1)}(\tau)\right] d\tau \\
&= \left[R^{(\kappa+1)}(0)\right] + \int_0^t r\left(\left[R^{(\kappa)}(\tau)\right],\tau\right) d\tau \ , \quad \text{or} \\
\left[R^{(\kappa+1)}(t)\right] &\subseteq \left[R^{(\kappa+1)}(0)\right] + t \cdot r\left(\left[R^{(\kappa)}([0\,;t])\right],[0\,;t]\right) \ , \quad 0 \leq t \leq T \ ,
\end{aligned}
\tag{9}
$$

is performed to determine the interval bounds of $\left[R^{(\kappa+1)}(t)\right]$. In the iteration formula (8), the right hand side of the set of differential equations is evaluated for the validated state enclosure $\left[x_{encl}^{(\kappa)}(t)\right]$ of the considered IVP in each iteration step κ which is obtained by

$$\left[x_{encl}^{(\kappa)}(t) \right] = x_{app}(t) + \left[R^{(\kappa)}(t) \right] . \tag{10}$$

∎

Proof. Using the Picard iteration (7), a bounding box $[B]$ of all states which are reachable in the time interval $t \in [0 ; T]$ can be determined according to Banach's fixed-point theorem. Substituting $[x_{encl}]$ for the bounding box $[B]$ on both sides of (7) leads to

$$\left[x_{encl}^{(\kappa+1)}([0 ; T]) \right] = [x^0] + [0 ; T] \cdot f\left(\left[x_{encl}^{(\kappa)}([0 ; T]) \right], [0 ; T] \right) . \tag{11}$$

Let the approximation error in (5) be defined by

$$[R([0 ; T])] := [R(0)] + [0 ; T] \cdot [\dot{R}([0 ; T])] , \tag{12}$$

where $[\dot{R}([0 ; T])]$ is a conservative interval enclosure of all possible time derivatives in the considered time interval. Then, the iteration formula (11) is equivalent to

$$x_{app}([0 ; T]) + \left[R^{(\kappa+1)}([0 ; T]) \right] = [x^0] + [0 ; T] \cdot f\left(\left[x_{encl}^{(\kappa)}([0 ; T]) \right], [0 ; T] \right) . \tag{13}$$

According to definition (12), $\left[\dot{R}^{(\kappa+1)}([0 ; T]) \right]$ is a guaranteed interval enclosure of all possible time derivatives of $\left[R^{(\kappa+1)}([0 ; T]) \right]$ in the time interval $[0 ; T]$. Analogously, $f\left(\left[x_{encl}^{(\kappa)}([0 ; T]) \right], [0 ; T] \right)$ includes the time derivative of the right hand side of (13). Therefore, differentiation with respect to time on both sides of (13) gives

$$\dot{x}_{app}([0 ; T]) + \left[\dot{R}^{(\kappa+1)}([0 ; T]) \right] = f\left(\left[x_{encl}^{(\kappa)}([0 ; T]) \right], [0 ; T] \right) . \tag{14}$$

Solving for $\left[\dot{R}^{(\kappa+1)} \right]$ leads directly to the iteration formula (8). Finally, evaluation of the sum of the approximate solution $x_{app}(t)$ and the bounds of the approximation error using outward rounding of the resulting interval provides a verified state enclosure of the solution of the considered IVP. ∎

The quality of the interval enclosures is influenced by two factors.

1. The initial approximation $x_{app}(t)$ in the complete time interval. Smaller deviations between the unknown exact solution and its initial approximation lead to tighter enclosures of the solution over a longer time span with less computational effort.
2. The time span. Tighter bounds can be obtained by subdivision of the time span $[0 ; T]$. This improves convergence of the iteration formula (8) and additionally leads to smaller interval bounds; see also *Step 3* in the following description of the algorithm.

In contrast to other validated techniques, we do not perform series expansions of the solution of the IVP, for which guaranteed error bounds of the discretization errors have to be determined; cf. interval Taylor series or interval Hermite-Obreschkoff methods implemented in VNODE. Hence, calculation of discretization error bounds via the local errors mentioned in the description of VNODE is not necessary in VALENCIA-IVP. The interval enclosure of $[R]$ can be obtained by applying the iteration formula (8). Note that it could also be obtained by applying any other validated ODE solver to the differential equation

$$\dot{R}(t) = -\dot{x}_{app}(t) + f\left(x_{app}(t) + R(t), t\right) \tag{15}$$

for the approximation errors which can be obtained by substituting (5) for $x(t)$ in (4).

Algorithm. In the following, the key components of VALENCIA-IVP are discussed.

Step 1: *Calculation of Reference Solutions*
As already pointed out, two possibilities for computation of reference solutions are implemented in VALENCIA-IVP, namely both calculation of analytical and numerical approximations of the solution of the considered IVP.

Analytical Reference Solution: An approximate reference solution of an appropriate linear IVP

$$\dot{x}_{app}(t) = f_{lin}\left(x_{app}(t)\right) \quad \text{with} \quad x^0_{app} = x_{app}(0) = \text{mid}\left([x^0]\right) = \frac{1}{2}\left(\underline{x}^0 + \bar{x}^0\right) \tag{16}$$

with the same dimension as the original system is calculated. Usually, linearization of the original system and replacement or neglection of nonlinear terms are used to solve the resulting system analytically. The reference solution can be improved by perturbation techniques, for which the state equation (4) is rewritten as a perturbed linear system

$$\dot{x}(t) = (1 - \varepsilon) \cdot f_{lin}\left(x(t)\right) + \varepsilon \cdot f\left(x(t), t\right) = f_\varepsilon\left(x(t), t, \varepsilon\right) \quad \text{with} \quad \varepsilon \in [0\,;\,1]\ . \tag{17}$$

The system (17) is linear for $\varepsilon = 0$ and is equal to the original nonlinear system for $\varepsilon = 1$ [Kha02]. For appropriately chosen, but yet unknown function vectors $[R(t)] \in \mathbb{R}^n$ for the approximation error, the solution of the IVP is enclosed by

$$[x_{encl}(t)] = \sum_{j=0}^{m} \left(\varepsilon^j y_{app,j}(t)\right) + [R(t)] = x_{app}(t) + [R(t)] \quad \text{with}$$

$$[\dot{x}_{encl}(t)] = \sum_{j=0}^{m} \left(\varepsilon^j \dot{y}_{app,j}(t)\right) + [\dot{R}(t)] = \dot{x}_{app}(t) + [\dot{R}(t)]\ . \tag{18}$$

The vectors $x(t)$ and $\dot{x}(t)$ in (17) are replaced by $[x_{encl}(t)]$ and $[\dot{x}_{encl}(t)]$ as defined in (18), where $[R(t)]$ and $[\dot{R}(t)]$ are set to zero while calculating the improved initial approximation. Then, sorting for identical powers of ε is performed on both sides of the expression. The resulting set of ODEs for $y_{app,j}(t)$ is solved analytically for the initial conditions $y^0_{app,0} = x^0_{app}$ and $y^0_{app,j} = 0$, for all $j \geq 1$ — again after linearization or replacement of nonlinear terms — in order to obtain an improved initial approximation $x_{app}(t)$ for $\varepsilon = 1$.

Numerical Reference Solution: Alternatively, a non-validated numerical approximation $\{x^N_i\}$, $i = 0, \ldots, L$ for the original IVP with the point interval $x^N_0 = \text{mid}\left([x^0]\right)$

as the initial condition can be calculated over the grid $\{t_i\}$ with $t_L = T$ by arbitrary non-validated IVP solvers. Since analytic expressions for $x_{app}(t)$ and its time derivative $\dot{x}_{app}(t)$ are required in the iteration scheme (8), analytical approximations should be calculated. This can be done by minimization of a distance measure

$$D = \sum_{i=1}^{L} d\left(x_i^N - x_{app}(t_i)\right) \overset{e.g.}{=} \sum_{i=1}^{L} \left\|x_i^N - x_{app}(t_i)\right\|_2^2 , \tag{19}$$

for all numerically determined points of the solution of the IVP.
A simple approximate solution determined by numerical methods is linear interpolation between grid points according to

$$x_{app}(t) = x_i^N + \frac{x_{i+1}^N - x_i^N}{t_{i+1} - t_i} \cdot (t - t_i) \quad \text{with}$$

$$\dot{x}_{app}(t) = \frac{x_{i+1}^N - x_i^N}{t_{i+1} - t_i} \quad \text{for} \quad t \in [t_i \,;\, t_{i+1}], \; i = 0, \ldots, L-1 \;. \tag{20}$$

The advantage of this method is that $x_{app}(t)$ is obtained with small computational effort. In the current C++ version of VALENCIA-IVP, this linear interpolation scheme is implemented together with computation of the numerical approximation $\{x_i^N\}$ by an explicit Euler method with constant step size.

However, arbitrary ODE solvers — also those using techniques for automatic step size control as well as solvers with embedded interpolants — can be applied to determine the numerical reference solutions after minor modifications of VALENCIA-IVP.

Analogously, higher-order interpolations between the numerically calculated grid-points expressed by parameterizable functions $x_{app}(t)$ can be included in the source code of this solver instead of linear interpolation. On the one hand, the deviation between the approximate and exact solutions of the IVP is reduced by these improved approximations. On the other hand, the dependency upon time of these higher-order interpolations is always nonlinear. Since the iteration formula (8), which is based on the nonlinear state equations, has to be evaluated for time intervals and not only for infinitesimally short points of time in the following *Steps 2–4*, the influence of overestimation is growing, if such interpolations are used. Due to these two effects, a compromise has to be found between improvement of the initial approximation and the computational effort, which is necessary to reduce the arising overestimation. For techniques aiming at the reduction of overestimation, see the discussion of advanced interval methods in *Step 4*.

Step 2:   Initialization of the Iteration Scheme
To start the iteration (8), initial interval approximations for $[R(t)]$ and $[\dot{R}(t)]$ are required. If possible, nonlinear terms originating from the state equation (4) are replaced by rough but still conservative bounds, e.g. $\sin(\cdot)$ and $\cos(\cdot)$ can be replaced by the interval $[-1\,;\,1]$. Afterwards, in the first iteration step $\kappa = 0$, $\left[\dot{R}^{(1)}(t)\right]$ is calculated. The iteration is continued, if $\left[\dot{R}^{(1)}(t)\right] \subseteq \left[\dot{R}^{(0)}(t)\right]$. Otherwise, the initial guesses for $[R(t)]$ and $[\dot{R}(t)]$ have to be modified. Note that the interval enclosure

$[R(0)]$ for the initial point of time has to be chosen such that all possible initial states are included, i.e., $[x^0] \subseteq x_{app}(0) + [R(0)]$.

Step 3: *Subdivision of the Time Span into Several Time Intervals*

If the time span $[0 \, ; \, T]$ is split into several shorter time intervals to improve convergence of the iteration and to reduce the width of the error bounds, again validated integration of $\left[R^{(\kappa+1)} \right]$ is necessary to obtain a guaranteed enclosure for the error term. As follows directly from (9), for both analytical and numerical reference solutions, the integration with respect to time is performed by

$$\left[R^{(\kappa+1)} (t_{i+1}) \right] = \left[R^{(\kappa+1)} (0) \right] + \sum_{j=0}^{i} (t_{j+1} - t_j) \cdot r \left(\left[R^{(\kappa)} \left([t_j \, ; \, t_{j+1}] \right) \right], [t_j \, ; \, t_{j+1}] \right)$$

(21)

for all $\{t_i\}$, $i = 0, \ldots, L-1$. For numerical reference solutions with linear interpolation (20) between the grid points, $\{t_i\}$ is determined by the non-validated ODE solver which has been applied in *Step 1*. Note that the grid on the time axis does not have to be equally spaced.

Step 4: *Calculation of the State Enclosures*

The state enclosure $[x(t)] \subseteq x_{app}(t) + [R(t)]$ determines whether improved initial approximations in *Step 1* and smaller time intervals in *Step 3* are necessary to reduce overestimation in the interval enclosures.

In the evaluation of (8), overestimation results from multiple occurrence of identical interval variables. This overestimation is reduced by mean-value rule evaluation of the right hand side of the iteration formula (8) as well as iterative improvement of the range of the expression on the right hand side including monotonicity tests [RKAH04, Kra05]. In VALENCIA-IVP, all partial derivatives required for these interval techniques are determined by algorithmic differentiation using FADBAD.

a) *Mean-Value Rule Evaluation of Iteration Formula*

Since natural interval evaluation of nonlinear expressions often leads to overestimation, the iteration formula (8) is evaluated using the mean-value rule

$$r(z) \in r(z_m) + \left. \frac{\partial r}{\partial z} \right|_{z=[z]} \cdot ([z] - z_m) \quad \text{for all} \quad z \in [z] \tag{22}$$

with the vector

$$[z] = \begin{bmatrix} [R(t_i)] \\ [t_i \, ; \, t_{i+1}] \end{bmatrix} \quad \text{and} \quad z_m = \text{mid}([z]) \tag{23}$$

containing all interval arguments of the right hand side of (8). These are the approximation errors R, the considered time interval, and all time-invariant uncertain parameters, if defined in the system model. To obtain the tightest possible enclosures, we use the intersection of the results of both natural interval extension and mean-value rule evaluation in all further computations.

b) *Monotonicity Test*

Additionally, VALENCIA-IVP performs a monotonicity test for further reduction of overestimation. In case of monotonicity of the component r_i, $i = 1, \ldots, n$, w.r.t. at least one z_j, $j = 1, \ldots, n+1$, i.e., if the lower bound of the interval evaluation of $\frac{\partial r_i}{\partial z_j}$ is strictly positive or if its upper bound is strictly negative, the interval $[z_j]$ can be replaced by one of the interval bounds as summarized in Tab. 1. For example, if $\inf\left(\frac{\partial r_i}{\partial z_j}\right) > 0$, $[z_j]$ can be replaced by $\underline{z}_j$ to compute the infimum of the range of r_i over $[z]$ and by $\overline{z}_j$ to compute its supremum. The range of r_i is then given by the interval hull of the results of both rows in Tab. 1.

Table 1. Replacement of arguments of the iteration formula in case of monotonicity

	$\inf\left(\frac{\partial r_i}{\partial z_j}\right) > 0$	$\sup\left(\frac{\partial r_i}{\partial z_j}\right) < 0$
$\inf\left\{ r_i(z)\big|_{z_j = \xi_j} \right\}$	$\xi_j = \underline{z}_j$	$\xi_j = \overline{z}_j$
$\sup\left\{ r_i(z)\big|_{z_j = \xi_j} \right\}$	$\xi_j = \overline{z}_j$	$\xi_j = \underline{z}_j$

c) Iterative Calculation of the Range

If the monotonicity test is not successful in at least one argument of r_i, all arguments of r_i with interval diameters which are significantly larger than zero can be split into several subintervals for which mean-value rule evaluation and monotonicity tests are applied again. Splitting is done at the interval midpoint of the component j_i^* of $[z]$ determined by

$$j_i^* = \underset{j=1,\ldots,n+1}{\arg\max} \left\{ \text{diag}\left\{ \text{diam}\left\{ \frac{\partial r_i}{\partial z}\bigg|_{z=[z]} \right\} \right\} \cdot \text{diam}\left\{ [z] \right\} \right\} . \tag{24}$$

This corresponds to the component of $[z]$ for which maximum reduction of overestimation is expected. Since only tight upper and lower bounds of r_i are desired, the splitting procedure is continued with the input intervals which lead to the smallest infimum/ largest supremum to improve the lower/ upper bounds of r_i. Splitting in VALENCIA-IVP is continued until a user-defined number of subintervals is reached or until r_i is monotonic for all input arguments. Finally, the *union* of all subintervals for r_i is determined to compute the improved enclosure of its range. For numerous practically relevant dynamical systems, only a small number of splittings is required to obtain good enclosures of the range if monotonicity is checked for each subinterval. Thus, compared to methods employing derivatives of high orders, the advantages of this procedure are the simplicity of implementation and often a smaller computational effort. Finally, no specific assumptions about differentiability of the state equations are necessary. After minor modifications of the selection criterion (24), iterative range computation is also applicable to systems with discontinuities and not highly differentiable state equations as e.g. idealization of friction in mechanical systems [RKAH06].

In [RHA06], the authors have demonstrated the applicability of VALENCIA-IVP to simulation of dynamical systems with both uncertain parameters and uncertain initial states. For a double pendulum with uncertainties in the initial angles, it has been shown that the achievable simulation quality is better than the results obtained by VNODE under the same conditions. Furthermore, the simulation quality of VALENCIA-IVP is comparable to COSY VI. However, the simulation times of the C++ implementation of VALENCIA-IVP are significantly smaller than the times required by COSY VI.

2.5 Other Validated IVP Solvers

COSY VI. The IVP solver COSY VI that is based on Taylor models has been presented in [BM98, MB04]. COSY VI seeks to improve conventional validated solvers with respect to modeling of the local functional behavior and control of the long-term growth of integration errors. The first task is performed using the Picard iteration (7) in combination with a fixed-point theorem. The long-term growth of integration errors is controlled by the so-called shrink wrapping method, which can be considered as a modified nonlinear version of the parallelepiped method.

VSPODE. This recently developed solver [LS06] is based on VNODE, but uses Taylor models as the underlying validated data type. It helps to obtain validated solutions of IVPs for ODEs with interval-valued parameters and initial values. The dependence of the solution on time is handled using interval Taylor series methods, as in VNODE, the dependence on the parameter vector and the initial state using Taylor models. This approach is reported to provide an efficient way to obtain a tight enclosure of all possible solutions to a parametric ODE system under uncertain conditions.

3 Strategies for Verification of MSS: A Case Study for MOBILE

To validate the dynamics of a mechanical system in the framework of an integrated tool, it is necessary to interface MSS and libraries implementing algorithms of validated analysis described in the previous Section. The realization of such an interface depends largely on the type of MSS we choose. From this point of view, all MSS can be roughly divided into "symbolic" and "numerical". The difference is that the former produces explicit expressions for the system model, and the latter does not. However, the results of the simulation for this model are as "exact" as those obtained by solving the expressions from the symbolic software on a computer.

The approaches to validation of both types of MSS are represented in Fig. 1. The common feature is the presence of a verification block consisting of a library for validated arithmetic such as PROFIL/BIAS or COSY INFINITY, optionally a tool for algorithmic (in the symbolic case, additionally automatic) differentiation such as FADBAD/TADIFF, and, finally, different solvers of (non)linear algebraic or differential systems of equations such as VNODE. In the symbolic case, the expressions for the model are produced using a computer algebra package (e.g. MAPLE) and solved directly by the tools from the verification block delivering the guaranteed results. In the numerical case, libraries have to be first adjusted[3] and then fused with the target MSS into its reliable version to guarantee the correctness of simulations.

---

[3] Here, "adjusted" means, for example, that IVP solvers which rely on symbolic representations of the state equations have to be adapted such that they work with representations provided by the numerical software.

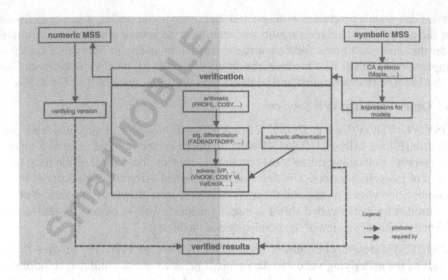

Fig. 1. Approaches to verification of "symbolic" and "numerical" MSS

In this Section, we focus on SMARTMOBILE, the validated MOBILE version described in more detail in [AKTT04, Aue07]. At first, the main features of MOBILE, the MSS of interest in this paper belonging to the numerical type in the above classification, are summarized. Then, we discuss the functionality and performance of the reliable version. Note that MOBILE is an open source tool. Since it is necessary to integrate the verification block from Fig. 1 into MSS in the numerical case, free access to the code of the modeling tool is crucial.

3.1 MOBILE

Let us consider the logical structure of MOBILE [Kec99]. Its central idea is the concept of a transmission element, which maps motion and loads between state objects (coordinate frames or variables) according to the formulas

$$
\begin{aligned}
q' &= \phi(q) \\
\dot{q}' &= \mathbf{J}_\phi \, \dot{q} \quad \text{with} \quad \mathbf{J}_\phi = \frac{\partial \phi}{\partial q} \\
\ddot{q}' &= \mathbf{J}_\phi \, \ddot{q} + \dot{\mathbf{J}}_\phi \, \dot{q} \\
Q &= \mathbf{J}_\phi^{\mathrm{T}} \, Q' \;.
\end{aligned}
\tag{25}
$$

All of the characteristics above are vectors, the dimension of which depends on the degrees of freedom of a multibody system. Here, q and q' are the generalized positions, $\dot{q}$ and $\dot{q}'$ the velocities, $\ddot{q}$ and $\ddot{q}'$ the accelerations, as well as Q and Q' the forces of the corresponding transmission element. The transmission of force is assumed to be ideal, i.e., power is neither generated nor consumed, for the sake of simplicity. The function ϕ is the mapping according to which motion is transmitted, and $\mathbf{J}_\phi$ is the corresponding Jacobian. Force is transmitted in the direction opposite to that of position, velocity, and acceleration.

Models in MOBILE are concatenations of transmission elements. The overall mapping of this concatenation from the original state into the final one is obtained by the composition of the corresponding mappings of the intermediate states. Concatenated elements can be considered as a single transmission element. This helps to solve the task of the global kinematics, that is, to obtain the positions, the orientations, the velocities, and the accelerations of all bodies of a mechanical system from the given generalized coordinates q and their time derivatives $\dot{q}$ and $\ddot{q}$.

All transmission elements are derived from the abstract class MoMap, which supplies their main functionality including the services doMotion() and doForce() for transmission of motion and force. Examples of transmission elements in MOBILE are such classes as MoRigidLink for modeling of rigid bodies or MoElementary-Joint for revolute and prismatic joints. Besides, there exist elements for modeling mass properties and applied forces. Special representations of the mapping (25) for these elements are to be found in [Kec93].

Transmission elements are assembled to chains implemented by the class MoMap-Chain. The methods doMotion() and doForce() can be used for a chain representing the system to determine the corresponding composite transmission function.

To model dynamics of a mechanical system, the equations of motion have to be built and solved. Their minimal form is given (based on the d'Alembert principle in the Lagrange form) by

$$M(q;t)\,\ddot{q} + b(q,\dot{q};t) = Q(q,\dot{q};t) \ , \tag{26}$$

where $M(q;t)$ is the generalized mass matrix, $b(q,\dot{q};t)$ the vector of generalized Coriolis and centrifugal forces, and $Q(q,\dot{q};t)$ the vector of applied forces.

In MOBILE, the kinetostatic method [Kec93] is employed to obtain the unknown mass matrix $M(q;t)$ and the force vector $\widehat{Q}(q,\dot{q};t) = b(q,\dot{q};t) - Q(q,\dot{q};t)$. Using the concatenation of motion and force transmission functions, one obtains the generalized forces of a multibody system in dependency on the motion and applied forces defined by q, $\dot{q}$, $\ddot{q}$. This process is called inverse dynamics φ_S^{D-1} of the system. The kinetostatic method is based on the knowledge that the transmission function φ_S^{D-1} produces the residue $\overline{Q}$ at the joints of the generalized coordinates which equal zero if the "real" derivatives $\ddot{q}$ of the mechanical state variables q, $\dot{q}$ are used as inputs of the inverse dynamics. That is, the relation (27) holds for the residue

$$-\varphi_S^{D-1} = -\overline{Q} = M(q;t)\,\ddot{q} + \widehat{Q}(q,\dot{q};t) \ , \tag{27}$$

whereas the equations of motion themselves have the form

$$M\,\ddot{q} + \widehat{Q} = 0 \ . \tag{28}$$

This fact provides an algorithm for the computation of M and $\widehat{Q}$.

After the introduction of a state vector $x = \left[q^T, \dot{q}^T\right]^T$, the state-space form of the state equations is obtained as

$$\dot{x} = \begin{bmatrix} \dot{q} \\ \ddot{q} \end{bmatrix} = \begin{bmatrix} \dot{q} \\ -M^{-1}\widehat{Q} \end{bmatrix} \ . \tag{29}$$

Finally, an IVP corresponding to (29) is solved by an appropriate integrator algorithm, e.g., Adams-Moulton-Bashforth's or Runge–Kutta's.

Equations of motion are generated by the class `MoEqmBuilder` which implements the kinetostatic method. The state-space form of these equations is obtained with the help of the class `MoMechanicalSystem` and subsequently solved by a `MoIntegrator`-derived class implementing one of the common algorithms for solving IVPs, e.g. `MoRungeKuttaIntegrator`.

One of the main features of MOBILE is the ability to model mechanical systems directly as executable programs. This allows the user to embed the resulting modules in existing libraries easily. Besides, the core of MOBILE is extendable owing to its open system design.

3.2 SMARTMOBILE

SMARTMOBILE produces guaranteed results by combining the theory from Section 2 with the modeling concepts from Section 3.1. Here, the numerical approach from Fig. 1 is implemented with the help of an interface based mainly on overloading: to bring validated arithmetics and algorithmic differentiation into MOBILE, we replace all the relevant occurrences of floating point data types with those of an appropriate validated one. Then, the validated algorithms, e.g. for solving linear systems of equations, are reimplemented to work with the new data type. The last and technically the most substantial step is the integration of a validated IVP solver, which has to use the same software for algorithmic differentiation and, as a rule, needs a thorough adjustment to MOBILE.

Before the actual reimplementation of MOBILE, we have to answer two fundamental designing questions: how to choose the basic validated data type and how to make the resulting tool independent of this basic data type. SMARTMOBILE solves the latter problem with the help of the template technique provided by C++ [VJ05]. The general idea is to systematically replace each MOBILE class containing a relevant member of the type `double` with its template equivalent. During this process, a type parameter is substituted for the floating point data type. Then, the so obtained template equivalent is substituted for all occurrences of the original class inside the other classes. The type parameter itself is specified by the user at the final stage of the system assembly according to the modeling task.

Some MOBILE classes along with their corresponding methods are *data type independent* and therefore easily convertible to templates. *Data type independent* means that no special validation algorithms aside from those overloaded by the basic data type are necessary for verification. In this case, a template equivalent can be implemented by simple data type substitution. Basic state objects and transmission elements from MOBILE fall into this category. However, there also exist data type dependent classes. These are, for example, `MoEqmBuilder` and `MoAdamsIntegrator`, classes for generating and solving equations of motion, respectively. They need a specialized treatment.

The proposed strategy was therefore to use pairs of classes in SMARTMOBILE: the first one represented the basic validated data type and was used instead of the placeholder of all the templates, and the second one implemented the simulation algorithm for this data type, for example, an IVP solver. This led us to the first question about the choice of the appropriate basic data type. In SMARTMOBILE, we provide the pair TMoInterval based on the data type INTERVAL from PROFIL/BIAS and TMoAWAIntegrator, an IVP solver based on VNODE.

INTERVAL alone is not sufficient for validation. It does not yet possess the facilities for obtaining Taylor coefficients of the solution to an IVP and the corresponding Jacobians of these coefficients necessary to solve the equations of motion in a validated way (cf. Section 2.4). Consequently, the next step is the introduction of algorithmic differentiation. It was decided to use the generic libraries FADBAD and TADIFF for that purpose. First, the data type TINTERVAL, which helps to automatically obtain Taylor coefficients of the algorithmic representation of a function f, is generated by TADIFF from the data type INTERVAL. Then, the forward mode of algorithmic differentiation is employed to get the Jacobian of f with the help of the data type FINTERVAL generated by FADBAD. At last, the library TADIFF is used once again to acquire the Jacobians of the Taylor coefficients with the help of the data type TFINTERVAL built on the basis of FINTERVAL.

FADBAD and TADIFF implement algorithmic differentiation through overloading. The general difficulty in solving IVPs (and consequently, simulating dynamics) with this implementation technique is as follows. Given a function to compute the right hand side of an ODE, we need two further functions, which differ from the first one in data types only, to obtain the Taylor coefficients and their Jacobians. Unfortunately, the common template solution to this problem cannot be employed in the context of MOBILE. Our solution is the class TMoInterval, containing members of all three types and overloading operators and elementary functions accordingly. A side effect of such code organization is that all required values can be obtained through a single call of the function responsible for the computation of the right hand side of the corresponding ODE.

The conversion to SMARTMOBILE is easy for the MOBILE user. The program for modeling and simulation of a multibody problem has to be altered only slightly for the user to obtain the corresponding validated one. First, the names of the transmission elements and state objects are changed from MoXXX to TMoXXX. Then, the template placeholder is defined and template syntax used as, for example, in TMoXXX <TMoInterval>. Finally, the appropriate IVP solver is called, for example, the supplied TMoAWAIntegrator instead of a non-validated integrator from MOBILE.

SMARTMOBILE successfully verifies the kinematics and dynamics for various classes of mechanical systems including non-autonomous and closed-loop ones (currently, with only one constraint). Its strength is the automatization and the generality of verified options for modeling and simulation (inside the class of systems allowed in MOBILE). However, depending on the considered system model, CPU times of the validated version are higher compared to the floating point one due to the use of validated arithmetic as well as inclusion of validated solvers for algebraic and ODE systems. Also, further strategies to improve the tightness of the result enclosures for complicated systems can be introduced. Nonetheless, a comparison of our approach to the symbolic one showed that although the symbolic strategy was better for obtaining dynamics of small systems, the numerical one proved to be either faster or less influenced by overestimation (but not both at the same time, cf. [Aue07]) for at least several complicated systems. Therefore, we feel encouraged to continue working on our tool and to further reduce overestimation and CPU time rather than to change to the symbolic approach.

4   Reducing the Overestimation in SMARTMOBILE

A primitive replacement of floating point computations by interval ones usually results in too wide intervals due to the dependency problem and especially the wrapping effect, which cannot be adequately handled by interval arithmetic alone. The treatment of the wrapping effect, which occurs when rotations are performed, is a key task in verification of mechanical systems (at least, in case of the numerical approach) since rotations are present in most models. This Section discusses several possible strategies which help to handle this problem.

4.1   Rotation Error Elimination

The wrapping effect can be minimized by using the rotation error elimination technique. The main idea of this technique presented in [Tra06] is that unnecessary rotations of interval values induced by the use of local coordinate systems during system's modeling are avoided with the help of an appropriate global coordinate system.

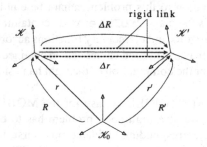

Fig. 2. Structure of a rigid link (dashed lines) in MOBILE

Consider this idea using the example of a rigid link connecting two bodies in the coordinate systems $\mathscr{K}$ and $\mathscr{K}'$ (cf. Fig. 2). For this transmission element, the transformation matrix R' and the translation vector r' between a reference coordinate system $\mathscr{K}_0$ and the new coordinate system $\mathscr{K}'$ are calculated according to

$$R' = R \cdot \Delta R \ ,$$
$$r' = \Delta R^{\mathrm{T}} \cdot r + \Delta r \ , \tag{30}$$

where R is the transformation matrix, and r is the translation vector between the reference coordinate system $\mathscr{K}_0$ and the old coordinate system $\mathscr{K}$. Further, ΔR is the constant rotation matrix, and Δr is the constant translation vector between $\mathscr{K}$ and $\mathscr{K}'$ [Kec93].

If the floating point values in this formula are naively replaced by their interval enclosures, the wrapping effect occurs inevitably with each coordinate transformation, because $[r]$ is multiplied by the transformation matrix $[\Delta R^{\mathrm{T}}]$. To reduce the wrapping effect, the interval translation vector $[r]$ is split into a floating point part r_{fp} and the correction interval $[\varepsilon]$ containing the information on the diameter of the interval enclosure in *the global coordinate system*, which is chosen to coincide with $\mathscr{K}_0$.

Let $p : \mathbb{IR}^n \mapsto \mathbb{R}^n$ map an interval vector $[v]$ to a floating point vector (e.g. the midpoint) and $c : \mathbb{IR}^n \mapsto \mathbb{IR}^n$ be defined as $c([v]) = [v] - p([v])$. Then the validated enclosure of R' as well as the new translation vector r'_{fp} and the new correction interval $[\varepsilon_{next}]$ are computed according to

$$
\begin{aligned}
[R'] &= [R] \cdot [\Delta R] \ , \\
r'_{fp} &= p([\Delta R^{\mathrm{T}}]r_{fp} + [\Delta r]) \ , \\
[\varepsilon_{next}] &= [\varepsilon] + \underbrace{[R] \cdot [\Delta R]}_{=[R']} \cdot c([\Delta R^{\mathrm{T}}]r_{fp} + [\Delta r]) \ .
\end{aligned}
\tag{31}
$$

That is, the rotation is applied only to the (floating point) value r_{fp}. All floating point errors are accounted for in $[\varepsilon_{next}]$. The overestimation that would arise from directly substituting an interval for r in the second part of (30) is avoided. Since r_{fp} is a floating point value, the term $c([\Delta R^{\mathrm{T}}]r_{fp} + [\Delta r])$ is small (assuming the matrices do not contain wide intervals). The position of $\mathscr{K}'$ can now be computed as

$$
[R'] \cdot r'_{fp} + [\varepsilon_{next}] \ .
\tag{32}
$$

The mapping $p([v])$ can be chosen arbitrarily. Here, the option $p([v]) \equiv 0$ seems to be the most advantageous as it eliminates one more rotation (cf. formulas (31) and (32)). In terms of implementation, the added information about the parameter error — the correction interval — in global coordinates requires a special treatment in all transmission elements of MOBILE, so that substantial changes are necessary to be able to use the rotation error elimination there.

Analogously to [Tra06], this technique was successfully applied to modeling of kinematics in SMARTMOBILE. However, it cannot be employed for modeling of dynamics just as easily, at least, not with the basic data type TMoInterval and the corresponding IVP solver TMoAWAIntegrator. The reason for this is the necessity to use midpoints of intervals inside MOBILE's transmission elements. These midpoints would have to be included into the algorithmic representation of the right side of the equations of motion in state-space form. That means that in order to "record" the program for algorithmic differentiation, the basic data type would have to support operations between graphs based on validated and non-validated data types and provide facilities for extracting a (non-validated) graph of midpoints from a given interval graph. These options are not supplied by FADBAD and TADIFF developers. It is possible, however, to do so in COSY INFINITY for RDA data types. The implementation of this error elimination technique in SMARTMOBILE for dynamics is an interesting topic for future research.

4.2   Uses and Limitations of Taylor Models in SMARTMOBILE: RDAInterval

Another possibility to deal with the dependency problem in SMARTMOBILE is to use a different kind of validated arithmetic as a basis for verified modeling. One of the options is the arithmetic based on Taylor models mentioned in Subsection 2.2. In order to use it in SMARTMOBILE, a new data type wrapper RDAInterval shown in Fig. 3 was introduced.

At first, several auxiliary members which serve as an interface to the C++ version of the Taylor model library COSY INFINITY are defined. It is necessary to introduce the number of monomials nm, the order of the Taylor model no, the maximal number of variables nv, the amount of memory space len required for this COSY INFINITY element, and the current number of variables varnum as static members of the class RDAInterval. The first three of them specify the parameters of a differential algebra which should be defined exactly once throughout the whole program. Furthermore, the amount of memory len for each instance should also remain the same. The member varnum is a counter of RDAInterval instances in the program which is not to exceed nv. The static method init(int, int, int) allows to set the parameters nm, no, and nv and to define the corresponding differential algebra with the help of the COSY INFINITY procedure daini(...).

The boolean member var shows whether the current instance of RDAInterval is really a Taylor model. Sometimes, if there are many variables in a program, several of them have to be defined as constant Taylor models, i.e., simple intervals. This is done partly to speed up Taylor model-based programs and partly to comply with memory restrictions imposed by COSY INFINITY (cf. page 21). The member Enclosure is a pointer at the Taylor model.

The constructor RDAInterval(...) helps to obtain a COSY INFINITY Taylor model from a given PROFIL/BIAS interval. If the Taylor model to be constructed is supposed to be a variable, the corresponding member Enclosure is defined as the midpoint of the given interval cons(mi) (the reference point) plus a standard COSY INFINITY Taylor model determined by the function rda(...) with the domain scaled to the diameter of the interval width(mi). The number of the variable should be specified while defining the standard Taylor model and incremented afterwards. If the current Taylor model is to be a constant, COSY INFINITY developers recommend the definition shown in the last non-empty line of the Fig. 3.

Consider the kinematics of a double pendulum (cf. Subsection 4.3) modeled in SMARTMOBILE first with the basic data type TMoInterval and then with RDAInterval. The respective positions of the pendulum's tip in a cartesian coordinate system are shown below up to the fourth digit after the decimal point (TMoInterval on the left, RDAInterval bounded by an interval on the right):

```
Position = [                 Position = [
[0          , 0      ];      [-0.19E-305,0.19E-305];
[ 1.1993 , 1.3853];         [ 1.2859    ,  1.2987 ];
[-0.1947 ,-0.0086]          [-0.1830    , -0.0204 ]
]                            ].
```

```
class RDAInterval {
private:
    static Cosy nm;    // the number of monomials
    static Cosy no;    // the order of TM
    static Cosy nv;    // the number of variables
    static int len;    // the length of the element
    static int varnum; // the number of the current variable
    bool var;          // do we have to use RDA?
    Cosy *Enclosure;   // the Taylor model
public:
    static void init(int, int, int);
    RDAInterval(const INTERVAL& x, bool var){
        if (this->var){
            *Enclosure=cons(x)+0.5*width(x)*rda(varnum,0,0,2);
            varnum++;
        } else *Enclosure=cons(x)+(0*rda(1,0,0,2)+(x-cons(x)));
    }
    ...
};
```

Fig. 3. RDAInterval — the Taylor model-based data type in SMARTMOBILE

The bounding algorithm for Taylor models specifies the smallest possible enclosure of zero for the first coordinate in case of Taylor models. The enclosures of the second and third coordinates are of about 93% and 13% tighter than for intervals.

As already mentioned, it is not always necessary to declare all variables in a program to be Taylor models. Rather, a "healthy" mixture of simple and RDA-enhanced intervals (united in the same data type RDAInterval) has to be deduced heuristically. At this point a question might arise why not simply declare *all* program variables as RDA intervals, which certainly cannot produce worse results. Aside from efficiency reasons, there is a more serious aspect to consider: memory restrictions imposed by the C++ version of the underlying library COSY INFINITY.

The version available to the authors can handle at most the differential algebra with twelve variables and the fifth order of Taylor series assuming that the maximum length of an RDA element equals 400. While it is relatively simple to increase the preset memory parameters controlling these characteristics in the FORTRAN version of COSY INFINITY, it cannot be as easily done in C++, which is confirmed by the COSY IN-FINITY developers. One reason is probably that the C++ version is an automatized translation from FORTRAN. Besides, COSY INFINITY programs with increased capacity are often slower both at compilation and running stages. Obviously, such a big tool as SMARTMOBILE is prone to use hundreds of variables, primary, auxiliary, or intermediate, so that these memory restrictions set back the development of programs based on RDAInterval there.

Until now we examined only the Taylor model-based simulation of kinematics, which was made possible by the data type RDAInterval. For dynamics, it is necessary to implement a specialized solver based on COSY VI that would work with this data type. This is still work in progress now, partly because of the above mentioned memory restrictions in COSY INFINITY. Another reason is that C++ implementations of COSY VI are still under development. To implement a version of COSY INFINITY and COSY VI more suitable for SMARTMOBILE is a demanding task which is nevertheless worth pursuing as the comparison for kinematics shows.

4.3   VALENCIA-IVP in SMARTMOBILE

To improve the quality of simulations, we have considered adjustments in MOBILE classes (Subsection 4.1) and alteration of the underlying validated arithmetics in SMARTMOBILE (Subsection 4.2) so far. Alternatively, other interval IVP solvers can be used, e.g., VALENCIA-IVP from Subsection 2.4. One of the advantages of this solver is that Taylor coefficients of the solution and their Jacobians do not need to be computed. Merely the Jacobian of the right hand side of the ODE is necessary.

To introduce VALENCIA-IVP into SMARTMOBILE, we implement a new basic data type which helps to obtain the numerical reference solution and the required Jacobian. It contains members of the types double (the reference solution), INTERVAL (the solution), and FINTERVAL (the Jacobian). Note that the Jacobian is always evaluated even if only double or INTERVAL values are required. Besides, we cannot avoid multiple function evaluations with the help of this new data type TMoFInterval as it was done in VNODE with TMoInterval.

The adjustment of VALENCIA-IVP to SMARTMOBILE resulted in the class TMoValenciaIntegrator. However, we are still working on transferring certain CPU time optimization strategies from the original version of VALENCIA-IVP into the SMARTMOBILE-compatible one at the moment. For example, the trick of evaluating only the necessary components of the set of state equations (and not all of them as usual) seems to be difficult to implement in TMoValenciaIntegrator because of the algorithmic form in which MOBILE produces the IVP to be solved.

Let us consider the example of the double pendulum with an uncertain initial angle of the first joint from [RHA06], where the authors modeled the dynamics with the help of the symbolic approach. Now, we study the dynamics of the double pendulum using a SMARTMOBILE model for the special case in which the lengths of both weightless arms of the pendulum are equal to 1 m and the two point masses amount to 1 kg each with the gravitational constant $g = 9.81 \frac{m}{s^2}$ (cf. Fig. 4). The initial values for angles (specified in rad) and angular velocities (in $\frac{rad}{s}$) are given as

$$\underline{\psi}^0 = \left[0.99 \frac{3\pi}{4} \quad -\frac{11\pi}{20} \quad 0.43 \quad 0.67 \right]^T ,$$

$$\overline{\psi}^0 = \left[1.01 \frac{3\pi}{4} \quad -\frac{11\pi}{20} \quad 0.43 \quad 0.67 \right]^T ,$$

where the initial angle of the first joint has an uncertainty of $\pm 1\%$ of its nominal value.

The results are summarized in Tab. 2. VALENCIA-IVP and VNODE denote here and in the following the VALENCIA-based and VNODE-based solvers in SMART-MOBILE, TMoValenciaIntegrator and TMoAWAIntegrator, resp. The abbreviations MVR and AIM indicate whether mean-value rule evaluation alone (MVR)

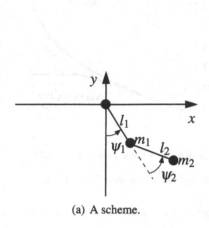

```
#define TMoInterval t;
TMoFrame<t> K0, K1, K2, K3, K4;
TMoAngularVariable<t> psi1, psi2;
// transmission elements
TMoVector<t> l1(0,0,-1), l2(0,0,-1) ;
TMoElementaryJoint<t> R1(K0,K1,psi1,xAxis) ;
TMoElementaryJoint<t> R2(K2,K3,psi2,xAxis) ;
TMoRigidLink<t> rod1(K1,K2,l1),rod2(K3,K4,l2) ;
t m1(1),m2(1) ;
TMoMassElement<t> Tip1(K2,m1),Tip2(K4,m2) ;
// the complete system
TMoMapChain<t> Pend;
Pend << R1<<rod1<<Tip1<<R2<<rod2<<Tip2 ;
// dynamics
TMoVariableList<t> q; q << psi1<<psi2 ;
TMoMechanicalSystem<t> S(q,Pend,K0,zAxis) ;
TMoAWAIntegrator I(S,0.0001,ITS_QR,15) ;
I.doMotion();
```

(a) A scheme.   (b) A model in SMARTMOBILE (abridged).

Fig. 4. The double pendulum

Table 2. Performance of `TMoValenciaIntegrator` in comparison to `TMoAWAInte-grator` for the double pendulum over the time interval $[0; 0.4]$

Solver	Break-down	CPU time in s
VALENCIA-IVP (MVR)	0.504	294
VALENCIA-IVP (AIM)	0.505	389
VNODE	0.424	1248

or in combination with monotonicity test and iterative calculation of range (AIM) as described in Step 4, Section 2.4, is used. The step size equals 10^{-4} in all cases. In case of `TMoAWAIntegrator` the QR-factorization algorithm with Taylor series of order 15 is chosen. We replace all floating point values of initial states, parameters, and the grid width on the time axis that are not exactly representable by machine numbers by their smallest possible interval enclosures in all computations.

The column "Break-down" of Tab. 2 contains the time after which the corresponding method no longer works. The last column indicates the CPU time (in seconds) which the solvers take to obtain the solution over the integration interval $[0; 0.4]$ on a Pentium 4, 3.0 GHz PC using CYGWIN. Additionally, the interval enclosures of the two angles ψ_1 and ψ_2 of the double pendulum are shown for identical time intervals in Fig. 5.

The state enclosures obtained with `TMoValenciaIntegrator` are tighter than those of `TMoAWAIntegrator` towards the end of the integration interval. Actually, the use of the former integrator reduces overestimation of the resulting interval boxes up to 839 times for this example, cf. Tab. 3. Here, the ratio between the pseudo volumes (product of the interval diameters of all components of a state vector) of the VNODE-based solver and the VALENCIA-based integrators is given for several reference points. However, the last row of Tab. 3 indicates that the use of the AIM strategies does not significantly improve the enclosures in the numerical case, in contrast to the symbolic approach from [RHA06]. Note also that even though the implementation of

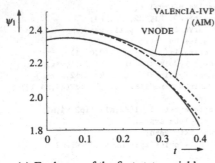 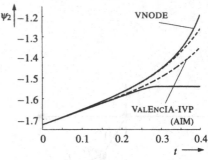

(a) Enclosure of the first state variable.   (b) Enclosure of the second state variable.

Fig. 5. Interval enclosures for the first and second state variable of the double pendulum

Table 3. Performance of `TMoValenciaIntegrator` in comparison to `TMoAWAInte-grator` for the double pendulum (cont'd)

Compared solvers	Reduction factor			
	$t = 0.1$	$t = 0.2$	$t = 0.3$	$t = 0.4$
VNODE vs. ValEncIA (AIM, MVR)	1.0817	1.7638	10.2574	839.564
ValEncIA: MVR vs. AIM	$1 + 8.2 \cdot 10^{-10}$	$1 + 5.0 \cdot 10^{-10}$	$1 + 4.2 \cdot 10^{-9}$	$1 + 2.8 \cdot 10^{-8}$

VALENCIA-IVP in SMARTMOBILE has not been optimized, the reduction of computing time in comparison to the VNODE-based integrator is considerable.

Here, we used constant step sizes in VNODE to free the comparison from the influence of the step size control strategy on the solution. If variable step sizes are allowed, the break-down time improves slightly to $t = 0.427$. However, the global error at $t = 0.424$ (the break-down time with the constant step size) amounts to 19.865 which is still larger than the error computed by VALENCIA-IVP at the same point. The real advantage of VNODE with variable step size is the reduction of CPU time. In this example, only 17 seconds are required to reach $t = 0.4$. In the following, we consider the constant step size strategy again.

Tab. 4 shows the results obtained for two further, more complex examples, a triple pendulum [AKTT04] (12 significant modeling elements, the resulting IVP has six state variables, initial uncertainty of order 10^{-2} in the first angle) and a one axis manipulator [ADL+05] (a closed-loop system with one constraint, 18 elements, two state variables, initial uncertainty of order 10^{-5} in the first state variable). Besides break-down and CPU times, Tab. 4 contains the column "Reduction factor" which indicates the reduction

Table 4. Performance of `TMoValenciaIntegrator` in comparison to `TMoAWAInte-grator` for the triple pendulum and the one axis manipulator

Example	Break-down		Reduction factor	CPU time in s	
	VALENCIA-IVP	VNODE		VALENCIA-IVP	VNODE
Triple pendulum	0.851	0.687	6604.977	386	1849
One axis manipulator	0.712	0.252	20.252	948	8854

ratio of the pseudo volumes between VNODE and VALENCIA-IVP (MVR) defined analogously to Tab. 3 at the reference points $t = 0.2$ for the one axis manipulator and $t = 0.6$ for the triple pendulum. For both examples, the CPU times are given over the integration interval $[0; 0.2]$.

The results for these examples confirm the general tendency: the use of TMoValenciaIntegrator in SMARTMOBILE reduces both overestimation and CPU times. However, the comparison of the break-down times of both solvers suggests that convergence of the iteration formula from VALENCIA-IVP should be further improved for state intervals with large diameters. Besides, we plan to improve the strategies for reduction of overestimation in VALENCIA-IVP.

5   Summary and Outlook

In this article, we explored the applicability of various validated concepts to modeling and simulation of mechanical systems in SMARTMOBILE with a special focus on overestimation reduction. For kinematics, the rotation error elimination technique and the employment of Taylor models were described. For dynamics, we showed by means of three characteristic examples with uncertain initial conditions that the use of the solver TMoValenciaIntegrator based on the newly developed solver VALENCIA-IVP in SMARTMOBILE reduced the pseudo volume of the state enclosures up to factors of 10^3 compared to the VNODE-based solver TMoAWAIntegrator with the constant step size strategy. Besides, under the same conditions, the former was four times faster than the latter for the considered tree-type multibody systems and even nine times faster for the closed-loop one.

The main direction for future development of SMARTMOBILE will be the implementation of Taylor model-based methods (e.g. COSY VI or VSPODE) and of the elimination technique for dynamic modeling. Both approaches will also be studied with respect to their performance on different modeling problems. This would help to provide users with the best solvers for their needs. Besides, an interesting point would be to study the influence of modeling on the solution, that is, how the performance of SMARTMOBILE depends on the number of transmission elements and variables for the same mechanical system.

As a further improvement of the efficiency of VALENCIA-IVP, an exclusion strategy for subintervals resulting from overestimation will be implemented which is based on a consistency test by backward integration of parts of the validated state enclosures determined by the present version of this program [RAH06]. Since these techniques involve simulation with several instead of only one interval box, efficient strategies for splitting and merging of interval boxes are required to achieve a good compromise between the increase of computational effort and the possible reduction of overestimation. These strategies have already been applied successfully outside of VALENCIA-IVP, see e.g. [RKAH07]. Furthermore, it will be explored how VALENCIA-IVP and other validated ODE solvers can be combined in order to benefit from the advantages of each technique, for example by the inclusion of an implicit integration technique to prevent unnecessary growth of interval bounds.

References

[ADL+05]   Auer, E., Dyllong, E., Luther, W., Stankovic, D., Traczinski, H.: Integration of Accurate Distance Algorithms into a Modeling Tool for Multibody Systems. In: Proc. of IMACS (2005)

[AKTT04]   Auer, E., Kecskeméthy, A., Tändl, M., Traczinski, H.: Interval Algorithms in Modeling of Multibody Systems. In: Alt, R., Frommer, A., Kearfott, R.B., Luther, W. (eds.) Dagstuhl Seminar 2003. LNCS, vol. 2991, pp. 132–159. Springer, Heidelberg (2004)

[Aue07]   Auer, E.: Interval Modeling of Dynamics for Multibody Systems. Journal of Computational and Applied Mathematics 199(2), 251–256 (2007)

[BM98]   Berz, M., Makino, K.: Verified Integration of ODEs and Flows Using Differential Algebraic Methods on High-Order Taylor Models. Reliable Computing 4, 361–369 (1998)

[BM02]   Berz, M., Makino, K.: COSY INFINITY Version 8.1. User's Guide and Reference Manual. Technical Report MSU HEP 20704, Michigan State University (2002)

[BS96]   Bendsten, C., Stauning, O.: FADBAD, a Flexible C++ Package for Automatic Differentiation Using the Forward and Backward Methods. Technical Report 1996-x5-94, Technical University of Denmark, Lyngby (1996)

[BS97]   Bendsten, C., Stauning, O.: TADIFF, a Flexible C++ Package for Automatic Differentiation Using Taylor Series. Technical Report 1997-x5-94, Technical University of Denmark, Lyngby (1997)

[DJvH02]   Deville, Y., Janssen, M., van Hentenryck, P.: Consistency Techniques for Ordinary Differential Equations. Constraint 7(3–4), 289–315 (2002)

[Gri00]   Griewank, A.: Evaluating derivatives: principles and techniques of algorithmic differentiation. SIAM, Philadelphia (2000)

[JKDW01]   Jaulin, L., Kieffer, M., Didrit, O., Walter, É.: Applied Interval Analysis. Springer, London (2001)

[Kec93]   Kecskeméthy, A.: Objektorientierte Modellierung der Dynamik von Mehrkörpersystemen mit Hilfe von Übertragungselementen (in German). In: Fortschrittberichte VDI, Reihe 20 Nr. 88, VDI-Verlag, Düsseldorf (1993)

[Kec99]   Kecskeméthy, A.: MOBILE Version 1.3. User's Guide (1999)

[Kha02]   Khalil, H.K.: Nonlinear Systems, 3rd edn. Prentice-Hall, Englewood Cliffs (2002)

[Knü94]   Knüppel, O.: PROFIL/BIAS—A Fast Interval Library. Computing 53, 277–287 (1994)

[Kra05]   Krasnochtanova, I.: Optimized Interval Algorithms for Simulation and Controller Design for Nonlinear Uncertain Systems Applied to Processes in Biological Wastewater Treatment, Master Thesis, University of Ulm (2005)

[Loh01]   Lohner, R.: On the Ubiquity of the Wrapping Effect in the Computation of the Error Bounds. In: Kulisch, U., Lohner, R., Facius, A. (eds.) Perspectives on Enclosure Methods, pp. 201–217. Springer, Wien, New York (2001)

[LS06]   Lin, Y., Stadtherr, M.A.: Validated Solution of Initial Value Problems for ODEs with Interval Parameters. In: NSF Workshop Proceeding on Reliable Engineering Computing, Savannah GA, February 22-24 (2006)

[MB04]   Makino, K., Berz, M.: Suppression of the Wrapping Effect by Taylor Model-Based Validated Integrators. Technical Report MSU HEP 40910, Michigan State University (2004)

[Ned99]   Nedialkov, N.S.: Computing Rigorous Bounds on the Solution of an Initial Value Problem for an Ordinary Differential Equation. PhD thesis, University of Toronto (1999)

[Ned02]   Nedialkov, N.S.: The Design and Implementation of an Object-Oriented Validated ODE Solver. Kluwer Academic Publishers, Dordrecht (2002)

[Neu02]   Neumaier, A.: Taylor Forms — Use and Limits. Reliable Computing 9, 43–79 (2002)

[NvM02]   Nedialkov, N.S., Mohrenschildt, M.v.: Rigorous Simulation of Hybrid Dynamic Systems with Symbolic and Interval Methods. In: Proc. of American Control Conference ACC, Anchorage, USA, pp. 140–147 (2002)

[RAH06]   Rauh, A., Auer, E., Hofer, E.P.: VALENCIA-IVP: A Comparison with Other Initial Value Problem Solvers. In: CD-Proc. of the 12th GAMM-IMACS International Symposium on Scientific Computing, Computer Arithmetic, and Validated Numerics SCAN 2006, Duisburg, Germany, IEEE Computer Society, Los Alamitos (2007)

[RHA06]   Rauh, A., Auer, E., Hofer, E.P.: A Novel Interval Method for Validating State Enclosures of the Solution of Initial Value Problems, Technical Report (2005), available online: http://vts.uni-ulm.de/doc.asp?id=6321

[Rih93]   Rihm, R.: Über Einschließungsverfahren für gewöhnliche Anfangswertprobleme und ihre Anwendung auf Differentialgleichungen mit unstetiger rechter Seite (in German). PhD thesis, University of Karlsruhe, Germany (1993)

[RKAH04]   Rauh, A., Kletting, M., Aschemann, H., Hofer, E.P.: Application of Interval Arithmetic Simulation Techniques to Wastewater Treatment Processes. In: Proc. of Modelling, Identification, and Control MIC 2004, Grindelwald, Switzerland, pp. 287–293 (2004)

[RKAH06]   Rauh, A., Kletting, M., Aschemann, H., Hofer, E.P.: Interval Methods for Simulation of Dynamical Systems with State-Dependent Switching Characteristics. In: Proc. of IEEE International Conference on Control Applications CCA 2006, Munich, Germany, pp. 355–360 (2006)

[RKAH07]   Rauh, A., Kletting, M., Aschemann, H., Hofer, E.P.: Reduction of Overestimation in Interval Arithmetic Simulation of Biological Wastewater Treatment Processes. Journal of Computational and Applied Mathematics 199(2), 207–212 (2007)

[RMB05]   Revol, N., Makino, K., Berz, M.: Taylor Models and Floating-Point Arithmetic: Proof that Arithmetic Operations are Validated in COSY. Journal of Logic and Algebraic Programming 64, 135–154 (2005)

[Rum99a]   Rump, S.M.: Interval Computations with INTLAB. Brazilian Electronic Journal on Mathematics of Computation 1 (1999)

[Rum99b]   Rump, S.M.: INTLAB — INTerval LABoratory. In: Csendes, T. (ed.) Developments in Reliable Computing, pp. 77–104. Kluwer Academic Publishers, Dordrecht (1999)

[Soc85]   IEEE Computer Society. IEEE Standard for Binary Floating-Point Arithmetic. Technical Report IEEE Std. 754-1985, American National Standards Institute (1985), http://standards.ieee.org

[Tra06]   Traczinski, H.: Integration von Algorithmen und Datentypen zur validierten Mehrkörpersimulation in MOBILE (in German). PhD thesis, University of Duisburg-Essen (2006)

[VJ05]   Vandevoorde, D., Josuttis, N.: C++ Templates. The Complete Guide. Addison-Wesley, Reading (2005)

Interval Subroutine Library Mission[*]

George F. Corliss[1], R. Baker Kearfott[2], Ned Nedialkov[3], John D. Pryce[4],
and Spencer Smith[5]

[1] Marquette University
[2] University of Louisiana at Lafayette
[3] McMaster University and Lawrence Livermore National Laboratory
[4] Cranfield University, RMCS Shrivenham
[5] McMaster University

Abstract. We propose the collection, standardization, and distribution
of a full-featured, production quality library for reliable scientific computing with routines using interval techniques for use by the wide community
of applications developers.

1 Vision – Why Are We Doing This?

The interval/reliable computing research community has long worked to attract
practicing scientists and engineers to use its results. We use any of the terms
interval, reliable, verified computation in the sense of producing rigorous bounds
on true results [1,2]. The Interval Subroutine Library (ISL) is a project to place
interval tools into the hands of people we believe will benefit from their use by
gathering and refining existing tools from many interval authors. We acknowledge that intervals carry a steep learning curve, and that they sometimes have
been over-promised. The winning strategy for widespread adoption of interval
technologies is the development of "killer applications" that are so much better
(in some sense) than current practice that practicing scientists and engineers
have no choice but to adopt the new technology.

The ISL team wants to see such killer applications appear, but producing
them is not our mission. The routine use of interval techniques by practicing
scientists and engineers is hampered by a lack of widely-used, comprehensive,
quality interval software that is available on all major platforms (Linux, Mac
OS, Unix, Windows). Once such software is available, use of interval techniques
is likely to grow along at least three paths: small-scale applications by scien-
tists/engineers in the course of their daily work; professionally built applications
in a specific area, such as global optimization or curve graphing; and the almost
invisible embedding of verified computing as a tool in commodity software such
as spreadsheets or scientific data analysis and document preparation.

ISL can provide the infrastructure for such developments. ISL targets application developers, those who are developing the significant applications. Interval-
based tools tailored for specific end-practitioner applications are developed by

[*] This work was supported in part by EPSRC Grant D033373/1.

P. Hertling et al. (Eds.): Real Number Algorithms, LNCS 5045, pp. 28–43, 2008.

applications developers with expertise in applications areas, but those developers are not finding interval tools they perceive as attractive for their applications. Currently, if we talk to a group of scientists or engineers about intervals and convince them of the value of interval techniques, when they ask, "Great! What software can I use?" there is a long, painful pause. We have many tools, packages, and research codes, but we have no CD that solves their problems with rigorous bounds.

Major obstacles to widespread adoption of interval techniques by developers of main-stream scientific and engineering applications include lack of speed and hardware support, lack of customer demand, lack of interoperability of interval tools, a steep learning curve, and many others. ISL, by itself, can solve **none** of those. ISL can help unify the work of the world-wide interval research community to ease the learning curve with easier to use, more portable software that interoperates better with other interval tools and with other common tools used in scientific computation. A central repository containing much of the high-quality work of the community helps attract customers, who, in turn, drive a broad-based demand for improved hardware support.

The ISL project itself does not author software; contributing authors do that. The goal of the ISL project is to gather and disseminate a library of high quality interval-based tools. The fundamental requirements is "Preserve containment." Routines are expected to return an enclosure of the correct mathematical result or provide a suitable indication of failure. The qualities of interest for the ISL project include

- correctness,
- comprehensiveness,
- reliability,
- performance,
- robustness,
- maintainability, and
- usability,
- portability.

To achieve these qualities, the ISL project encourages its contributing authors to use sound software engineering principles, including documentation, good architecture, thorough testing, and coding standards. The documentation produced and the process of assembling the library also support the goal of achieving high quality. Documentation should be complete, consistent, correct, usable, verifiable, maintainable, and reusable. The development process should have the qualities of productivity, timeliness, and transparency.

The authors are embarking upon a plan for the cooperative development of such a library. This paper lays out the broad scope of the project.

1.1   Short-Term Goals

By the end of 2007, we expect to offer interval Basic Linear Algebra Subroutines (BLAS) levels 1 and 2 and a collection of problem-solving packages, mostly chosen from existing software, including linear systems, optimization, and differential equations. The collection may include utilities for automatic differentiation, Taylor models, and constraint propagation. For our plan to achieve this, see §3.1.

1.2 Long-Term Goals

In perhaps 6 to 8 years, we hope to offer a library of interval tools with coverage comparable to early releases of the IMSL or NAG libraries, SLATEC [3], the popular *Numerical Recipes* books [4,5], the GNU Scientific Library(GSL) [6], or other comparable libraries. The library will be freely available, and we shall also encourage its appearance in commercial products. The library should be used by a significant number of applications widely used in their respective domains. For our plan to achieve this, see §3.2.

1.3 History

To provide a comprehensive interval problem-solving library is an old idea, much of whose history in this section was kindly provided by an anonymous referee. In 1976, Ulrich Kulisch and H.W. Wippermann proposed developing language support and problem-solving routines in interval arithmetic. They raised funds and jointly with Nixdorf developed what finally became PASCAL–XSC. The language, the compiler, and the subroutine library are still available from www.xsc.de. The Russian translation of the language and the corresponding toolbox volume just appeared in its third edition.

When IBM became aware of this development, a close cooperation with the Kulisch institute at Karlsruhe was started in 1980. They jointly developed and implemented a Fortran extension called ACRITH–XSC, together with a large number of problem-solving routines with automatic result verification corresponding to the PASCAL–XSC development.

As a next step, a C++ class library called C–XSC was developed at the Kulisch institute at Karlsruhe in the early 1990s, with books [7] and a toolbox volume with problem-solving routines [8]. The software was recently updated, extended, and adapted to new versions of C and C++ at the institute of W. Krämer at Wuppertal. The software is publicly available, including source code, from www.math.uni-wuppertal.de/wrswt/xsc/cxsc_new.html. C–XSC runs on many platforms, and comes with a large number of elementary and special functions for real, complex, real interval, and complex interval data. They all come with proven error bounds. The problem-solving routines that come with C–XSC cover much of the problem space envisioned by the present ISL project.

The filib++ package [9,10], from Wuppertal, offers fast computation of guaranteed bounds for interval versions of a comprehensive set of elementary functions, and it adds an extended mode for exception-free computation mode relying on containment sets [11]. Filib++ uses templates and traits classes to obtain an efficient, easily extendable, and portable C++ library. Filib++ also comes with an extensive set of problem-solving routines. It is available from www.math.uni-wuppertal.de/wrswt/software/filib.html.

Another library providing guaranteed bounds is INTLAB [12] (based on Matlab) from the institute of S.M. Rump at Hamburg. Other, Fortran based libraries are offered by R.B. Kearfott in Louisiana and W. Walter at Dresden.

The development of killer applications to promote interval technologies is surely possible with existing interval environments, but they have not yet happened. By building on the work of our predecessors, ISL, if successful, will increase the penetration of interval techniques into the mainstream of scientific and engineering computation and lowering the barriers to the development of that killer application.

2   Product – What Will We Deliver?

To meet the needs of a wide community of applications developers in a broad cross-section of applications areas, we need a portable, comprehensive, production-quality library of interval tools solving most of the standard problems of scientific computation. The library should be available in a downloadable or CD form. The library should also have a clear licensing structure that protects authors, while still encouraging commercialization.

2.1   Contents

We envision a hierarchical library, with units organized into chapters roughly as suggested by Figure 1:

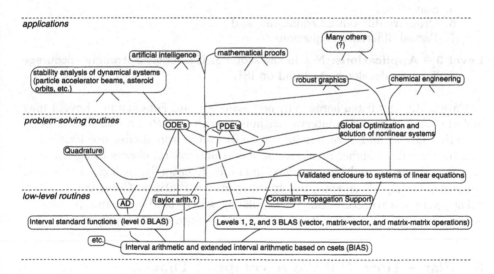

Fig. 1. A tentative hierarchical structure

Level 0 – Basic Interval Arithmetic: Interval arithmetic including
1. Constructors,
2. Arithmetic operations,
3. Comparison operators,
4. Input/output, and
5. Elementary functions.

Level 0 should be consistent with a C++ interval arithmetic standard, such as that proposed by proposed by Brönnimann, Melquiond, and Pion (BMP) [13], should a suitable standard be adopted. Level 0 should have a large overlap with the functionality provided by many current interval arithmetic packages, including PROFIL/BIAS [14,15], `filib++` [9,10], Boost [16], Gaol [17], C–XSC [7,8,18,19,20], and the Sun C++ compiler [21].

Level 1 – Utilities: Level 1 units are called by several units in Level 2 to provide capabilities including

1. Error handler,
2. Additional (non-basic) interval arithmetic features,
3. Vectors and matrix classes,
4. Level 1 and level 2 BLAS,
5. Automatic differentiation,
6. Taylor model arithmetic, and
7. Constraint propagation.

Level 2 – Problem-solving routines: Chapter contents will initially mirror the contents of many numerical analysis texts, and will grow with time.

1. Linear systems, eventually including sparse and eigensystems,
2. Nonlinear systems,
3. Optimization,
4. Quadrature,
5. Statistics,
6. Ordinary differential equations, and
7. Partial differential equations.

Level 3 – Applications: Not in the scope of ISL, but we strongly encourage applications developers to build on ISL.

Capabilities not listed here are by no means left out. For example, Level 1 may include multiple precision interval arithmetic with an API close to that of Level 0, so that Level 2 and Level 3 units can easily switch from double precision intervals to intervals of higher precision, or vector and matrix classes using elliptical representations for multi-dimensional intervals. Eventually, each chapter should contain a variety of general- and special-purpose routines. Categorizing interval software in a structure roughly paralleling widely-used approximate libraries encourages interval researchers to consider gaps in interval coverage.

3 Plan – How Will We Accomplish That?

We have both short-term (two years) and long-term (3–10 years) plans.

3.1 Short-Term Plan: Gather, Organize, and Disseminate

For perhaps two years, this is primarily a library project in the sense of identifying, collecting, organizing, and making available work that already exists.

Step 1 – Language standardization. ISL is a C++ library. Brönnimann et al. have proposed to add intervals to the C++ language standard [13]. The ISL team is working for the strengthening and the adoption of this proposal. The BMP proposal can become the basis for our ISL Level 0 BIAS well before it is approved. Several existing implementations of intervals in C++ are reasonably close to the proposed standard, so multiple (almost) reference implementations are available.

Step 2 – Pilot inclusion into ISL. Select about three existing packages for initial inclusion into ISL. This gives us a chance to prototype policies, procedures, and practices for incorporating existing work. See the discussion of some issues in §3.3–3.9. In particular, this paper is **not** a call for participation, as we are still working to refine how ISL will work.

Step 3 – Invite participation. Once some of the issues of policies, procedures, and practices for incorporating existing work have been refined, we will invite 6–10 researchers to submit their work for inclusion in ISL. At this stage, the number of packages we will invite remains modest, as we develop experience, participation, and visibility.

We hope for a very preliminary release including parts of Steps 1 and 2 by the end of 2006 and a release including about five ISL Level 2 problem-solving routines by the end of 2007.

3.2   Long-Term Plan

After we gain experience and visibility from the short-term "gather, organize, and make available" activities, we expect to expand the scope of the library by inviting contributions from the interval community. Work will be managed along the model of many successful open source projects. We anticipate releases each 1–2 years. We will continue development of a free version of ISL, while seeking a commercial partner such as NAG, Sun, IBM, Intel, or Microsoft.

Next, we turn our attention in subsections 3.3–3.9 to some of the issues that must be settled to ensure the success of ISL.

3.3   Language and Environment

We do not wish to ignite religious warfare, but we must choose an appropriate computer language for ISL. In the short-term "gather, organize, and make available" activities, we can include packages in any language. Most existing interval software is in Matlab or some dialect of Fortran or C++. It is attractive to suppose we can support all languages, but with finite resources, our goals are served better by focusing on one language. **ISL is in C++** because there appears to be more existing interval software in C++ than in other candidates. Inter-language interoperability depends on support from the languages themselves.

3.4 Organizational Structure

Quality, comprehensive libraries are not compiled by a single person or small group of people over a short time. There are many models we can follow of software development by large, loosely-coupled teams over several years, including the LAPACK project [22], PETSc [23,24,25], and many open source projects such as the GSL [6].

The ISL project is coordinated by a steering committee, currently, the authors of this paper. We meet occasionally as a group, and subsets meet as possible at conferences. The steering committee sets directions and policies, such as those outlined in this paper. In the long-term steady state, the role of the steering committee is somewhat like that of the editorial board of a major journal, overseeing the work of authors, referees, and the publication process.

3.5 Adding Value

There are several good packages for interval arithmetic corresponding to our proposed Level 0 BIAS, there are many interval-based problem-solving routines corresponding to our Levels 1 and 2, and there are a few comprehensive projects such as Karlsruhe XSC Toolbox books [8] and Neumaier's COCONUT [26]. Kreinovich does an admirable job of capturing and maintaining pointers to many interval projects at [27]. What value does ISL add?

We return to our initial premise. Although many interval tools are available, there is no single source, a web site or a CD offering a standardized, portable, peer-reviewed suite of tools that install and work together. In the long term, we envision a comprehensive, universally used library. This is in contrast to offering general languages, such as in the COCONUT project or General Algebraic Modeling System (GAMS), or offering graphical user interfaces, such as in commercial packages such as Maple. We view the effort as promoting standardization, portability, and reuse. In the short term, ISL works with contributing authors to gather existing interval tools, standardize their installation and interfaces, perform peer review acceptance and comparative testing, provide examples, and make these tools available from a single source.

3.6 Quality Assurance

All interval code has to have the quality of correctness. The code must obey the rule "Preserve containment."

Contributions are peer refereed. To be considered for inclusion in ISL, normally we expect the algorithm to have been the subject of at least one peer refereed journal paper. Codes, testing, and documentation are also refereed, similar to the standards for an Association for Computing Machinery (ACM) Algorithm. We intend that publications and programs associated with ISL be held to the highest academic and software engineering standards.

We strongly recommend that contributing authors follow modern software engineering practices, where this term encompasses methods proposed by Parnas [28], Literate Programming [29,30], agile methods such as Extreme Programming [31] and Test-Driven Development [32], and others. Generally, we favor the more formal methods because specifications for, say, a linear equation solver, are not expected to change significantly while development is being done.

The ideal contribution to ISL consists of the following parts.

A specification of the software requirements, including the mathematical statement of the problem and information on the required inputs and possible output values. With respect to the inputs, the specification indicates clearly any constraints that exist on the data. Where necessary, a flag shall be specified whose values indicate the reason for failure when a solution cannot be determined. All contributions to ISL share the goal of achieving the qualities listed in the introduction, especially the requirement, "Containment is preserved." However, it is difficult to write validatable specifications of non-functional requirements. For instance, validating correctness is challenging for scientific computing problems, because formal proofs of correctness are difficult and often overly conservative, although non-formal proofs with rigor comparable to proofs in the mathematical literature **are** often appropriate. Therefore, rather than specify the requirements, the approach is taken of describing the final software product, typically including statements of the form, "It finds an enclosure of correct solution if the input lies in set Y." This description is given in the software validation report, discussed below.

A design specification. The ideal specification includes an API or function signature plus semantics, often modeled on specifications for corresponding packages for approximate solutions.

A software validation report. The software validation report is about a combination of observed scope, tightness and speed (plus maybe memory load), and the observed interplay among them. It characterizes the problems that are successfully solved.

The contributing authors are asked to provide the evidence that the software meets the stated requirements and to describe the level to which the software meets the software quality goals. The software validation can consist of informal and formal analysis, testing and a summary of important software metrics. Techniques for informal and formal analysis include code walkthroughs, code reviews, and inspection. Techniques such as literate programming can be employed so that confidence can be built on the correctness of the code, in a similar sense to how confidence is built by mathematicians inspecting a mathematical proof. The summary of testing also builds confidence by showing the test cases that were passed and that any user can download and run for themselves. The descriptions set expectations for the behavior of the program in similar situations. The descriptions can be tested for lies; for instance, the validation report might assert, "the software was run over a given range of inputs on machines x and y and the program terminated in 5 seconds or less with an enclosure of the correct answer in 87% of cases, terminated in 5 seconds or less with a failure indicator in 10% of cases, and had not terminated in 5 seconds in the remaining 3% of cases."

More detailed quality assurance policies and procedures are being developed based on our experiences in Step 2 of the Short Term Plan outlined in §3.1.

3.7 Licensing

Especially in view of Sun being granted several interval-related patent applications, the interval community is increasingly aware of the importance of the protection of intellectual property. ISL needs a carefully considered license which

- protects rights and reputations of authors,
- provides for free distribution, and
- encourages commercialization.

To help us in that, we are gathering and evaluating examples including

- GNU Lesser General Public License (LGPL), Modified Berkeley, Boost, MIT, and other Open Source licenses;
- licenses of various interval packages; and
- intellectual property policies of some (possibly) participating universities.

3.8 Publications

Since many potential participants in the ISL project are academics, the project will not succeed without clear publication opportunities:

- Continuing publication of incremental and innovative development of interval software;
- Identification of omissions in coverage (holes) as development targets;
- Comparative testing of similar packages in the spirit of Enright and Hull [33] or Mazzia, Iavernaro, and Magherini [34] for approximate ODE solvers or Pryce [35,36] for Sturm-Liouville solvers; and
- Suites of test problems for interval problem-solving routines, e.g., Corliss and Yu for interval arithmetic operations and elementary functions [37]. Interval test suites should include many problems from existing test suites for approximate solvers and also problems intended to test existence and containment properties.
- Software engineering publications related to the development of scientific software with respect to appropriate process models, methodologies and documentation. Software engineering has mostly ignored scientific software and placed most of the emphasis on research on safety critical systems and information systems. There is room to make contributions by looking at the issues that are specific to scientific software.

For academic researchers, release of software to ISL in addition to journal publications offers an external, peer-reviewed process for recognizing research contributions and wider dissemination than links from the authors' web site.

While valuing the role of *Reliable Computing* as **the** core journal in this research field, we encourage contributing authors to publish in a wide variety of journals, especially journals in applications areas, to help bring the message of intervals to as wide an audience as possible. Intervals are much closer to the main stream of scientific computing than many of us realize, as new applications and researchers using interval techniques are published regularly. We help more people learn about intervals by publishing in the outlets they read.

In the long-term steady state, having code accepted for inclusion in ISL may be viewed as equivalent to a journal paper, probably contributing more to the overall advance of the infrastructure of science than many journal papers.

3.9 Funding

While it would be welcome if someone wanted to provide large funding, that is unlikely. If we look at the models of LAPACK or most open source projects, there may be modest funding somewhere for overall leadership and organization, but the developers are on their own to secure their own funding. One would hope that contributing to a large, well-organized, well-publicized international effort might help many interval researchers get our own work funded.

Similarly, it would be welcome if a large software company provided the leadership and modest funding for the champion to lead an open source project. In other fields of study, with more obvious customers, several firms have made significant contributions to various open source projects.

4   Partners – How Can You Help?

Clearly, the long-term goal is ambitious, requiring the work of many people over many years. This section outlines our vision of an ideal partnership of contributing authors, chapter architects, referees, and ISL steering committee.

4.1   Contributing Author

A contributing author contributes any of

- Code unit for the library to solve some well-defined problem of scientific computing, e.g., constraint propagation, linear systems, optimization;
- Functionality or performance improvements, corrections, or extensions to an existing unit of the library;
- Test suites;
- Documentation; and
- API architecture for a chapter of the library.

For a new unit for the library, a contributing author should submit

- User Guide: installation, requirements, examples;
- Maintenance documentation: system architecture, detailed design, test plan and report;
- Source code; and
- Journal article (with quality of TOMS article and algorithm).

A typical interaction might be

1. Contributing author contacts (or is contacted by) the steering committee.
2. They discuss a suitable problem of scientific computing, specifications, licensing, etc.
3. Contributing author submits a suitable research publication to a journal.
4. Contributing author submits to ISL source; installation instructions; documentation of the problem, description of algorithms, examples of use, references, etc.; acceptance and other tests; copies of journal papers, etc.
5. ISL or chapter editor sends submitted materials to referees.
6. Usual iterations with editors, referees, and authors.
7. ISL accepts or declines the submission.
8. After acceptance, ISL maintains discussions with the contributing author to ensure that updates to the original work are reflected in the library.

ISL should be more than a listing of web links to contributing authors' pages. That requires some process, at least semi-formal. The short-term "gather, organize, and make available" phase of the project will be used to find an appropriate balance of a formal process to ensure quality and a light-weight process all can use effectively. For example, there is no need to duplicate refereeing work already performed for journal publication.

4.2   Chapter Architect/Editor

In the short term, the ISL steering committee are the architects of the library and the editors for contributed units. As the scope grows, each chapter of the library (see §2.1) has an architect/editor responsible for

- External architecture of the chapter, problem coverage, consistent API;
- Internal architecture, shared utilities; and
- Collaboration with contributing authors for this chapter.

4.3   Referee

The referee contributes to the overall quality of the library by providing an external assessment. The referee reviews materials submitted by contributing authors including source, installation instructions, documentation of the problem, description of algorithms, examples of use, references, acceptance and other tests. The referee is assessing the library materials as they affect application developers who use the library, rather than the more academic concerns of a traditional journal referee. We anticipate that some referees are anonymous, and some are collegial.

The ISL refereeing process may be modeled on the process for refereeing Association for Computing Machinery (ACM) Algorithms. We encourage ISL contributing authors to submit their work as ACM Algorithms, in which case, the ISL refereeing is sharply truncated.

We anticipate that some refereeing work leads to publishable careful comparative testing of similar packages and compilation of sets of standard test problems for interval problem-solving routines.

4.4 Applications Development

The goal of the ISL project is to get quality, portable, uniform, comprehensive interval tools into use by developers of applications used by practicing scientists and engineers. Our target audience includes

- Developers in the interval community, our contributing authors. For example, authors of interval differential equations or optimization solvers benefit from automatic differentiation, constraint propagation, and linear solvers in ISL.
- Scientists from applications areas developing more reliable software than that currently available.
- Small commercial companies seeking the competitive advantage of high reliability software in their market niche.
- Large commercial companies who develop market-leading software packages in industry segments, such as chemical engineering, structural engineering, financial modeling, chip and circuit design, supply chain planning, industrial process engineering, etc.

ISL targets the developers of software for use in these areas.

5 Will ISL Succeed?

It is natural to ask why this ISL project might be more successful than its predecessors described in §1.3. We seek not to disparage the work of others but to highlight opportunities open to the community as a whole to increase the penetration of interval techniques into the common practice of scientific and engineering computation.

Stand on the shoulders of giants – ISL should incorporate as much as possible existing work in the underlying packages for interval arithmetic and the higher-level problem-solving routines.

C++ Standard – If intervals are standardized in C++, intervals become available in the daily software toolbox of most developers of scientific and engineering software, without even a need to download. A standard makes it much easier for several packages to inter-operate, removing the current impediments of slight inconsistencies in the interval arithmetic provided by for example by C–XSC, filib++, PROFIL/BIAS, and Sun.

Interoperability – Interval problem-solving routines available with C–XSC, filib++, PROFIL/BIAS, and Sun each have strengths. If we can leverage a C++ interval standard for the underlying interval operations and migrate existing problem-solving routines toward more consistent interfaces, it becomes easier for developers to use the best routines from several authors.

Hardware support – Broader adoption of interval software is much more attractive if interval arithmetic is supported directly by hardware, a position advanced by Professor Kulisch for many years, most recently in [38].

Customers – Major hardware vendors will support intervals when there is customer demand, and customers do not demand what they cannot experience.

More widely available, portable, easy to use interval software facilitates experiencing the certainty afforded by interval techniques by a broader audience.

Open source – The success of many broad-based open source software development projects demonstrates an alternative to the Not Invented Here model of development in single institutes by leveraging the strengths of many researchers, rather than putting the resources of the research community into competing projects.

World-wide – ISL should involve researchers from Japan, China, India, Brazil, and others as well as traditional seats in Europe, Canada, and US.

Connections with point-based algorithms – Approximate algorithms using pure floating-point arithmetic are becoming more reliable, e.g., Baron for global optimization. A wider network of developers can leverage connections with point numerical analysis and applications developers. Advances from approximate algorithms can improve the speed of interval algorithms, and interval insights can improve the reliability of approximate algorithms at critical steps.

Ease the steep learning curve – Compared with approximate algorithms, most interval algorithms require a deeper understanding of the problem, of the method of solution being used, and of the coding of the problem. We reduce that gap by very careful attention to ease of use and by promoting more consistent interfaces across a wider variety of problem-solving routines.

"Deeper" analysis – We have problem-solving routines covering a wide variety of problems, but there are opportunities to improve the set of problems each can handle and their scalability to larger problems.

"Broader" analysis – We have solved the general case for linear and nonlinear systems, differential equations, and optimization repeatedly. A centralized repository makes it easier to recognize the need for special-purpose solvers, e.g., sparse systems, linear ordinary differential equations, linear programming problems, and many more.

Structure for academic recognition for software development – Most members of the interval community are academic researchers, rewarded for publishing, not for coding. ISL may come to provide in the long term a structure for recognition of peer-reviewed software development, leading to increased funding opportunities.

It will take a long time. It will take lots of people. And it will take money.

6   Conclusions

Initially, ISL intends to be a single source for as large a body of existing interval routines as resources allow. In the longer term, ISL offers a quality, portable, uniform, comprehensive, problem-solving library. Eventually, we aspire to be seamlessly integrated with tools and libraries for approximate computation.

Acknowledgement

An initial draft of this paper was prepared during a visit of the first author to Tibor Csendes at the Department of Applied Informatics, University of Szeged, Hungary, January 16 - 20, 2006. We thank Tibor for his kind hospitality.

We thank anonymous referees for contributing significantly, especially for providing much of the content of §1.3.

References

1. Moore, R.E.: Methods and Applications of Interval Analysis. SIAM, Philadelphia (1979)
2. Neumaier, A.: Interval Methods for Systems of Equations. Cambridge University Press, Cambridge (1990)
3. Fong, K., Jefferson, T., Suyehiro, T., Walton, L.: Guide to the SLATEC common mathematical library. Technical report (1990), netlib.org, http://www.netlib.org/slatec/
4. Press, W.H., Flannery, B.P., Teukolsky, S.A., Vetterling, W.T.: Numerical Recipes in Fortran: The Art of Scientific Computing, 2nd edn. Cambridge University Press, Cambridge (1992); Also available for Fortran 90, C, and C++
5. Press, W.H., Teukolsky, S.A., Vetterling, W.T., Flannery, B.P.: Numerical Recipes in C++: The Art of Scientific Computing, 2nd edn. Cambridge University Press, Cambridge (2002)
6. GSL: GNU Scientific Library (1996 - June 2004), http://www.gnu.org/software/gsl/
7. Klatte, R., Kulisch, U., Wiethoff, A., Lawo, C., Rauch, M.: C-XSC – A C++ Library for Extended Scientific Computing. Springer, Heidelberg (1993)
8. Hammer, R., Hocks, M., Kulisch, U., Ratz, D.: Numerical Toolbox for Verified Computing I — Basic Numerical Problems. Springer, Heidelberg (1993)
9. Lerch, M., Tischler, G., von Gudenberg, J.W., Hofschuster, W., Krämer, W.: filib++, a fast interval library supporting containment computations. ACM Transactions on Mathematical Software 32(2) (2006)
10. Lerch, M., Tischler, G., Wolff von Gudenberg, J., Hofschuster, W., Krämer, W.: The interval library filib++ 2.0 - design, features and sample programs (preprint 2001/4), Universität Wuppertal, Wuppertal, Germany (2001)
11. Pryce, J.D., Corliss, G.F.: Interval arithmetic with containment sets. Computing 78(3), 251–276 (2006)
12. Rump, S.M.: INTLAB interval toolbox, version 5.2 (1999–2006), http://www.ti3.tu-harburg.de/intlab.ps.gz
13. Brönnimann, H., Melquiond, G., Pion, S.: A proposal to add interval arithmetic to the C++ standard library. Technical Report N1843-05-0103, CIS Department, Polytechnic University, New York, and Laboratoire de l'Informatique du Parallélisme, École Normale Supérieure de Lyon, and INRIA Sophia Antipolis (2005–2006)
14. Knüppel, O.: PROFIL/BIAS – A fast interval library. Computing 53(3–4), 277–287 (1994), http://www.ti3.tu-harburg.de/profil_e
15. Knüppel, O.: PROFIL/BIAS v 2.0. Bericht 99.1, Technische Universität Hamburg-Harburg, Harburg, Germany (1999)

16. Brönnimann, H., Melquiond, G., Pion, S.: The Boost interval arithmetic library (2006), http://www.cs.utep.edu/interval-comp/main.html
17. Goualard, F.: Gaol, not just another interval library (2006), http://www.sourceforge.net/projects/gaol/
18. Hofschuster, W.: C–XSC – A C++ Class Library web page (2004) http://www.math.uni-wuppertal.de/wrswt/xsc/cxsc.html
19. Hofschuster, W., Krämer, W.: C–XSC 2.0: A C++ library for extended scientific computing. In: Alt, R., Frommer, A., Kearfott, R.B., Luther, W. (eds.) Dagstuhl Seminar 2003. LNCS, vol. 2991, pp. 15–35. Springer, Heidelberg (2004)
20. Hofschuster, W., Krämer, W., Wedner, S., Wiethoff, A.: C–XSC 2.0: A C++ library for extended scientific computing. Preprint BUGHW–WRSWT 2001/1, Universität Wuppertal (2001)
21. Sun Microsystems.: C++ interval arithmetic programming reference (2004–2006) http://docs.sun.com/db/doc/806-7998
22. Anderson, E., Bai, Z., Bischof, C., Blackford, S., Demmel, J., Dongarra, J., Du Croz, J., Greenbaum, A., Hammarling, S., McKenney, A., Sorensen, D.: LA-PACK User's Guide, 3rd edn. SIAM, Philadelphia (1999); Certain derivative work portions have been copyrighted by the Numerical Algorithms Group Ltd. http://www.netlib.org/lapack/, http://www.nacse.org/demos/lapack/.
23. Balay, S., Buschelman, K., Eijkhout, V., Gropp, W.D., Kaushik, D., Knepley, M.G., McInnes, L.C., Smith, B.F., Zhang, H.: PETSc users manual. Technical Report ANL-95/11 - Revision 2.1.5, Argonne National Laboratory (2004)
24. Balay, S., Buschelman, K., Gropp, W.D., Kaushik, D., Knepley, M.G., McInnes, L.C., Smith, B.F., Zhang, H.: PETSc Web page (2001), http://www.mcs.anl.gov/petsc
25. Balay, S., Gropp, W.D., McInnes, L.C., Smith, B.F.: Efficient management of parallelism in object oriented numerical software libraries. In: Arge, E., Bruaset, A.M., Langtangen, H.P. (eds.) Modern Software Tools in Scientific Computing, pp. 163–202. Birkhäuser Press (1997)
26. Neumaier, A.: COCONUT Web page (2001-2003), http://www.mat.univie.ac.at/~neum/glopt/coconut
27. Kreinovich, V.: Interval Computations (2006), http://www.cs.utep.edu/interval-comp/main.html
28. Parnas, D.L.: Software Fundamentals: Collected Papers by David L. Parnas. Addison-Wesley, Reading (2001)
29. Knuth, D.E.: Literate programming. The Computer Journal 27(2), 97–111 (1984)
30. LiterateProgramming: Literate Programming Web page (2000–2005), http://www.literateprogramming.com/
31. Burke, E.M., Coyner, B.M.: Java Extreme Programming Cookbook. O'Reilly, Sebastopol (2003)
32. Beck, K.: Test-Driven Development: By Example. Addison-Wesley, Reading (2003)
33. Hull, T., Enright, W., Fellen, B., Sedgwick, A.: Comparing numerical methods for ordinary differential equations. SIAM J. Numer. Anal. 9, 603–637 (1972)
34. Mazzia, F., Iavernaro, F., Magherini, C.: Test set for IVP solvers, release 2.2 (2003), http://pitagora.dm.uniba.it/~testset/
35. Pryce, J.D.: A test package for Sturm-Liouville solvers. ACM Trans. Math. Software 25(1), 21–57 (1999)

36. Pryce, J.D.: Algorithm 789: SLTSTPAK, a test package for Sturm-Liouville solvers. ACM Trans. Math. Software 25(1), 58–69 (1999)
37. Corliss, G.F., Yu, J.: Testing COSY's interval and Taylor model arithmetic. In: Alt, R., Frommer, A., Kearfott, R.B., Luther, W. (eds.) EDBT 2004. LNCS, vol. 2992, pp. 91–105. Springer, Heidelberg (2004)
38. Kirchner, R., Kulisch, U.W.: Hardware support for interval arithmetic. Reliable Computing 12(3), 225–237 (2006)

Convex Polyhedral Enclosures of Interval-Based Hierarchical Object Representations

Eva Dyllong

University of Duisburg-Essen, Faculty of Engineering, Department of Computer
Science, Lotharstrasse 65, D-47048 Duisburg, Germany

Abstract. In this paper, we discuss approaches to constructing convex
polyhedral enclosures of interval-based hierarchical structures. Hierarchical object representations are the data structures most frequently used
for reconstructing real scenes. This object modelling does not depend on
the nature of a real solid but only on the chosen maximum level of the
hierarchical structure. This is a useful property for objects with complex
shapes that are difficult to describe via exact mathematical formulas.
We focus on reliable object modeling using an interval-based octree data
structure. To obtain a convex polyhedral enclosure of an octree, we seek
feasible ways to limit the number of considered points. For this purpose,
we use the concept of extreme vertices of the tree nodes. Accurate algorithms for constructing the convex hull of these vertices yield a convex
polyhedron as an adaptive and reliable object enclosure at each level of
the tree.

1 Introduction

The three major approaches to modeling a solid object suitable for computer
processing are boundary representations (B-Rep), constructive solid geometry
(CSG) and spatial enumeration methods. Octrees fall in the third category. They
provide a common technique for reconstructing a scene relying on a hierarchical
representation of objects using axis-aligned bounding boxes.

1.1 Relevant Properties of Hierarchical Object Representation

The octree-based object representation does not depend on the nature of the
real solid. This is a useful property for objects with complex structures that
are difficult to describe using exact mathematical expressions, including those
with internal cavities. An adaptive enclosure of a real solid that depends only
on the chosen maximum level of the tree and an efficient execution of Boolean
operations are additional advantages of the octree data structure. Because of
their adaptive depth, octrees offer a faithful enclosure of a real object and can
be used to describe a virtual environment constructed from camera data.

On the other hand, the applicability of this data structure to objects in the
context of robotic simulation has its limits. One of the reasons for this is that
in a dynamically changing environment a lot of arbitrary rigid motion transformations are needed, which are in general computationally difficult to realize for

P. Hertling et al. (Eds.): Real Number Algorithms, LNCS 5045, pp. 44–56, 2008.

octrees. That is because, whenever an octree moves, it must be recalculated to reflect its new position. This update is not as straightforward as for other object representations, such as boundary representations. Otherwise, unaligned octrees could be applied to model moving objects. Moreover, to prevent accumulation errors and to avoid the wrapping effect, the input octree should always be used, and only the motion matrix should be changed.

Furthermore, the distance computations between two objects in the scene, which are frequently used in many applications, become difficult and slow for unaligned octrees during the simulation. Using convex hull of an octree allows one to apply algorithms for computing the distance between convex polyhedra and to speed up the distance computation as well as to reduce the wrapping effect which occurs due to rotated boxes which are covered by axis-aligned ones. This beneficial effect has been shown in a recent diploma thesis [Grim06] that compares the efficiency of distance algorithms for octrees and their polyhedral enclosures.

1.2   Purpose and Outline of This Paper

As mentioned above, for a moving object it is advantageous to use a polyhedral instead of hierarchical representation. However the second one is the preferred underlying data structure subsequent to the reconstruction of a scene. For this reason, we focus on computing adaptive and reliable polyhedral enclosures at each level of the tree. This approach yields a form of a polyhedral hierarchical object representation as a result. Moreover, using interval-based hierarchical structures we make sure that all object points are enclosed, which is an important premise in the field of motion planning, particularly in regard to collision testing. To obtain a convex polyhedral enclosure of an octree, we conveniently determinate the extreme vertices of the tree nodes and afterwards construct the convex hull of these vertices.

This paper is organized as follows: Section 2 gives the underlying definition of an interval-based hierarchical object representation. The concept of extreme vertices is presented in Section 3. Section 4 deals with our reliable algorithm for computing convex polyhedral enclosures of hierarchical structures. Details on the reliable implementation of the algorithms presented are given in Section 5. Section 6 offers some concluding remarks.

2   Interval-Based Hierarchical Object Representation

An octree is an efficient data structure used to represent spatial data hierarchically in a tree-based structure with eight child nodes. The idea is to recursively subdivide a cube including objects of three-dimensional space into eight mutually disjoint voxels until the required closeness to the object is reached [Same90]. Each node is checked to see whether it is full (black), partially empty (gray) or empty (white) of solid material. If the nodes are empty or full, they do not need to be subdivided any further. In the case of partial emptiness, the nodes need to be subdivided to create a higher level of the octree. The subdivision process

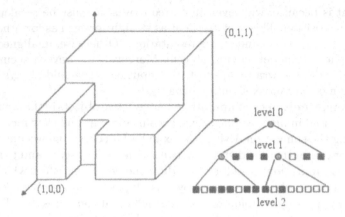

Fig. 1. An object representation using the octree data structure

is repeated until all the nodes are either full or empty or until the maximum resolution level has been reached. To obtain an outer or an inner hierarchically structured approximation of the object, partially occupied leaves are filled or emptied. An object representation using the octree data structure is illustrated in Figure 1.

Since an explicit pointer-based octree storage is expensive in terms of memory requirements, we use a more compact linear encoding, the depth-first (DF-) representation of an octree. In this representation an octree is stored by listing consecutively the octree nodes encountered on performing a preorder traversal of the octree. The symbols used are G (gray node), B (black node), and W (white node). Since there are only three different characters, two bits per node are sufficient for storing the octree. As an example, the octree in Figure 1 yields the DF-representation: GGBWBBBWBBBBBGBWBWWWWWWBB.

An octree node belonging to an arbitrary hierarchy level geometrically defines an axis-aligned cube. In the case of an axis-aligned octree, all cubes have as vertices machine numbers that are multiples of powers of two. But this needn't to be true after a rotation. Due to rounding errors, the vertices must be replaced by small intervals with machine numbers, or each side of the cube has to be stored as an interval to obtain a reliable enclosure of a given real object. Furthermore, interval arithmetic is used to carry out elementary operations on the cubes [Moor66].

Interval arithmetic is the most common technique providing reliable solutions for many numerical problems. In interval arithmetic numbers are replaced by intervals, representing the imprecision associated with each number. Basic arithmetical operations, such as the sum, difference, product or inverse of intervals, or even elementary functions, like sine, cosine, etc. are well defined in interval arithmetic. Unfortunately, the loss of dependencies between variables is an often criticized drawback of this arithmetic. It results in an overestimation of intervals that increases with the number of interval evaluations.

We reduce the problem of overestimation by isolating the basic arithmetical expressions that have to be implemented in such a way that they yield a result that is as close as it can be to the best possible machine representation. Since scalar products occur frequently and are important basic operations in many geometric computations, it is advantageous to perform the scalar product calculation with the same precision as the basic arithmetical operations. By using the exact scalar product, we can delay the onset of qualitative errors and improve the robustness of the implementation [Dyll04].

3   Concept of Extreme Points

The convex hull of a geometric object is the smallest convex set containing the object. In the case of a finite set of points, the convex hull can be identified with the smallest polyhedron that contains the points. The vertices of this polyhedron are called the extreme points of the convex hull.

A simple and straightforward way of determining the convex hull of an octree is to build the set of all vertices of cubes that define the octree nodes containing the object and to construct the convex hull from these points. However, this is not an efficient method. If the maximum hierarchy level of the octree changes, the points defining the vertices vary, and a completely new set of points has to be investigated. Furthermore, there are several vertices in the tested set that belong to the interior of the object and thus are not relevant to the computation of the convex hull.

Since only points on the boundary are relevant to the computation of the convex hull, we can reduce the number of the points by investigating only those vertices that belong to exactly one of the black or gray cubes. In the following, a vertex is said to be an extreme vertex of cubes if none of the adjacent cubes belong to the object. We use the extreme vertices on the boundary to obtain a convex polyhedral enclosure of the given octree-based object representation. Furthermore, a suitable update of the extreme vertices contributes to performance enhancement of this approach whenever the maximum level has changed.

While the level increases, only gray cubes containing extreme vertices take part in the update process. Figure 2 shows examples of how the update process is performed. For the sake of clarity, a two-dimensional case has been chosen. The rectangle with the black mark on the left shows a gray node of a quadtree at level k containing an extreme vertex. On the opposite side are listed twenty of the sixty-two conceivable cases of the four subrectangles at level $k + 1$. The arrows illustrate the update of the extreme vertices.

Extreme vertices that belong to black nodes have been retained unchanged at all higher levels. The updates of a gray rectangle with an extreme vertex in the upper right, upper left or bottom left corner work in an analogous way. In the case of an octree, we use a transformation of the octree into a bintree and perform the update of the black marks successively in x-, y- and z-direction.

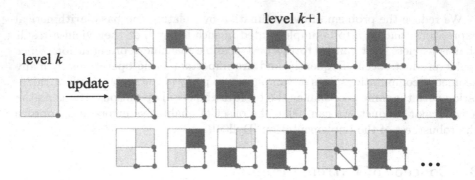

Fig. 2. Update of extreme vertices in 2D

4   Convex Polyhedral Enclosures

The paradigm of our algorithm for constructing the hull in three dimensions is the same as in two dimensions (cf. [DyLu06]). This is because of the natural correspondence between the underlying nodes: rectangles to cubes, and the portability of the concept of extreme vertices to higher dimensions. Hence, to make clear the idea, we first describe in detail the determination of the convex hull of a hierarchical data structure by considering a quadtree Q and afterwards examine an octree in order to explain some apparent differences.

4.1   Accurate Convex Hull Algorithm for a Quadtree

Let us consider an axis-aligned quadtree Q at level k. In the preprocessing phase of the algorithm, all rectangles that cannot contain an extreme vertex point will be removed. For this purpose the rectangle hull $[x_{\min}^{(k)}, x_{\max}^{(k)}] \times [y_{\min}^{(k)}, y_{\max}^{(k)}]$ of Q at level k, $k = 0, 1, \ldots$, is constructed, that is, the smallest axis-aligned rectangle that covers Q at level k, with the left-, right-, bottom- and topmost corners of Q (see Figure 3).

Let $S_k^{(i)} = [x_L^{(i)}, x_R^{(i)}] \times [y_L^{(i)}, y_R^{(i)}]$, $i = 1, \ldots, n(n \leq 4^k)$ be the black or gray nodes at level k with machine-representable corners. If $P_R^{\min}$ and $P_R^{\max}$ both belong to gray nodes at level $k - 1$ (see Figure 3), we update $x_{\max}^{(k)}$ as follows:

$$x_{\max}^{(k)} = \max\{x_R^{(i)} | i = 1, \ldots, n \wedge S_k^{(i)} \in S_{k-1}^{(j)} \text{ with } x_R^{(j)} = x_{\max}^{(k-1)}\}.$$

The values $x_{\min}^{(k)}$, $y_{\min}^{(k)}$ and $y_{\max}^{(k)}$ are updated respectively. We consider the axis-aligned subrectangle R_1 spanned by $P_B^{\max} = (x_B^{\max}, y_B^{\max})$ and $P_R^{\min} = (x_R^{\min}, y_R^{\min})$ at level k and repeat the following update process:

1. Update $P_R^{\min}$ by searching all $S_k^{(i)} \in S_{k-1}^{(j)}$ with $x_R^{(j)} = x_{\max}^{(k-1)}$ and $P_B^{\max}$ by searching all $S_k^{(i)} \in S_{k-1}^{(j)}$ with $y_L^{(j)} = y_{\min}^{(k-1)}$; $i \in \{1, \ldots, n\}$. This phase takes time $O(d \cdot 2^{(d-1)k})$, where d denotes the dimension.

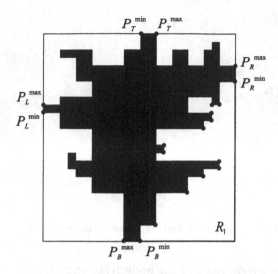

Fig. 3. Rectangle hull of Q

2. Update the list L_1 of extreme vertices of the subrectangle R_1 at level k by searching all $S_k^{(i)} \in S_{k-1}^{(j)}$ with $S_{k-1}^{(j)}$ containing extreme vertices at level $k-1$ (see Figure 2 for details). The nodes with the black marks are nodes with the (possible) extreme points of the convex hull at levels $k-1$ and k respectively. Thus the time complexity is bounded by $2^d \cdot d^k$.

3. Sort the list L_1 with respect to y_L, i.e. $L_1 = \{(x_R^{(j)}, y_L^{(j)}), j = 1, \ldots, m, m \leq n\}$ with $y_L^{(j)} \leq y_L^{(j+1)}$ (cp. the modified Graham scan algorithm in [RaRo03]). Set $x_{curr} = x_B^{\max}$, $M_1 = \{P_B^{\max}\}$. For $j = 1, \ldots, m$ do: if $x_{curr} < x_R^{(j)}$, put $x_{curr} := x_R^{(j)}$ and insert $(x_R^{(j)}, y_L^{(j)})$ into the list M_1. Finally, insert $P_R^{\min}$ into M_1 (see Figure 3). Sorting the list has time complexity $O(m \log m)$.

4. Eliminate the points of M_1 that are not extreme points at level k by using the left-turn test together with interval arithmetic and the exact scalar product (see Section 5.3). This can be done in linear time.

The remaining subrectangles R_2 spanned by $P_R^{\max}$ and $P_T^{\max}$, R_3 spanned by $P_L^{\max}$ and $P_T^{\min}$, and R_4 spanned by $P_L^{\min}$ and $P_B^{\min}$ are handled similarly and yield lists M_2, M_3 and M_4. The composite list M with the sublists M_1, M_2, M_3 and M_4 results in the convex hull of Q at level k. If all corners are machine numbers, the computation is rounding-error free. The computed hull is the smallest possible machine-representable convex hull of the given quadtree. Note that for the update of the convex hull at level k, instead of the list M the sublists L_1, L_2, L_3 and L_4 of (possible) extreme points at level $k - 1$ are used. That is because the list M at level $k - 1$ does not necessarily contain all black marks relevant for the convex hull at level k.

Apart from the smallest possible machine-representable convex polyhedral enclosure of the given object, this approach yields as a by-product the convex

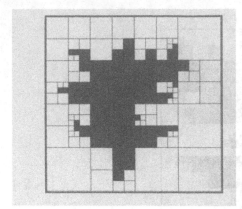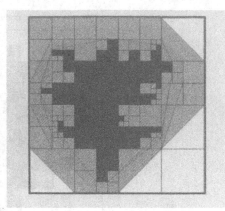

Fig. 4. A quadtree at level 5 and its convex hulls at levels 2, 3, 4 and 5

hulls at all approximation levels of the underlying quadtree from the first to the maximum level (see Figure 4).

4.2   Accurate Convex Hull Algorithm for an Octree

In case of an octree the preprocessing phase can be done analogously starting with the smallest axis-aligned parallelepiped covering the octree and finishing with a set M of the extreme points of the convex hull we are searching for.

At the start of the update process, instead of the four rectangular regions $R_1, \ldots, R_4$ eight box regions located with respect to each of the eight corners of the three dimensional extension $[x_{\min}^{(k)}, x_{\max}^{(k)}] \times [y_{\min}^{(k)}, y_{\max}^{(k)}] \times [z_{\min}^{(k)}, z_{\max}^{(k)}]$ of the rectangle hull at level k, $k = 0, 1, \ldots$, have to be investigated.

Step 2 of this process can be provided by transforming the octree into a bintree and updating the black marks successively in the three directions. While performing the update, each black mark at level k may create at most three extreme vertices at level $k + 1$.

To eliminate the extreme vertices that are not extreme points of the convex hull in step 4, the QuickHull algorithm was used. QuickHull is a variation on the incremental algorithm and can be generalized like the incremental algorithm to a higher number of dimensions. At each step of the algorithm, one triangle of the current hull is selected and an associated point that is furthest away from the triangle is found and added to the hull as follows: We delete the selected triangle and any other triangles that can see this point. Then we add all the triangles that connect the new point and an edge of the hole in the hull that results from the deleted triangles. We repeat this step for each edge of the hole to finish connecting the new point to the remainder of the old hull. Points that can not see any of the new triangle are inside the hull and can be discarded. We select a new triangle and repeat the steps above until there are no points outside

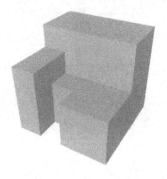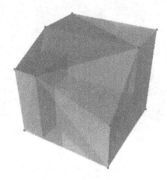

Fig. 5. The octree described in Figure 1 and its convex hull

the hull, that is, points that are visible to at least one triangle of the new hull; for details see [BDH96]. In the worst case the QuickHull algorithm is $O(n^2)$, but in practice it is not worse than $O(n \log n)$, where n denotes the number of points.

We use the orientation test described in Section 5 to test the visibility of a point. The furthest points can also be found by evaluating an arithmetic expression that can be implemented in such a way that the exact scalar product is used, thereby delivering highly accurate results.

Figure 5 shows as an example of an octree and its convex hull at level 2.

5   Reliable Implementation

In this section we show how our algorithm for constructiong convex polyhedral enclosures can be implemented to perform the construction reliably. An octree after a rotation has vertices which are interval values. This provides for an interval-based implementation. The algorithm itself is not interval-based but the realization makes it that. Due to the fact that the special features of interval arithmetics have been considered while developing the algorithm, the processing of the implementation appears nearly straightforward.

Furthermore, it is to mention that owing to several import filters and inclusion tests like the interval version of the criterion from Walach and Zeheb a number of reliable construction feasibilities for the user are available.

5.1   Software Environment

We have implemented the hull algorithms described above in C++ programming language using the interval arithmetic of C-XSC (a C++ class library for eXtended Scientific Computation) together with the C++ toolbox for verified computing [HHKR95], [Zhan05].

Its wide range of tools for scientific computation such as interval and multiple-precision arithmetics, rounding control or the exact scalar product makes C-XSC

 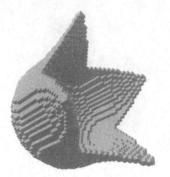

Fig. 6. A surface defined by equation (1) and plotted with Maple

Fig. 7. An octree at level 5 defined by (2)

well suited for reliable programming with result verification. Furthermore, the C++ toolbox contains several problem-solving methods for some standard problems of numerical analysis, such as evaluations of polynomials or arithmetic expressions, and nonlinear systems of equations, which are used as problem solvers in our implementation.

We used the library OpenGL and the graphical widget toolkit Qt for visualization.

5.2   Constructions of an Octree

Our implementation provides several ways of constructing a quadtree or an octree. A hierarchical object representation can be imported from a ASCII file containing the DF-representation of the tree or containing another type of hierarchical representations, e.g. subpavings from SIVIA (Set Inversion Via Interval Analysis) [JKDW01]. Moreover, an axis-aligned tree can be created manually node by node. An interval polynomial expression can be put as input, too.

Let us consider as an example an object whose surface can be described by points fulfilling the following multivariate polynomial equation

$$x^2 - y^3 + z^2 - 1 = 0, \tag{1}$$

where $x \in [-2, 2]$ and $y \in [-2, 2]$. Figure 6 shows the surface plotted with the computer algebra system Maple.

To model inaccuracies of the input data coming for instance from measuring sensors during the reconstruction the interval data type can be used. An octree at level 5 computed for the interval polynomial expression

$$x^2 - y^3 + z^2 = [0.95, 1.01] \tag{2}$$

if using our implementation is illustrated in Figure 7.

To reliably decide which parts of the scene belong to an object a common inclusion test applying interval arithmetic is used. Alternatively, it is possible to test the sign of the polynomial in the interval examined. For our example the sign of the following multivariate polynomial

$$P(x, y, z) = -x^2 + y^3 - z^2 + [0.95, 1.01]$$

can be investigated.

For this purpose an interval version of the criterion from Walach and Zeheb [WaZe80] for testing the sign of a multivariate polynomial in a box is implemented. This positivity test has the advantage that the procedure can be iterated in such a way that in each step the dimension decreases. For details of the transformation of this criterion to interval valued polynomials see [FaLu00].

Theorem 1 (Walach-Zeheb Test). *Let* $Q = [\underline{x}_1, \overline{x}_1] \times \ldots \times [\underline{x}_n, \overline{x}_n] \subset \Re^n$ *be a box and* $P : Q \to \Re$ *a multivariate polynomial. Then* $P(x) > 0$ *for all* $x \in Q$, *if and only if the following conditions hold:*

(i) it holds $P|_{x_i = \underline{x}_i} > 0$ *on the whole hyperplane* $x_i = \underline{x}_i$ *and* $P|_{x_i = \overline{x}_i} > 0$ *on the whole hyperplane* $x_i = \overline{x}_i$ *for all* $i = 1, \ldots, n$.
(ii) the system of n *equation in* n *variables,*

$$P(x) = 0$$
$$\frac{\partial P(x)}{\partial x_i} = 0 \quad \text{for all} \quad i = 1, \ldots, n - 1$$

does not have any solution in Q.

The implementation of the sign test for polynomials uses a solver from the C++ toolbox to compute enclosures for all solutions of systems of nonlinear equations given by continuously differentiable functions based on Hansen and Sengupta's method and a modification of Ratz's method [HHKR95].

5.3   Orientation Tests

For a reliable implementation of our hull algorithm, the computation of geometric predicates – in particular orientation tests – are required with guaranteed accuracy. However, a common floating point implementation of these tests in two or three dimensions may lead to inconsistent results due to an accumulation of roundoff errors. This occurs because the result of both predicates frequently depends on the sign of a small determinant.

Let $\mathbf{a} = (a_1, a_2)$, $\mathbf{b} = (b_1, b_2)$ and $\mathbf{c} = (c_1, c_2)$ be three points in the plane. If $l_{\vec{a}b}$ is the line through $\mathbf{a}$ and $\mathbf{b}$, then $\mathbf{c}$ is on the left side of $l_{\vec{a}b}$ as long as the determinant D holds (see Figure 8)

$$D = \begin{vmatrix} a_1 & a_2 & 1 \\ b_1 & b_2 & 1 \\ c_1 & c_2 & 1 \end{vmatrix} > 0.$$

Expanding the determinant, we obtain the sum

$$D = a_1b_2 + a_2c_1 + b_1c_2 - a_1c_2 - a_2b_1 - b_2c_1.$$

In case the points are vertices of an axis-aligned quadtree, the coordinates of the points are exactly represented machine numbers and the sum can be computed as the scalar product of $\mathbf{v} = (a_1, a_2, b_1, a_1, a_2, b_2)$ and $\mathbf{w} = (b_2, c_1, c_2, -c_2, -b_1, -c_1)$. Hence, we can implement the sign of $D = \mathbf{x} \cdot \mathbf{y}$ exactly using the exact scalar product in C-SXC.

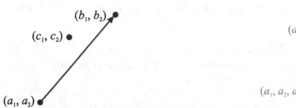

Fig. 8. The left-turn test   **Fig. 9.** The visibility test

Let $\mathbf{a}$, $\mathbf{b}$, $\mathbf{c}$ and $\mathbf{d}$ be four points in three dimensions. The orientation or visibility test yields a positive value if $\mathbf{d}$ lies below the oriented plane passing through the points $\mathbf{a}$, $\mathbf{b}$, and $\mathbf{c}$. Oriented plane means that the points appear in counterclockwise order when viewed from above the plane (see Figure 9). The test may be done using the sign of the following determinant

$$D = \begin{vmatrix} a_1 & a_2 & a_3 & 1 \\ b_1 & b_2 & b_3 & 1 \\ c_1 & c_2 & c_3 & 1 \\ d_1 & d_2 & d_3 & 1 \end{vmatrix}.$$

D can be computed as a sum of 24 monomials of the form $\pm x_1 y_2 z_3$. In order to get the sign correctly, we need to evaluate the determinant with relative error less than one. An evaluation of this dot product expression can be carried out with maximum accuracy using multiple-precision arithmetics provided in C-XSC.

5.4   Time Comparison of Distance Routines

The implemented tool has been upgraded with accurate distance routines for interval-based hierarchical structures. In addition, the computing time required for the distance computation between two hierarchical structures with the time needed for distance computation between their polyhedral enclosures has been compared [Grim06]. Owing to the reduction of points that define the object, a noticeable saving of the computing time as a result of building its polyhedral enclosure was expected. In case of the octree described in Figure 1 for example, there are 13 cube nodes each with 8 vertices and 12 rectangle facets to be considered during the distance computation instead of 11 vertices and 17 triangular facets in case of its convex enclosure. For the quadtree shown in Figure 4 there are 33 square nodes each with four vertices which would be relevant to the computation in comparision to 12 vertices of the corresponding polygon. Hence, the time reduction in case of moving objects is all the more significant and could be detected on several examples as a speed up of the computation by a factor of ten or greater es reported in [Grim06] .

6   Summary and Future Works

In this paper an adaptive algorithm for reliable computation of convex polyhedral enclosures of an interval-based hierarchical object representation is discussed. An efficiently construction of convex polyhedral enclosures affects advantageously further treatment or processing of the scene. Our approach uses the concept of extreme vertices of the nodes on the boundary and yields the smallest possible machine-representable convex polyhedron containing the octree representation of the given object at each level of the tree. Details of the reliable implementation of the convex hull algorithm are provided.

For the future we plan to transform the paradigm of our hull algorithm to higher dimensions. For this reason we seek feasible ways of combining the concept of extreme points with a method based on an interval linear solver to obtain a reliable polyhedral enclosure for higher dimensional hierarchical structures. The experimental data presented in [KrLa04] suggest that *Gaussian* elimination with *total pivoting* can be used as a solver that is robust to the dimension involved and to uncertainties in the coordinates of the points.

References

[BDH96]   Barber, C.B., Dobkin, D.P., Huhdanpaa, H.: The Quickhull algorithm for convex hull. ACM Transactions on Mathematical Software 22(4), 469–483 (1996)

[Dyll04]   Dyllong, E.: Akkurate Abstandsalgorithmen mit Ergebnisverifikation. Ph.D. thesis, University of Duisburg-Essen, VDI Reihe 20, Nr. 390, Düsseldorf (2004)

[DyLu06]   Dyllong, E., Luther, W.: Verified convex hull and distance computation for octree-encoded objects. Journal of Computational and Applied Mathematics (2006)

[FaLu00]   Fausten, D., Luther, W.: Verifizierte Lösungen von nichtlinearen polynomialen Gleichungssystemen. Technical Report SM-DU-477, University of Duisburg (2000)

[Grim06]   Grimm, C.: Verläßliche Abstandsalgorithmen für intervallbasierte Octreemodelle und ihre konvexen Einschlüsse – Ein Effizienzvergleich. Diploma thesis, University of Duisburg (2006)

[HHKR95]   Hammer, R., Hocks, M., Kulisch, U., Ratz, D.: C++ Toolbox for Verified Computing. Basic Numerical Problems. Springer, Berlin (1995)

[KrLa04]   Krivsky, S., Lang, B.: Using Interval Arithmetic for Determining the Structure of Convex Hulls. Numerical Algorithms 37, 233–240 (2004)

[Moor66]   Moore, R.: Interval Analysis. Prentice Hall, Englewood Cliffs (1966)
[RaRo03]   Ratschek, H., Rokne, J.: Geometric Computations with Interval and New Robust Methods. Horwood Publishing, Chichester (2003)
[Same90]   Samet, H.: The Design and Analysis of Spatial Data Structures. Addison-Wesley Publishing Company, Reading (1990)
[Zhan05]   Zhang, Min.: Konvexe Einschlüsse von hierarchischen intervallbasierten Modellen. Diploma thesis, University of Duisburg-Essen (2005)
[WaZe80]   Walach, E., Zeheb, E.: Sign Test of Multivariate Real Polynomials. IEEE Trans. on Circuits and Systems 27(7), 619–625 (1980)
[JKDW01]   Jaulin, J., Kieffer, M., Didrit, O., Walter, E.: Applied Interval Analysis. Springer, London (2001)

Real Algebraic Numbers:
Complexity Analysis and Experimentation

Ioannis Z. Emiris[1], Bernard Mourrain[2], and Elias P. Tsigaridas[1]

[1] Department of Informatics and Telecommunications
National Kapodistrian University of Athens, Hellas
{emiris,et}@di.uoa.gr
[2] GALAAD, Inria
Sophia-Antipolis, France
mourrain@sophia.inria.fr

Abstract. We present algorithmic, complexity and implementation results concerning real root isolation of a polynomial of degree d, with
integer coefficients of bit size $\leq \tau$, using Sturm (-Habicht) sequences and
the Bernstein subdivision solver. In particular, we unify and simplify the
analysis of both methods and we give an asymptotic complexity bound of
$\widetilde{\mathcal{O}}_B(d^4\tau^2)$. This matches the best known bounds for binary subdivision
solvers. Moreover, we generalize this to cover the non square-free polynomials and show that within the same complexity we can also compute the
multiplicities of the roots. We also consider algorithms for sign evaluation, comparison of real algebraic numbers and simultaneous inequalities,
and we improve the known bounds at least by a factor of d.

Finally, we present our C++ implementation in SYNAPS and some preliminary experiments on various data sets.

1 Introduction

The representation and manipulation of shapes is important in many applications, e.g. CAGD, non linear computational geometry, robotics and molecular
biology. The usual underlying models for these shapes are parametrized patches
of rational surfaces, BSplines, natural quadrics and implicit algebraic curves or
surfaces. Geometric processing on these objects, e.g. computing boundary representations [8], arrangements [18,38], Voronoi diagram of curved objects [21],
requires the intensive use of polynomial solvers and computations with algebraic
numbers. In such applications, a geometric model may involve several thousands
of algebraic primitives. Their manipulations involve the computation of intersection points, of predicates on these intersection points (such as the comparison of
coordinates), of the sign of polynomial expressions at these points (such as the
sign of a polynomial which defines the boundary of an object). The coordinates
of these intersection points, which are the solutions of polynomial equations, are
real algebraic numbers that we need to manipulate efficiently.

The objective of this paper is to give an overview of effective computations with
real algebraic numbers, which unify, simplify and improve previous approaches.

P. Hertling et al. (Eds.): Real Number Algorithms, LNCS 5045, pp. 57–82, 2008.
© Springer-Verlag Berlin Heidelberg 2008

We will tackle both complexity analysis and practical issues. We consider two algorithms for real root isolation of univariate integer polynomials, one based on Sturm sequences and one based on Descartes' rule of sign and we will put both under the general concept of a subdivision solver. We will also analyse algorithms for sign evaluation, comparison of real algebraic numbers and the problem of simultaneous inequalities.

Our aim is to provide better insights on these algorithms and better bounds on their complexity. For the analysis we consider the bit complexity model which is more realistic than the arithmetic one in the problems we are interested in. Our algorithms are essentially output sensitive, since they depend not only on the input bit size, but also on the actual separation bound, as we will see.

Notation. In what follows $\mathcal{O}_B$ means bit complexity and the $\widetilde{\mathcal{O}}_B$-notation means that we are ignoring logarithmic factors. For a polynomial $f \in \mathbb{Z}[X]$, $\deg(f)$ denotes its degree. By $\mathcal{L}(f)$ we denote an upper bound on the bit size of the coefficients of f (including a bit for the sign). For $a \in \mathbb{Q}$, $\mathcal{L}(a)$ is the maximum bit size of the numerator and the denominator. Let $\mathsf{M}(\tau)$ denote the bit complexity of multiplying two integers of bit size at most τ and $\mathsf{M}(d, \tau)$ denote the bit complexity of multiplying two univariate polynomials of degrees bounded by d and coefficient bit size at most τ. Using FFT, $\mathsf{M}(\tau) = \mathcal{O}_B(\tau \log^{c_1} \tau)$ and $\mathsf{M}(d, \tau) = \mathcal{O}_B(d\tau \log^{c_2}(d\tau))$ for suitable constants c_1, c_2.

Prior Works. Various algorithms exist for polynomial real root isolation, but most of them focus on square-free polynomials. There is a huge bibliography on the problem and the references cited in this paper are only the tip of the iceberg of the existing bibliography.

Collins and Akritas [10] introduced a subdivision-based real root isolation algorithm that relies on Descartes' rule of sign (we call it *Descartes solver* from now on) and derived a bound of $\widetilde{\mathcal{O}}_B(d^6 \tau^2)$. Johnson [27] improved the bound to $\widetilde{\mathcal{O}}_B(d^5 \tau^2)$ without using fast Taylor shifts [48], whereas a gap in his proof was corrected by Krandick [29]. The latter also presented a different way of traversing the subdivision tree. Rouillier and Zimmermann (c.f. [43] and references therein) presented a unified approach with optimal memory management for various variants of the Descartes solver.

An algorithm (we call it *Bernstein solver* from now on) that is based on a combination of Descartes' rule and on the properties of the Bernstein basis was first introduced by Lane and Riesenfeld [31] and a bound on its complexity was first obtained by Mourrain et al [39]. The interested reader may also refer to Mourrain et al [37] for a variant with optimal memory management and the connection to Descartes solver. In the same context, Eigenwillig et al [15] proposed a randomized algorithm for square-free polynomials with bit stream coefficients. The complexity of all these algorithms was bounded by $\widetilde{\mathcal{O}}_B(d^6 \tau^2)$. Recently, it was proven that the number of steps of both Descartes and Bernstein solver is $\widetilde{\mathcal{O}}_B(d\tau)$ [16,46], which is the crucial step for obtaining a $\widetilde{\mathcal{O}}_B(d^4 \tau^2)$ bound for both solvers [16], provided the polynomials are square-free.

Another algorithm, also based on Descartes' rule of sign but which does not rely on subdivision schemes, is the CF algorithm [1,47]. It exploits the continued fraction expansion of the real roots of a polynomial in order to isolate them. The expected bit complexity of this algorithm is $\widetilde{\mathcal{O}}_B(d^4\tau^2)$ [47].

If we restrict ourselves to real root isolation using Sturm (or Sturm-Habicht) sequences (we call it *Sturm solver* from now on) the first complete complexity analysis is probably due to Heindel [26]; see also [11], who obtained a complexity of $\widetilde{\mathcal{O}}_B(d^7\tau^3)$. Du et al [14], giving an amortized-like argument for the number of subdivisions, obtained a complexity of $\widetilde{\mathcal{O}}_B(d^4\tau^2)$, for square-free polynomials.

Another family of solvers (that we call numerical), compute an approximation of all the roots (real and complex) of a polynomial up to a desired accuracy (see e.g. [45,40]). They are based on the construction of balanced splitting circles in the complex plane and achieve the quasi-optimal complexity bound $\widetilde{\mathcal{O}}_B(d^3\tau)$, if we want to isolate the roots. However, performance in practice does not always agree with that predicted by asymptotic analysis. Let us also mention the Aberth solver [5,6], which has unknown (bit) complexity, but is very efficient in practice.

For sign evaluation and comparison as well as computations with real algebraic numbers the reader may refer to [42]. In [19] for degree ≤ 4, it is proved that these operations can be performed in $\mathcal{O}(1)$, or $\widetilde{\mathcal{O}}_B(\tau)$. For the problem of simultaneous inequalities (SI), we are interested in computing the (number of) real roots of a polynomial f, such that n other polynomials achieve specific sign conditions, where the degree of all the polynomials is bounded by d and their bit size by τ. Ben-Or, Kozen and Reif [3] presented the BKR algorithm for SI and Canny [7] improved it in the univariate case (by a factor) achieving $\mathcal{O}(n(m\,d\log(m)\log^2(d) + m^{2.376}))$ arithmetic complexity, where m is the number of real roots of f. Coste and Roy [12] introduced Thom's encoding for the real roots of a polynomial and SI in this encoding (see also [44]). Their approach is purely symbolic and works over arbitrary real closed fields. They state a complexity of $\widetilde{\mathcal{O}}_B(N^8)$, using fast multiplication algorithms but not fast computations and evaluation of polynomial sequences, where $N \geq n, d, \tau$. Basu, Pollack and Roy [2] presented an algorithm for SI where the real algebraic numbers are in isolating interval representation, with complexity $\widetilde{\mathcal{O}}_B(nd^6\tau^2)$ or $\widetilde{\mathcal{O}}_B(N^9)$, that uses repeated refinements of the isolating intervals and does not assume fast multiplication algorithms.

Results. For the problem of real root isolation of a univariate polynomial, we consider the general concept of a subdivision solver, Fig. 1, that mimics the binary search algorithm. The Sturm and Bernstein solvers count differently the number of real roots of a polynomial in an interval; see Cor. 2.1 and Prop. 3.1, respectively. Moreover, the Sturm solver counts exactly the number of real roots while the Bernstein solver provides an overestimation. However, exploiting the fact that Descartes' rule of sign (Prop. 3.1) can not overestimate the number of real roots by more than the degree of the polynomial, both solvers can be put under the general concept of the subdivision solver of Fig. 1. With exactly the same arguments we can prove that they perform the same number of steps.

The analysis that we present, Prop. 5.2, unifies and simplifies significantly the previous approaches and applies as is to the Descartes solver, as well.

For the Sturm solver we present an algorithm with complexity $\widetilde{\mathcal{O}}_B(d^4\tau^2)$, that improves the result of Du et al [14], see also [13], by extending it to non square-free polynomials. We simplify significantly the proof (Th. 5.1) and unify it with the Bernstein approach. We also show that computing the multiplicities of the roots can be achieved within the same complexity bound.

For the Bernstein solver, we simplify the proof of Eigenwillig et al [16, 46] for the number of subdivisions by considering the subdivision tree at an earlier level and by using Th. 4.1 exactly as stated in [27, 30]. Thus, we arrive at the same bound for the Bernstein subdivision method (Th. 5.1) as in [16], but for polynomials which are not necessarily square-free.

Real root isolation is an important ingredient for the construction of algebraic numbers represented in isolating interval representation, see Def. 6.1. We analyze the complexity of comparison, sign evaluation and simultaneous inequalities (Sec. 6). Even though the algorithms for these operations are not new [2, 19, 42, 51], the results from real solving and optimal algorithms for polynomial remainder sequences, allow us to improve the complexity of all the algorithms, at least by a factor of d (Cor. 6.1, 6.2). For SI we prove a bound (Cor. 6.3) of $\widetilde{\mathcal{O}}_B(d^4\tau \max\{n, \tau\})$.

These algebraic operations ought to have efficient and generic implementations so that they can be used by other scientific communities. We present a package of SYNAPS [36] that provides these functionalities on real algebraic numbers and exploits various algorithmic and implementation techniques. Experimental results (Sec. 7) illustrate the advantages of the software and our implementation of various algorithms for real root isolation.

Our results extend directly to the bivariate case, i.e. real solving a polynomial system, sign evaluation of a bivariate polynomial evaluated over two algebraic numbers and SI in two variables. However, due to reasons of space, we cannot present these results here. The reader may refer to [22, 20].

Outline. In Sec. 2, we recall the main ingredients of the Sturm solver and analyse them in detail. Sec. 3 presents the ingredients of the Bernstein solver and their complexity. In Sec. 4, we present the general scheme for two subdivision algorithms based on Sturm-Habicht sequences and on the Bernstein basis representation, for real root isolation and computation of the multiplicities. The following section is devoted to the complexity analysis of both methods. Sec. 6 is devoted to operations with real algebraic numbers, i.e. comparison, sign evaluation and SI. Sec. 7 illustrates our implementation in SYNAPS and experiments concerning real root isolation on various data sets. Finally, we sketch our current and future work in Sec. 8.

2 Preliminaries for Sturm–Habicht Sequences

We recall here the main ingredients related to Sturm(-Habicht) sequence computations and their bit complexity.

Let $f = \sum_{k=0}^{p} f_k x^k, g = \sum_{k=0}^{q} g_k x^k \in \mathbb{Z}[x]$ where $\deg(f) = p \geq q = \deg(g)$ and $\mathcal{L}(f) = \mathcal{L}(g) = \tau$. We denote by $\mathrm{rem}(f, g)$ and $\mathrm{quo}(f, g)$ the remainder and the quotient, respectively, of the Euclidean division of f by g, in $\mathbb{Q}[x]$.

Definition 2.1. [32] *The* signed polynomial remainder sequence *of f and g, denoted by* **SPRS** (f, g)*, is the polynomial sequence*

$$R_0 = f, R_1 = g, R_2 = -\mathrm{rem}(f, g), \ldots, R_k = -\mathrm{rem}(R_{k-2}, R_{k-1}),$$

with k the minimum index such that $\mathrm{rem}(R_{k-1}, R_k) = 0$. The quotient sequence *of f and g is the polynomial sequence $\{Q_i\}_{0 \leq i \leq k-1}$, where $Q_i = \mathrm{quo}(R_i, R_{i+1})$ and the* quotient boot *is $(Q_0, Q_1, \ldots, Q_{k-1}, R_k)$.*

There is a huge bibliography on signed polynomial remainder sequences (c.f. [2, 49, 51] and references therein). Gathen and Lucking [50] presented a unified approach to subresultants, while El Kahoui [17] studied the subresultants in arbitrary commutative rings. For the Sturm-Habicht (or Sylvester-Habicht) sequences the reader may refer to González-Vega et al [24], see also [2, 32, 33].

In this paper we consider the Sturm-Habicht sequence of f and g, i.e. **StHa**(f, g), which contains polynomials proportional to the polynomials in **SPRS** (f, g). Sturm-Habicht sequences achieve better bounds on the bit size of the coefficients.

Let M_j be the matrix which has as rows the coefficient vectors of the polynomials $f x^{q-1-j}$, $f x^{q-2-j}$, $\ldots, f x, f$, $g, g x, \ldots, g x^{p-2-j}, g x^{p-1-j}$ with respect to the monomial basis $x^{p+q-1-j}$, $x^{p+q-2-j}, \ldots, x, 1$. The dimension of M_j is $(p + q - 2j) \times (p + q - j)$. For $l = 0, \ldots, p + q - 1 - j$ let M_j^l be the square matrix of dimension $(p + q - 2j) \times (p + q - 2j)$ obtained by taking the first $p + q - 1 - 2j$ columns and the $l + (p + q - 2j)$ column of M_j.

Definition 2.2. *The Sturm-Habicht sequence of f and g, is the sequence*

$$\mathbf{StHa}(f, g) = (H_p = H_p(f, g), \ldots, H_0 = H_0(f, g)),$$

where $H_p = f, H_{p-1} = g, H_j = \delta_j \sum_{l=0}^{j} \det(M_j^l) x^l$ and $\delta_j = (-1)^{(p+q-j)(p+q-j-1)/2}$. The sequence of principal Sturm-Habicht coefficients $(h_p = h_p(f, g), \ldots, h_0(f, g))$ is defined as $h_p = 1$ and h_j is the coefficient of x^j in the polynomial H_j, for $0 \leq j \leq p$. When $h_j = 0$ for some j then the sequence is called defective, *otherwise* non-defective.

If **StHa**(f, g) is non-defective then it coincides up to sign with the classical subresultant sequence introduced by Collins [9], see also [51]. However, in the defective case, one can have better control on the bit size of the coefficients in the sequence, see e.g. [32, 33].

Theorem 2.1. [2, 41, 33, 49] *There is an algorithm that computes* **StHa**(f, g) *in $\mathcal{O}_B(pq \, \mathsf{M}(p\tau))$, or $\widetilde{\mathcal{O}}_B(p^2 q \tau)$. Moreover, $\mathcal{L}(H_j(f, g)) = \mathcal{O}(p\tau)$.*

Let the quotient boot that corresponds to **StHa**(f, g), be **StHaQ**$(f, g) = (Q_0, Q_1, \ldots, Q_{k-1}, H_k)$. The number of coefficients in **StHaQ**(f, g) is $\mathcal{O}(q)$ and their bit size is $\mathcal{O}(p\tau)$ [2, 41].

Theorem 2.2. [2, 32, 41, 49] *The quotient boot, the resultant and the gcd of f and g, can be computed in $\mathcal{O}_B(q \log(q) \, \mathsf{M} \, (p\tau))$ or $\widetilde{\mathcal{O}}_B(pq\tau)$.*

Theorem 2.3. [32, 41] *There is an algorithm that computes the evaluation of $\mathbf{StHa}(f, g)$ over a number a, where $\mathsf{a} \in \mathbb{Q} \cup \{\pm\infty\}$ and $\mathcal{L}(\mathsf{a}) = \sigma$ with complexity $\mathcal{O}_B(q \log(q) \, \mathsf{M} \, (\max\{p\tau, q\sigma\}))$ or $\mathcal{O}_B(q\mathsf{M} \, (\max (p\tau, q\sigma)))$ if $\mathbf{StHaQ}(f, g)$ is already computed. In both cases the complexity is $\widetilde{\mathcal{O}}_B(q \max\{p\tau, q\sigma\})$.*

In many cases, e.g. real root isolation, sign evaluation, comparison of real algebraic numbers, we need the evaluation of $\mathbf{StHa}(f, f')$ over a rational number of bit size $\mathcal{O}(p\tau)$. If we perform the evaluation by Horner's rule then for every polynomial in sequence, we must perform $\Omega(p)$ multiplications between numbers of bit size $\mathcal{O}(p\tau)$ and $\mathcal{O}(p^2\tau)$, thus the overall complexity is $\mathcal{O}_B(p^3\mathsf{M} \, (p\tau))$.

However, when we compute the complete $\mathbf{StHa}(f, f')$ in $\mathcal{O}_B(p^2\mathsf{M} \, (p\tau))$ (Th. 2.1), the quotient boot is computed implicitly [41, 2]. Thus, we can use the quotient boot in order to perform the evaluation even if we have already computed all the polynomials in the sequence. Notice also that the computation should be started by the last element of the quotient boot so as to avoid the costly evaluation of the first two polynomials in the sequence using Horner's scheme.

Even though this approach is optimal, it involves big constants in its complexity, thus it is not efficient in practice when the length of the sequence is not sufficiently big or when the sequence is defective, see e.g. [13]. Moreover, special techniques should be used for its implementation to avoid costly operations with rational numbers. So, as it is always the case with optimal algebraic algorithms, the implementation is far from a trivial task.

Theorem 2.4. [2] *The square-free part of f, i.e. f_{red}, can be computed from $\mathbf{StHa}(f, f')$, in $\mathcal{O}_B(p \log(p) \, \mathsf{M} \, (p\tau))$ or $\widetilde{\mathcal{O}}_B(p^2\tau)$, and $\mathcal{L}(f_{red}) = \mathcal{O}(p + \tau)$.*

Let $W_{(f,g)}(\mathsf{a})$ denote the number of modified sign changes of the evaluation of $\mathbf{StHa}(f, g)$ over a. Notice that $W_{(f,g)}(\mathsf{a})$ does not refer to the usual counting of sign variations, since special care should be taken for the presence of consecutive zeros [2, 24, 32].

Theorem 2.5. [2, 51, 42] *Let $f, g \in \mathbb{Z}[x]$ be relatively prime polynomials, where f is square-free and f' is the derivative of f and its leading coefficient $f_d > 0$. If $\mathsf{a} < \mathsf{b}$ are both non-roots of f and γ ranges over the roots of f in (a, b), then $W_{(f,g)}(\mathsf{a}) - W_{(f,g)}(\mathsf{b}) = \sum_{\gamma} \text{sign} \, (f'(\gamma)g(\gamma))$.*

Corollary 2.1. *If $g = f'$ then $\mathbf{StHa}(f, f')$ is a Sturm sequence and Th. 2.5 counts the number of distinct real roots of f in (a, b).*

For the Sturm solver $V(f, [\mathsf{a}, \mathsf{b}])$ will denote $V(f, [\mathsf{a}, \mathsf{b}]) = W_{(f,f')}(\mathsf{a}) - W_{(f,f')}(\mathsf{b})$.

3 Preliminaries for the Bernstein Basis Representation

In this section we present the main ingredients needed for the representation of polynomials in the Bernstein basis.

Let $\mathbb{R}[x]_d$ be the set of real polynomials of degree d. For $a < b \in \mathbb{R}$, we denote by $B_d^i(x; a, b) = \binom{d}{i} \frac{(x-a)^i(b-x)^{d-i}}{(b-a)^d}$ $(i = 0, \ldots, d)$ the Bernstein basis of $\mathbb{R}[x]_d$ on an interval $[a, b]$.

For any polynomial $f = \sum_{i=0}^d b_i B_d^i(x; a, b) \in \mathbb{R}[x]_d$ represented in the Bernstein basis, the coefficients $\boldsymbol{b} = (b_i)_{i=0,\ldots,d}$ are called *control coefficients* of f. We denote by $V(f, [a, b]) \equiv V(\boldsymbol{b})$, the number of sign changes in this sequence $\boldsymbol{b}$ (after removing zero coefficients).

The following theorem, which is a direct consequence of Descartes' rule, allows us to bound the number of real roots of f on the interval $[a, b]$

Proposition 3.1. [2] *The number N of real roots of f on (a, b) is bounded by $V(f, [a, b])$. Moreover $N \equiv V(f, [a, b]) \mod 2$.*

Given a polynomial f represented in the Bernstein basis on an interval $[a, b]$, de Casteljau's algorithm (see e.g. [2,37]), allows us to compute its representation in the Bernstein bases on the two sub-intervals, $I_L = [a, (1-t)a+tb]$ and $I_R = [(1-t)a+tb, b]$, where $0 \le t \le 1$. Namely, $\boldsymbol{b}_L = (b_0^i)_{i=0,\ldots,d}$ (resp. $\boldsymbol{b}_R = (b_i^{d-i})_{i=0,\ldots,d}$) are the control coefficients of f on I_L (resp. I_R), where $b_i^0 = b_i, i = 0, \ldots, d$, and

$$b_i^r = (1 - t) b_i^{r-1} + t b_{i+1}^{r-1}(t), \quad 0 \le i \le d - r, \ 0 \le r \le d. \tag{1}$$

In order to analyse the complexity of the de Casteljau algorithm we recall some polynomial transformations related to the Bernstein representation, see [37] for more details. Let $\mathbb{R}[x, y]_{[d]}$ be the set of homogeneous polynomials of degree d in (x, y). For any $f \in \mathbb{R}[x]_d$, we denote by $\bar{f}$ the homogenisation of f in degree d. For $\lambda \ne 0, \mu \in \mathbb{R}$, consider the following maps, $\mathbb{R}^2 \to \mathbb{R}^2$:

- $\rho : (x, y) \mapsto (y, x)$,
- $H_\lambda : (x, y) \mapsto (\lambda x, y)$, $H_\lambda' : (x, y) \mapsto (x, \lambda y)$,
- $T_\mu : (x, y) \mapsto (x - \mu y, y)$, $T_\mu' : (x, y) \mapsto (x, y - \mu x)$.

The composition of the previous maps with $\bar{f}$ induces invertible transformations on the set of polynomials of degree d. The corresponding maps for non-homogeneous polynomials, which we denote using the same names, are: $\forall f \in \mathbb{R}[x]_d$, $\rho(f) = x^d f(1/x)$, $H_\lambda(f) = f(\lambda x)$, $H_\lambda'(f) = f(\lambda^{-1}x)$, $T_\mu(f) = f(x - \mu)$ and $T_\mu'(f) = (1 - \mu x)^d f(\frac{x}{1-\mu x})$.

For any polynomial, $f(x) = \sum_{i=0}^d b_i B_d^i(x; a, b)$, we have

$$\rho \circ T_1 \circ \rho \circ H_{b-a} \circ T_{-a}(f) = \sum_{i=0}^d \binom{d}{i} b_i x^i.$$

Now consider another interval $[c, e]$. The representation of f in the Bernstein basis on $[c, e]$ is $f(x) = \sum_{i=0}^d b_i' B_d^i(x; c, e)$. The map which transforms f from its Bernstein representation on $[a, b]$ to its Bernstein representation on $[c, e]$, i.e. from $\sum_{i=0}^d \binom{d}{i} b_i x^i$ to $\sum_{i=0}^d \binom{d}{i} b_i' x^i$ is

$$\rho \circ T_1 \circ \rho \circ H_{e-c} \circ T_{-c} \circ T_a \circ H_{\frac{1}{b-a}} \circ \rho \circ T_{-1} \circ \rho = T_1' \circ H_{e-c} \circ T_{a-c} \circ H_{\frac{1}{b-a}} \circ T_{-1}'. \tag{2}$$

If we consider $[a, b] = [0, 1]$ and $[c, e] = [0, \frac{1}{2}]$ then map (2) becomes: $\rho \circ T_{-1} \circ \rho \circ H_{\frac{1}{2}} \circ \rho \circ T_1 \circ \rho$. And after simplifications, we obtain

$$\Delta_L : \overline{f} \mapsto \overline{f}(x + \frac{y}{2}, \frac{y}{2}) = \overline{f} \circ T_{-1} \circ H'_{\frac{1}{2}}. \tag{3}$$

If we consider the symmetric case, i.e. $[a, b] = [0, 1]$ and $[c, e] = [\frac{1}{2}, 1]$, then map (2) becomes: $\rho \circ T_{-1} \circ \rho \circ H_{\frac{1}{2}} \circ T_{-\frac{1}{2}} \circ \rho \circ T_1 \circ \rho$. It corresponds to the following map on the homogeneous polynomials:

$$\Delta_R : \overline{f} \mapsto \overline{f}(\frac{x}{2}, \frac{x}{2} + y) = \overline{f} \circ T'_{-1} \circ H_{\frac{1}{2}}.$$

In both cases, multiplication by 2^d yields the maps

$$\overline{\Delta}_R : \overline{f} \mapsto \overline{f}(x, x + 2y) \quad \text{and} \quad \overline{\Delta}_L : \overline{f} \mapsto \overline{f}(2x + y, y), \tag{4}$$

that operate operate on polynomials with integer coefficients.

Proposition 3.2. *Let $(b_i)_{i=0,\ldots,d} \in \mathbb{Z}^{d+1}$ be the coefficients of a polynomial f in the Bernstein basis on the interval $[a, b]$, and let ν be a bound on the bit size of coefficients. The complexity of computing the Bernstein coefficients of f on the two sub-intervals $[a, \frac{a+b}{2}]$ and $[\frac{a+b}{2}, b]$ is bounded by $\widetilde{\mathcal{O}}_B(d(d+\nu))$ and their bit size is bounded by $d + \nu$.*

Proof. Using the de Casteljau scheme, Eq. (1) using $t = \frac{1}{2}$, we prove by induction that the coefficients $b_i^r = \frac{(b_i^{r-1} + b_{i+1}^{r-1})}{2}$ are of the form $\frac{\overline{b}_i^r}{2^r}$, where $\overline{b}_i^r \in \mathbb{Z}$ is of bit size $\leq \nu + r$. Reducing to the same denominator 2^d, we obtain integer coefficients of bit size $\leq \nu + d$.

We denote by ν' the bit size of the coefficients $\left(\binom{d}{i} b_i\right)_{i=0,\ldots,d}$ where $(b_i)_{i=0,\ldots,d}$ are the coefficients of f in the Bernstein basis $(B_d^i(x; a, b))_{i=0,\ldots,d}$. Notice that $\nu' \leq \nu + d$.

In order to compute the coefficients of f on $[a, \frac{a+b}{2}]$ and $[\frac{a+b}{2}, b]$, we apply the same operations as when we compute the coefficients of a polynomial in the Bernstein basis on $[0, \frac{1}{2}]$ and $[\frac{1}{2}, 1]$, given its coefficients in the Bernstein basis on $[0, 1]$.

According to (3), applying de Casteljau's algorithm corresponds first to multiply by the binomial coefficients, then to shift, $y \mapsto x + y$, then to scale one variable of the homogeneous polynomial $\overline{f}$ by $\frac{1}{2}$, and finally to divide by the binomial coefficients[1].

Since the bit size of the binomial coefficients is bounded by d (their sum is 2^d), multiplying the b_i by them costs at most $\widetilde{\mathcal{O}}_B(d(\nu + d))$. Shifting by 1 a polynomial of degree d with coefficients of bit size $\leq \nu + d$ requires $\widetilde{\mathcal{O}}_B(d(d+\nu))$ bit operations [49, 48] and the resulting polynomial has (coefficient) bit size $\mathcal{O}(\nu + d)$. Consequently, scaling the variable of the (resulting) polynomial by $\frac{1}{2}$ and computing the quotient by the binomial coefficients costs $\widetilde{\mathcal{O}}_B(d(\nu + d))$.

---

[1] There is no need for the division step if we have to apply repeatedly the shift operation.

Therefore, the complexity of computing the Bernstein coefficients of f on the sub-interval $[a, \frac{a+b}{2}]$ is bounded by $\widetilde{\mathcal{O}}_B(d(\nu + d))$. By symmetry, inverting the order of the coefficients of f, we obtain the same bound for the coefficients of f on $[\frac{a+b}{2}, b]$, which ends the proof. $\qquad\qquad\square$

4 Subdivision Solver

Let $f = \sum_{i=0}^{d} a_i x^i \in \mathbb{Z}[x]$, such that $\deg(f) = d$ and $\mathcal{L}(f) = \tau$, and let f_{red} be its square-free part. We want to isolate the real roots of f, i.e. to compute intervals with rational endpoints that contain one and only one root of f, as well as the multiplicity of every real root.

In Fig. 1, we present the general scheme of the subdivision solver that we consider, augmented appropriately so that it also outputs the multiplicities. It uses an external function $V(f, I)$, which bounds the number of roots of f in the interval I. A real root isolation algorithm can be put under the subdivision solver concept of Fig. 1 if it can provide the $V(f, I)$ function. In the case of the Sturm solver, $V(f, I)$ returns the exact number (counted without multiplicities) of the real roots of f in I (Cor. 2.1). In the case of Bernstein solver, $V(f, I)$ is equal to the number of real roots of f in I (counted with multiplicities) modulo 2 (Prop. 3.1).

Separation Bounds. An important quantity for the analysis of the subdivision solvers is a bound on the minimal distance, $\mathrm{sep}(f)$, between the roots of a univariate polynomial f (also called *separation bound*), or more generally on the product of distances between roots. We recall here classical results, slightly adapted to our context. For the separation bound it is known, e.g. [2,34,51], that

---

ALGORITHM: Real Root Isolation
Input: A polynomial $f \in \mathbb{Z}[x]$, such that $\deg(f) = d$ and $\mathcal{L}(f) = \tau$.
Output: A list of intervals with rational endpoints, which contain one and only one real root of f and the multiplicity of every real root.

1. Compute the square-free part of f, i.e. f_{red}
2. Compute an interval $I_0 = (-B, B)$ with rational endpoints that contains all the real roots. Initialize a queue Q with I_0.
3. While Q is not empty do
    a) Pop an interval I from Q and compute $v := V(f, I)$.
    b) If $v = 0$, discard I.
    c) If $v = 1$, output I.
    d) If $v \geq 2$, split I into I_L and I_R and push them to Q.
4. Determine the multiplicities of the real roots, using the square-free factorization of f.

Fig. 1. Real root isolation subdivision algorithm

$\text{sep}(f) \geq d^{-\frac{d+2}{2}}(d+1)^{\frac{1-d}{2}}2^{\tau(1-d)}$, thus $\log(\text{sep}(f)) = \mathcal{O}(d\tau)$. The latter provides a bound on the bit size of the endpoints of the isolating intervals.

Recall, that Mahler's measure, see e.g. [34,51,2], of a polynomial f is $\mathcal{M}(f) = |a_d| \prod_{i=1}^{d} \max\{1, |\gamma_i|\}$, where a_d is the leading coefficient and γ_i are all the roots of f. We know that $\mathcal{M}(f) < 2^\tau \sqrt{d+1}$ [2,34]. Thus, the following inequality [2,34] holds:

$$\mathcal{M}(f_{red}) \leq \mathcal{M}(f) < 2^\tau \sqrt{d+1}. \tag{5}$$

For the minimum distance between two consecutive real roots of a square-free polynomial, Davenport-Mahler bound is known [13] (see also [27,30]). The conditions for this bound to hold were generalized by Du et al [14]. A similar bound, with less strict hypotheses, also appeared in [35]. Using (5) we can provide a bound similar to [27] for non square-free polynomials.

Theorem 4.1 (Davenport-Mahler bound revisited). *Let* $A = \{\alpha_1, \ldots, \alpha_k\}$ *and* $B = \{\beta_1, \ldots, \beta_k\}$ *be subsets of* distinct *(complex) roots of* f *(not necessarily square-free) such that* $\beta_i \notin \{\alpha_1, \ldots, \alpha_i\}$ *and* $|\beta_i| \leq |\alpha_i|$*, for all* $i \in \{1, \ldots, k\}$*. Then*

$$\prod_{i=1}^{k} |\alpha_i - \beta_i| \geq \mathcal{M}(f)^{-d+1} d^{-\frac{d}{2}} \left(\frac{\sqrt{3}}{d}\right)^k.$$

The bound also holds when $\alpha_1 > \beta_1 = \alpha_2 > \beta_2 = \ldots \alpha_k > \beta_k := \alpha_{k+1}$*, are distinct real roots of* f*.*

Proof. Use [27] and (5). □

5   Complexity Analysis of Real Root Isolation

In this section, we bound the number of bit operations for isolating the real roots of a polynomial using the Sturm and the Bernstein solver. In what follows we present in detail the complexity of each step of the subdivision algorithm (see Fig. 1).

We consider the tree associated with a run of the subdivision algorithm on a polynomial f. Each node represents an interval. The root of the tree corresponds to the initial interval $I_0 = [a, b]$. The algorithm splits every interval which is not a leaf of the tree to two equal sub-intervals. The depth of a node of the tree (associated with an interval I) is $\log_2(|I_0|/|I|)$. This is also the number of subdivisions performed to obtain the sub-interval I from I_0. The number of steps (subdivisions) that the algorithm performs equals the total number of nodes of the subdivision tree, or in other words equals the number of intervals (sub-intervals of I_0) that are tested. Notice that the arguments are independent of the subdivision solver, Sturm or Bernstein in this paper, that we use.

5.1   Square-Free Factorisation [Step 1]

The computation of f_{red} can be done in $\widetilde{\mathcal{O}}_B(d^2\tau)$ (Th. 2.4). Notice that $\mathcal{L}(f_{red}) = \mathcal{O}(d+\tau)$. In order to simplify notation, we assume that $d = \mathcal{O}(\tau)$, thus

$\mathcal{L}(f_{red}) = \mathcal{O}(\tau)$. Notice also that after this computation, the Sturm-Habicht sequence $\mathbf{StHa}(f)$ is available. We do not need the complete sequence but only the quotient boot, thus this computation can be done in $\widetilde{\mathcal{O}}_B(d^2\tau)$ (Th. 2.2). However, we may also assume that the complete sequence is computed, with complexity $\widetilde{\mathcal{O}}_B(d^3\tau)$ (Th. 2.1), since this step is not the bottleneck of the algorithm.

5.2 Root Bounds and Initialization [Step 2]

The Cauchy bound states that if α is a real root of f then $|\alpha| \le B = 1 + \max\left(\left|\frac{a_{d-1}}{a_d}\right|, \ldots, \left|\frac{a_0}{a_d}\right|\right) \le 2^\tau$. Various upper bounds are known for the absolute value of the real roots [2, 51, 49]. However, asymptotically the bit size of all the bounds is the same, i.e. $B \le 2^\tau$. Thus, we can take $I_0 = [a, b]$, with $a \le -2^\tau$ and $b \ge 2^\tau$.

For the Sturm solver, before starting the main loop, we may have to compute the Sturm-Habicht sequence of f, which costs $\widetilde{\mathcal{O}}_B(d^3\tau)$ (Th. 2.1).

For the Bernstein solver, we have to represent f_{red} in the Bernstein basis on $[a, b]$. This can be done in $\mathcal{O}(d^2)$ arithmetic operations and it produces coefficients of size $\mathcal{O}(d(d+\tau))$. The cost of this transformation is bounded by $\widetilde{\mathcal{O}}_B(d^3(d+\tau))$.

In both methods, the initialization step can be done in $\widetilde{\mathcal{O}}_B(d^3(d+\tau))$.

5.3 Computing $V(f, I)$ and Splitting [Steps 3.a-d]

Suppose that the algorithm is at depth h of the subdivision tree. The tested interval, say I, has endpoints of bit size bounded by $\tau + h$, since each subdivision step increases the bit size by one.

Using the Sturm solver, we compute $V(f, I)$, Cor. 2.1, by evaluating $\mathbf{StHa}(f)$ over the endpoints of I. The cost of the evaluation is $\widetilde{\mathcal{O}}_B(d^2(\tau + h))$ (Th. 2.3). Then we split I, i.e. compute the middle point of it, in $\mathcal{O}_B(\tau + h)$.

Using the Bernstein solver, we compute $V(f, I)$ by counting the number of sign variations in the control coefficients of f in I. This can be done in $\mathcal{O}(d)$ operations. We denote by $\tau_0 = \mathcal{O}(d(d+\tau))$ (Sec. 5.2) a bound on the bit size of the coefficients of f in the Bernstein basis on the interval I_0. By Prop. 3.2, since we performed h subdivisions so far, starting from a polynomial with coefficients of bit size τ_0, the coefficients of f on I are of bit size $\tau_0 + dh$ and the complexity of the splitting operation is in $\widetilde{\mathcal{O}}_B(d(d + \tau_0 + dh)) = \widetilde{\mathcal{O}}_B(d^2(d + \tau + h))$ (Sec. 5.2).

Consequently, for both solvers, steps 3.a-d can be performed with complexity $\widetilde{\mathcal{O}}_B(d^2(d + \tau + h))$.

5.4 Subdivision Tree Analysis [Step 3]

In this section, we analyse the total number of subdivisions.

A bound on the number of the subdivision steps was derived in [30, Th. 5.5, 5.6], where in Rem. 5.7 the authors state: "The theorem (5.6) implies the dominance relations $hk \preceq n\log(nd)$ and $h \preceq n\log(nd)$ which can be used in an

asymptotic analysis of Algorithm 1 when the ring S of the coefficients is $\mathbb{Z}$",
where k is the number of internal nodes of depth h in the recursion tree of the
subdivision algorithm based on Descartes' rule, n is the degree and d is the
Euclidean norm of the polynomial. In [46, Th. 5], a $\mathcal{O}(d\tau + d\log d)$ bound is
derived and, later on, [16] proved optimality under the mild assumption that
$\tau = \Omega(\log d)$.

Our arguments for bounding the number of the subdivision steps are a combination and/or simplification of the arguments in [30,14,46]. Our proof (prop. 5.2)
is simpler than the one in [16,46] since the handling of the subdivision tree stops
at an earlier level and we use Th. 4.1 (as stated in [27] and [30]) without any
modifications. We also simplify substantially the proof of [14], for Sturm solver.

We denote by $\mathcal{I}$ the set of intervals which are the parents of two leaves in the
subdivision tree in Sturm (resp. Bernstein) solver. By construction, for $I \in \mathcal{I}$,
$V(f, I) \geq 2$ but for the two sub-intervals I_L, I_R of I, $V(f, I_L)$ and $V(f, I_R)$ are
in $\{0, 1\}$, since these intervals are leaves of the subdivision tree. Moreover, for
the Sturm solver, it holds that $V(f, I) = 2$ and $V(f, I_L) = V(f, I_R) = 1$.

Notice that $|\mathcal{I}|$ is less than $V(f, I_0)$, since at each subdivision the sum of the
variations of f on all the intervals cannot increase, for both methods (see [39,37]
for the Bernstein solver). In particular, we have $|\mathcal{I}| \leq d$.

Let α_I (or β_I) be a, possibly complex, root of f whose real part is in I.

Proposition 5.1. *If $I \in \mathcal{I}$ then there exist two distinct (complex) roots $\alpha_I \neq \beta_I$
of f such that $|\alpha_I - \beta_I| < 2|I|$.*

Proof. Consider an interval $I \in \mathcal{I}$ which contains two leaves I_L, I_R of the subdivision tree. We have the following possibilities for the sign variation of f on the
two sub-intervals I_L, I_R:

$(1, 1)$: for both solvers, there are two distinct real roots $\alpha \in I_L, \beta \in I_R$ in I and
$|\alpha - \beta| \leq |I|$. This is the only case, which can happen in Sturm's solver.
$(0, 0)$: this may happen only in the Bernstein solver. Since the sign variations
are $V(f, I) \geq 2$, the first circle theorem [37,2,30] implies that there exist two
complex conjugate roots $\beta, \overline{\beta}$ in the disc $D(m(I), \frac{|I|}{2})$. Therefore, $|\beta - \overline{\beta}| \leq |I|$.
$(1, 0)$ **or** $(0, 1)$: this may also happen only in the Bernstein solver. There is
a real root α in I. Since $V(f, I) \geq 2$, the second circle theorem [37, 2, 30]
implies that there exists two complex conjugate roots $\beta, \overline{\beta}$ in the union of
the discs $D(m(I) \pm \frac{1}{2\sqrt{3}}\mathbf{i}|I|, \frac{1}{\sqrt{3}}|I|)$, which is contained in a disc of diameter
$2|I|$. Therefore $|\beta - \alpha| < 2|I|$.

Thus the proposition holds.   □

In addition, we can prove the following result.

Lemma 5.1. *Let $\{\alpha_I, \beta_I\} \cap \{\alpha_{I'}, \beta_{I'}\} \neq \emptyset$, then $I \cap I' \neq \emptyset$.*

Proof. For the Sturm solver, this property is clear since $\alpha_I, \beta_I \in I$.

Let us consider the Bernstein subdivision method. We suppose that $I \cap I' = \emptyset$.
Without loss of generality, we can assume in the proof that $I \neq I'$, $|I| \geq |I'|$,
and that $\forall x \in I, \forall y \in I', x \leq y$.

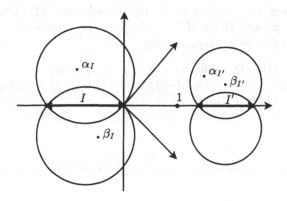

Fig. 2. Two disjoint intervals I and I' and the associated two circles to each of them

Since the intervals are obtained by binary subdivision, we can assume that the distance between I and I' is at least $|I'|$. Using scaling and translation, we can assume that the right endpoint of I is 0 and that $I' = [1+u, 2+u]$ ($u \geq 0$). The tangents to the larger circles, containing I and the roots α_I and β_I at $(0,0)$, are $\frac{\sqrt{3}}{2}x \pm \frac{y}{2} = 0$ (see Fig. 2). We denote by R_I the union of the corresponding discs, so that $\alpha_I, \beta_I \in R_I$.

The center of the discs, whose union $R_{I'}$ contains the roots $\alpha_{I'}$ and $\beta_{I'}$, are $(\frac{3}{2} + u, \pm\frac{\sqrt{3}}{6})$ and their radius is $\frac{\sqrt{3}}{3}$ (see Fig. 2). A direct computation of the distance between these centers and the two tangent lines shows that $R_I \cap R_{I'} = \emptyset$. Consequently, if $I \cap I' = \emptyset$, then we conclude that $\{\alpha_I, \beta_I\} \cap \{\alpha_{I'}, \beta_{I'}\} = \emptyset$. □

Let us number the intervals of $\mathcal{I}$ in increasing order and denote by $\mathcal{I}'$ the subset with an even index and by $\mathcal{I}''$ the subset with an odd index. By Lem. 5.1, the pairs $\{\alpha_I, \beta_I\}$ for $I \in \mathcal{I}'$ (resp. $\mathcal{I}''$) are disjoint. Thus, by Th. 4.1 (exchanging the role of α_I and β_I if necessary), we have

$$\prod_{I \in \mathcal{J}} |\alpha_I - \beta_I| \geq \mathcal{M}(f)^{-d+1} d^{-\frac{d}{2} - |\mathcal{J}|} \sqrt{3}^{|\mathcal{J}|}, \tag{6}$$

for $\mathcal{J} = \mathcal{I}'$ or $\mathcal{J} = \mathcal{I}''$. This is the key argument for the following result:

Proposition 5.2. *The number of subdivisions in both methods is in* $\mathcal{O}(d\tau + d\log(d))$.

Proof. The number N of subdivisions equals the number of internal nodes in the subdivision tree. It is less than the sum of the depth of I, for $I \in \mathcal{I}$:

$$N \leq \sum_{I \in \mathcal{I}} \log\frac{|b-a|}{|I|}$$
$$\leq |\mathcal{I}| \log|b - a| - \sum_{I \in \mathcal{I}} \log|I|$$
$$\leq |\mathcal{I}| \log|b - a| + |\mathcal{I}| - \sum_{I \in \mathcal{I}} \log|\alpha_I - \beta_I| \text{ (Prop. 5.1).}$$

By (6), we have $-\sum_{I \in \mathcal{I}'} \log|\alpha_I - \beta_I| \leq (d-1)\log(\mathcal{M}(f)) + (\frac{d}{2} + |\mathcal{I}'|)\log d - |\mathcal{I}'|\log\sqrt{3}$. A similar bound applies for $\mathcal{I}''$.

As we can take $\mathsf{a} = -2^\tau, \mathsf{b} = 2^\tau$ (by Cauchy bound), $\log\mathcal{M}(f) \leq \tau + \frac{1}{2}\log(d+1)$ (Eq. 5) and $|\mathcal{I}'| + |\mathcal{I}''| = |\mathcal{I}| \leq d$, the number of internal nodes, N, in the subdivision tree is bounded by

$$\begin{aligned} N &< |\mathcal{I}| + |\mathcal{I}|\log|\mathsf{b} - \mathsf{a}| - \sum_{I\in\mathcal{I}}\log|\alpha_I - \beta_I| \\ &\leq d + d(\tau+1) + (d-1)(2\tau + \log(d+1)) + 2\,d\log d \\ &= \mathcal{O}\left(d\tau + d\log d\right). \end{aligned}$$

□

Remark 5.1. The constant in this bound on the number of subdivisions can be divided by 2, in Sturm method, by applying directly Th. 4.1 to α_I, β_I for $I \in \mathcal{I}$.

5.5   Multiplicities [Step 4]

In order to compute the multiplicities of the roots, we compute the square-free factorization, i.e. a sequence of square-free coprime polynomials $(g_1, g_2, \ldots, g_m)$ with $f = g_1 g_2^2 \cdots g_m^m$ and $g_m \neq 1$. The algorithm of Yun [49] computes the square-free factorization in $\widetilde{\mathcal{O}}_B(d^2\tau)$. To be more specific the cost is twice the cost of the computation of $\mathbf{StHa}(f, f')$ [23]. Moreover, $\deg(g_i) = \delta_i \leq d$ and $\mathcal{L}(g_i) = O(d\tau)$ by Mignotte's bound [34], where $1 \leq i \leq m$.

At every isolating interval, one and only one g_i must have opposite signs at its endpoints, since g_i are square-free and pairwise coprime. If a g_i changes sign at an interval then the multiplicity of the real root that the interval contains is i. Each g_i can be evaluated over an isolating point in $\widetilde{\mathcal{O}}_B(\delta_i^2 d\tau)$, using Horner's rule. We can evaluate it over all the isolating points (there are at most $d+1$), in $\widetilde{\mathcal{O}}_B(\delta_i d^2\tau)$ [49, 51]. Since $\sum_{i=1}^m \delta_i \leq d$ the overall cost is $\widetilde{\mathcal{O}}_B(d^3\tau)$.

5.6   Complexity of Real Root Isolation

In this section we prove that the bit complexity bound for the two subdivision solvers is $\widetilde{\mathcal{O}}_B(d^4\tau^2)$:

Theorem 5.1. *Let $f \in \mathbb{Z}[x]$, with $\deg(f) = d$ and $\mathcal{L}(f) = \tau$, not necessarily square-free. We can isolate the real roots of f and determine their multiplicities using Sturm or Bernstein solver in $\widetilde{\mathcal{O}}_B(d^4\tau^2)$. Moreover, the endpoints of the isolating intervals have bit size bounded by $\mathcal{O}(d\tau)$.*

Proof. In order to isolate the real roots of f, we first compute its square-free part (step 1). This can be done in $\widetilde{\mathcal{O}}_B(d^2\tau)$ arithmetic operations and yields a polynomial f_{red}, which coefficients are of bit size bounded by $\mathcal{O}(d+\tau)$ (see Sec. 5.1). This step is not necessary for the Sturm solver.

The initialisation step costs $\widetilde{\mathcal{O}}_B(d^3(d+\tau))$ (Sec. 5.2).

Then we run the main loop of the subdivision algorithm. The cost of a subdivision step at level h is $\widetilde{\mathcal{O}}_B(d^2(d+\tau+h))$ (Sec. 5.3).

By Prop. 5.2, the number of subdivisions and the depth h of any node of the subdivision tree is $\widetilde{\mathcal{O}}(d\tau)$. Therefore, the overall complexity of both subdivision solvers is $\widetilde{\mathcal{O}}_B(d^2(d+\tau+d\tau)\,d\tau) = \widetilde{\mathcal{O}}_B(d^4\tau^2)$.   □

Remark 5.2. In Sec. 5.1 we assumed that $d = \mathcal{O}(\tau)$. If we drop this assumption then the complexity of real root isolation is $\widetilde{\mathcal{O}}_B(d^6 + d^5\tau + d^4\tau^2)$. If $N = \max\{d, \tau\}$ then in both cases the complexity bound is $\widetilde{\mathcal{O}}_B(N^6)$.

6   Real Algebraic Numbers

The real algebraic numbers, i.e. those real numbers that satisfy a polynomial equation with integer coefficients, form a real closed field denoted by $\mathbb{R}_{alg} = \overline{\mathbb{Q}}$. From all integer polynomials that have an algebraic number α as root, the primitive one (the gcd of the coefficients is 1) with the minimum degree is called *minimal*. The minimal polynomial is unique (up to a sign), primitive and irreducible [51]. Since we use Sturm-Habicht sequences, it suffices to deal with algebraic numbers, as roots of any square-free polynomial and not as roots of their minimal ones. In order to represent a real algebraic number we choose the *isolating interval representation*.

Definition 6.1. *The isolating-interval representation of real algebraic number* $\alpha \in \mathbb{R}_{alg}$ *is* $\alpha \cong (f(x), I)$*, where* $f(x) \in \mathbb{Z}[x]$ *is square-free and* $f(\alpha) = 0$*,* $I = [a, b]$*,* $a, b, \in \mathbb{Q}$*,* $\alpha \in I$ *and* f *has no other root in* I*.*

Using the results of Sec. 2 and 3 we can compute the isolating interval representation of all the real roots of a polynomial f, with $\deg(f) = d$ and $\mathcal{L}(f) = \tau$, in $\widetilde{\mathcal{O}}_B(d^4\tau^2)$ and the endpoints of the isolating intervals have bit size $\mathcal{O}(d\tau)$.

Comparison and sign evaluation. We can use Sturm-Habicht sequences in order to find the sign of a univariate polynomial, evaluated over a real algebraic number and to compare two algebraic numbers. We improve existing bounds by one factor.

Corollary 6.1. *Let* $g(x) \in \mathbb{Z}[x]$*, where* $\deg(g) \leq d$ *and* $\mathcal{L}(g) = \tau$*, and a real algebraic number* $\alpha \cong (f, [a, b])$*. We can compute* $\mathrm{sign}(g(\alpha))$ *in* $\widetilde{\mathcal{O}}_B(d^3\tau)$*.*

Proof. By Th. 2.5

$$W_{(f,g)}(a) - W_{(f,g)}(b) = \mathrm{sign}(g(\alpha)) \cdot \mathrm{sign}(f'(\alpha)),$$

and thus

$$\mathrm{sign}(g(\alpha)) = \left(W_{(f,g)}(a) - W_{(f,g)}(b)\right) \cdot \mathrm{sign}(f'(\alpha)).$$

Thus we need to perform two evaluations of **StHa**(f, g) over the endpoints of the isolating interval of α. The complexity of each is $\widetilde{\mathcal{O}}_B(d^3\tau)$ (Th. 2.3). The sign of $f'(\alpha)$ can be computed as

$$\mathrm{sign}(f'(\alpha)) = \mathrm{sign}(f(b)) - \mathrm{sign}(f(a)).$$

Notice that $f(a)f(b) < 0$, since γ is the unique real root of f in $[a, b]$. It reasonable to assume that the signs of f over the endpoints of the isolating interval are known. If this is not the case then we can evaluate f over them, using Horner's rule, with complexity $\widetilde{\mathcal{O}}_B(d^3\tau)$, since we need d multiplications of numbers of bit size $\mathcal{O}(d^2\tau)$.

We conclude that the overall complexity of the operation is $\widetilde{\mathcal{O}}_B(d^3\tau)$. $\qquad\square$

Corollary 6.2. *We can compare two real algebraic numbers in isolating interval representation in $\widetilde{\mathcal{O}}_B(d^3\tau)$.*

Proof. Let two algebraic numbers $\gamma_1 \cong (f_1(x), I_1)$ and $\gamma_2 \cong (f_2(x), I_2)$ where $I_1 = [a_1, b_1]$, $I_2 = [a_2, b_2]$. Let $J = I_1 \cap I_2$. When $J = \emptyset$, or only one of γ_1 and γ_2 belong to J, we can easily order the 2 algebraic numbers. If $\gamma_1, \gamma_2 \in J$, then $\gamma_1 \geq \gamma_2 \Leftrightarrow f_2(\gamma_1) \cdot f_2'(\gamma_2) \geq 0$. We obtain the sign of $f_2(\gamma_2)$, using Cor. 6.1, thus the complexity of comparison is $\widetilde{\mathcal{O}}_B(d^3\tau)$. $\qquad\square$

Simultaneous inequalities Let $f, A_1, \ldots, A_{n_1}, B_1, \ldots, B_{n_2}, C_1, \ldots, C_{n_3} \in \mathbb{Z}[x]$, with degree bounded by d and coefficient bit size bounded by τ. We wish to compute the number of and the real roots, γ, of f such that $A_i(\gamma) > 0$, $B_j(\gamma) < 0$ and $C_k(\gamma) = 0$ and $1 \leq i \leq n_1, 1 \leq j \leq n_2, 1 \leq k \leq n_3$. Let $n = n_1 + n_2 + n_3$.

Corollary 6.3. *There is an algorithm that solves the problem of simultaneous inequalities (SI) in $\widetilde{\mathcal{O}}_B(d^4\tau \max\{n, \tau\})$.*

Proof. First we compute the isolating interval representation of all the real roots of f in $\widetilde{\mathcal{O}}_B(d^4\tau^2)$ (Th. 5.1). There are at most d. For every real root γ of f, for every polynomial A_i, B_j, C_k we compute $\text{sign}(A_i(\gamma))$, $\text{sign}(B_j(\gamma))$ and $\text{sign}(C_k(\gamma))$. Sign determination costs $\widetilde{\mathcal{O}}_B(d^3\tau)$ (Cor. 6.1) and in the worst case we must compute n of them. Thus the overall cost is $\widetilde{\mathcal{O}}_B(\max\{nd^4\tau, d^4\tau^2\})$. $\qquad\square$

This improves the known bounds by one or two factors in the bit complexity model.

7 Implementation and Experiments

In this section, we describe the package for algebraic numbers available in the library SYNAPS [36]. The purpose of this package is to provide a set of tools, for the manipulation of algebraic numbers, needed in applications such as Geometric modeling, and non linear computational geometry. In the problems encountered in these domains, the degree of the involved polynomials is not necessarily very high (≤ 100), but geometric operations require an intensive use of algebraic solvers.

For this reason, in this section we focus on univariate equations of *small degree* in opposition with the first sections, but the input bit size may be beyond machine precision. We analyse the behavior of our solvers, in this range of problems which appear in our geometric applications and for which the asymptotic bounds may not be pertinent indicators. We do not consider large degree problems, where memory management issues might influence the solving strategy.

In SYNAPS, there are several solver classes; their interface is as follows:

```
template < class T > struct SOLVER {
  typedef NumberTraits<T>::RT    RT;
  typedef NumberTraits<T>::FT    FT;
  typedef NumberTraits<T>::FIT   FIT;

  typedef UPolDse<T>             Poly;
  typedef root_of<T, Poly>       RO_t;
  ... };
```

where `RT` is the ring number type (typically $\mathbb{Z}$), `FT` is the field number type (typically $\mathbb{Q}$), `FIT` is the interval type, `Poly` is the univariate polynomial, `RO_t` is the type for real algebraic numbers, etc.

Algebraic numbers are of the form:

```
template <class T, class UPOL=UPolDse<T> >
struct root_of {
  NumberTraits<T>::Interval_t interval_;
  UPOL polynomial_;
  ... };
```

parametrized by the type of coefficients and univariate polynomials. This allows flexibility and an easy parametrisation of the code.

In order to construct a real algebraic number the user may select from several different univariate solvers, that we are going to describe hereafter. The other functionalities that we provide are the comparison, `bool compare(const RO_t& a, const RO_t& b)` and the evaluation of signs `int sign_at(const Poly& P, const RO_t& a)`, based on interval evaluation and if necessary on the computation of Sturm-Habicht sequences. This involves several additional functions for computing subresultant sequences with various methods (Euclidean, Subresultants, Sturm-Habicht, etc), for computing the GCD, the square-free part, etc. We also provide the four operations, i.e. $\{+, -, *, /\}$, of `RO_t` with `RT`'s (integer type) and `FT`'s (rational type).

Perhaps the most important operation is the construction of real algebraic numbers, i.e. real root isolation of univariate polynomials. Several subdivision solvers have been tested for the construction of these algebraic numbers. We report here on the following solvers:

(S_1) `solve( f,IslSturm<ZZ>());`
(S_2) `solve( f,IslBzInteger<QQ>());`
(S_3) `solve( f,IslBzBdgSturm<QQ>());`

These solvers take as input, polynomials with integer or rational coefficients and output intervals with rational endpoints. All use the same initial interval.

S_1 (`IslSturmQQ` in the plots) is based on the construction of the Sturm-Habicht sequence and subdivisions, using rational numbers or large integers provided by the library GMP.

S_2 (`IslBzIntegerZZ` in the plots) is an implementation of the Bernstein subdivision solver, using integer coefficients. The polynomial is converted to the

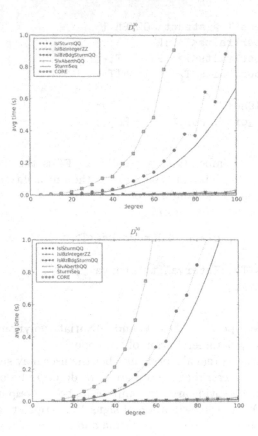

Fig. 3. Random polynomials of bit size 30 (top) and 50 (bottom) bits

Bernstein representation on the initial interval, using rational arithmetic. Then, the coefficients are reduced to the same denominator, and the numerators are taken. Finally, the integer version of de Casteljau's algorithm, see Eq. (4), is applied at each subdivision step.

S_3 (IslBzBdgSturmQQ in the plots) is a combination of two solvers. In a first part, the polynomial is converted to the Bernstein representation on the initial interval, using rational arithmetic and its coefficients are rounded to double intervals. The Bernstein subdivision solver is applied on this interval representation and stops when it certifies the isolation of a root or when it is not possible to decide the existence and uniqueness of a root from the "sign" $(-, +, ?)$ of the interval coefficients. In this case, in order to complete the isolation process, the solver S_2 is used on the intervals which are suspect.

We also compare with the time needed for computing the square-free factorization of the tested polynomial (SturmSeq in the plots). Our implementation is based on Yun's algorithm and Sturm-Habicht sequences.

We test against CORE [28] (CORE in the plots), and MPSOLVE, a numerical solver based on Aberth's method [5] and implemented by G. Fiorentino and D.

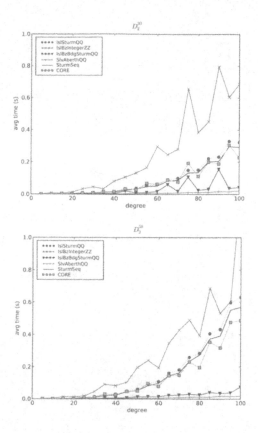

Fig. 4. Random polynomials with multiple roots of bit size 30 (top) and 50 (bottom) bits

Bini [6] (`SlvAberthQQ` in the plots), that are open source tools with real solving capabilities. CORE implements the Sturm solver. In order to isolate the real roots of a polynomial first it computes its square-free part and then performs isolation. MPSOLVE implements an iterative method for approximating the roots of a polynomial and the implementation is based on multiprecision floats. We set the output precision of the algorithm to 30 digits. The code of MPSOLVE is integrated in SYNAPS and called similarly to the other solver (S_1, S_2 and S_3), i.e `solve( f, Aberth<RR>())`.

Other libraries such as [25], or EXACUS with Leda [4], or RS [43], have not been tested, due to accessibility obstacles. Namely LEDA is a commercial software and RS, neither open source, is accessible as a binary code through its MAPLE (v. 9.5) interface, which we did not have at the time of the experiments. For experiments against these libraries and the package of Rioboo [42] in AXIOM, for degree ≤ 4, the reader may refer to [19].

Our data are polynomials of degree $d \in \{5, \ldots, 100\}$ and coefficient bit size $\tau \in \{10, 20, 30, \ 40, 50\}$ with various attributes. Namely D_1^τ denotes random

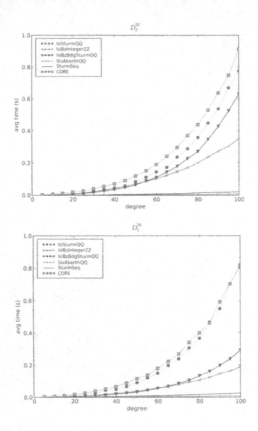

Fig. 5. Mignotte polynomials of bit size 30 (top) and 50 (bottom) bits

polynomials with few real roots and D_2^τ random polynomials with multiple real roots. D_3^τ denotes polynomials with d (multiple) integer real roots and D_4^τ polynomials with d (multiple) rational real roots. D_5^τ denotes Mignotte polynomials, i.e. $a\left(x^d - 2(Kx - 1)^2\right)$, where $K \in [5..30]$, D_6^τ polynomials that are the product of two Mignotte polynomials and D_7^τ Mignotte polynomials with multiple roots.

For reasons of space, we present timings, which are the average of 100 runs, only for D_1^{30}, D_1^{50}, D_2^{30}, D_2^{50}, D_5^{30}, D_5^{50}, D_7^{30} and D_7^{50}. The experiments were performed on an Pentium (2.6 GHz), using g++ 3.4.4 (Suse 10). We have to emphasize that we do not consider experimentation as a competition, but rather as a starting point for improving existing implementations.

For polynomials with few, distinct and well separated real roots, this is the case for D_1 and D_2 (see Fig. 3 and 4), S_1 is clearly the worst choice, since the huge time for the computation of the Sturm sequence dominates the time for its evaluation. In such data sets, S_2 or even approximate solvers are the solvers of choice, since not much output precision is needed in order to isolate the roots.

When there are roots that are very close and/or there are multiple roots, this is the case for D_5 and D_7 (see Fig. 5 and 6), then the computation time of the

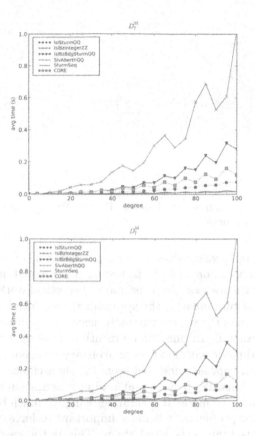

Fig. 6. Mignotte-like polynomials with multiple roots of bit size 30 (top) and 50 (bottom) bits

Sturm-Habicht sequence is negligible (for the experiments that we performed). In such cases a combined solver is the solver of choice, since it isolates the well separated roots and also provides good initial intervals for S_2, if needed. Special notice should be given to the bad behavior of the Bernstein solver (S_2) in the presence of multiple roots. The expected similar asymptotic behavior of Sturm and Bernstein subdivision solvers can be guessed on these experimentations, though the degree is not very high (≤ 100). This applies in the worst case (Mignotte-like polynomials), whereas for random polynomials, the asymptotics of the two solvers seem to be different.

We have to mention that CORE does not compute the multiplicities of the real roots. In addition, for the Aberth solver (SlvAberthQQ in the plots), even though we specified its parameters in order to search for real roots only and detect multiplicities, since it is a numerical solver it must be given an output precision. In order to be sure in advance that we isolate all real roots, the output precision should be equal to the (theoretical) separation bound. In almost all cases, MPSOLVE is the fastest implementation.

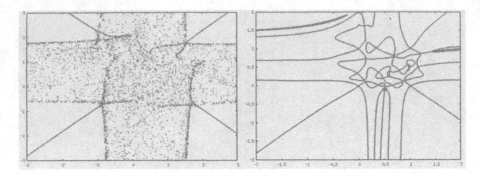

Fig. 7. Left: Approximation using doubles. Right: Approximation using Bernstein solver and interval arithmetic.

For the exact solvers, we consider solver S_3, which is a combination of solvers, as the most promising option. It is the fastest on random instances and comparable to S_2 on all other instances. However, more theoretical work is needed so that we can provide some guarantee for the approximations. Another important issue is the implementation of the square-free factorization, which seems to be the bottleneck for all the exact algorithms. As we mentioned before, our implementation depends on the arithmetic with integers of arbitrary precision, provided by GMP. This implementation approach does not seem the right choice for the square-free factorization algorithms. We believe that an implementation based on machine arithmetic combined with modular techniques will give much better results.

In some geometric problems, it is more important to have controlled approximation of the roots than to isolated them. This is the case in the following example where we want to draw a curve defined by an implicit equation. In this specific problem, the polynomial $f(x,y)$ is of degree 43 in each variable with coefficients of bit size 50 [8]. In order to get a picture of the implicit curve in the box $[a,b] \times [c,d]$, we solve the univariate polynomials $f(a + k\frac{(b-a)}{N}, y)_{k=0,...,N-1}$ ($N = 200$) and then exchange the role of x and y. The subdivision is stopped, when the precision of 10^{-4} is reached, without checking the existence and uniqueness of the roots in the computed intervals.

Two types of solvers have been tested:

- The first one ($\texttt{SlvBzStd<double>}$) is a direct implementation of the Bernstein solver with $\texttt{double}$ arithmetic. It produces the left part of Fig. 7. We see that in some regions, the solver is more sensible to numerical errors, and behaves almost "like a random generator of points".
- The second solver ($\texttt{SlvBzBdg<QQ>}$), similar to S_3, uses exact (rational) arithmetic to convert the input polynomial to its Bernstein representation. Then it normalises the coefficients and rounds up and down the rational numbers to the closest $\texttt{double}$ numbers[2]. Then the main subdivision loop is performed

---

[2] For that purpose, one can use for instance the function $\texttt{get\_double}$ of MPFR ($\texttt{http://www.mpfr.org/}$) with correct rounding mode.

on double interval arithmetic, extending the sign count to this context. If all the interval coefficients contain 0, we recompute the representation of the initial polynomial (using exact rational arithmetic) and run again the rounding and subdivision steps with double arithmetic, until we get the required precision. This produces the right part of Fig. 7.

The Bernstein solver based on interval arithmetic and using this symbolic-numeric strategy can be applied efficiently (even for input polynomials with large coefficient size) to geometric problems, where (controlled) approximate results are sufficient. Its main advantage is that it exploits the performance of machine precision arithmetic for the main loop of the algorithm and the approximation properties of the Bernstein representation. Notice that the size of the problem is prohibitive for exact subdivision based solvers.

8   Current and Future Work

Our experiments imply that the combination of symbolic and numeric techniques leads to very promising implementations. Along these lines, we plan to improve the existing implementation of solvers, so that we can approximate with guarantees the roots of a polynomial with exact coefficients. The applications of Bernstein methods to polynomials with approximate coefficients is also under investigation. We are also extending our package in SYNAPS so that it can handle computations in an extension field.

There are a lot of open questions concerning exact algorithms for real root isolation. Just to mention few of them: Is there any class of polynomials, with few real roots, such that the Bernstein solver performs $\mathcal{O}(d\tau)$ subdivision steps but the Sturm solver performs only a constant number? Does the $\widetilde{\mathcal{O}}_B(d^4\tau^2)$ hold for complex root isolation? What is the expected complexity of the exact subdivision solvers?

The most important open problem for a theoretical and hopefully practical point of view is the following: Is there any exact (subdivision based) solver with complexity $\widetilde{\mathcal{O}}_B(d^3\tau)$, similar to the numerical solvers?

Acknowledgments

All authors acknowledge partial support by IST Programme of the EU as a Shared-cost RTD (FET Open) Project under Contract No IST-006413-2 (ACS - Algorithms for Complex Shapes).

References

1. Akritas, A.: An implementation of Vincent's theorem. Numerische Mathematik 36, 53–62 (1980)
2. Basu, S., Pollack, R., Roy, M.-F.: Algorithms in Real Algebraic Geometry. Algorithms and Computation in Mathematics, vol. 10. Springer, Heidelberg (2003)

3. Ben-Or, M., Kozen, D., Reif, J.H.: The complexity of elementary algebra and geometry. J. Comput. Syst. Sci. 32, 251–264 (1986)

4. Berberich, E., Eigenwillig, A., Hemmer, M., Hert, S., Kettner, L., Mehlhorn, K., Reichel, J., Schmitt, S., Schömer, E., Wolpert, N.: EXACUS: Efficient and Exact Algorithms for Curves and Surfaces. In: Brodal, G.S., Leonardi, S. (eds.) ESA 2005. LNCS, vol. 3669, pp. 155–166. Springer, Heidelberg (2005)

5. Bini, D.: Numerical computation of polynomial zeros by means of Aberth's method. Numerical Algorithms 13(3–4), 179–200 (1996)

6. Bini, D., Fiorentino, G.: Design, analysis, and implementation of a multiprecision polynomial rootfinder. Numerical Algorithms, 127–173 (2000)

7. Canny, J.: Improved algorithms for sign determination and existential quantifier elimination. The Computer Journal 36(5), 409–418 (1993)

8. Cazals, F., Faugère, J.-C., Pouget, M., Rouillier, F.: The implicit structure of ridges of a smooth parametric surface. Technical Report 5608, INRIA (2005)

9. Collins, G.: Subresultants and reduced polynomial remainder sequences. J. ACM 14, 128–142 (1967)

10. Collins, G., Akritas, A.: Polynomial real root isolation using Descartes' rule of signs. In: SYMSAC 1976, pp. 272–275. ACM Press, New York (1976)

11. Collins, G., Loos, R.: Real zeros of polynomials. In: Buchberger, B., Collins, G., Loos, R. (eds.) Computer Algebra: Symbolic and Algebraic Computation, 2nd edn., pp. 83–94. Springer, Wien (1982)

12. Coste, M., Roy, M.F.: Thom's lemma, the coding of real algebraic numbers and the computation of the topology of semi-algebraic sets. J. Symb. Comput. 5(1/2), 121–129 (1988)

13. Davenport, J.H.: Cylindrical algebraic decomposition. Technical Report 88–10, School of Mathematical Sciences, University of Bath, England (1988), http://www.bath.ac.uk/masjhd/

14. Du, Z., Sharma, V., Yap, C.K.: Amortized bound for root isolation via Sturm sequences. In: Wang, D., Zhi, L. (eds.) Int. Workshop on Symbolic Numeric Computing, School of Science, Beihang University, Beijing, China, pp. 81–93 (2005)

15. Eigenwillig, A., Kettner, L., Krandick, W., Mehlhorn, K., Schmitt, S., Wolpert, N.: A Descartes Algorithm for Polynomials with Bit-Stream Coefficients. In: Ganzha, V.G., Mayr, E.W., Vorozhtsov, E.V. (eds.) CASC 2005. LNCS, vol. 3718, pp. 138–149. Springer, Heidelberg (2005)

16. Eigenwillig, A., Sharma, V., Yap, C.K.: Almost tight recursion tree bounds for the Descartes method. In: ISSAC 2006: Proceedings of the 2006 International Symposium on Symbolic and Algebraic Computation, pp. 71–78. ACM Press, New York (2006)

17. El Kahoui, M.: An elementary approach to subresultants theory. J. Symb. Comput. 35(3), 281–292 (2003)

18. Emiris, I., Kakargias, A., Teillaud, M., Tsigaridas, E., Pion, S.: Towards an open curved kernel. In: Proc. Annual ACM Symp. on Computational Geometry, pp. 438–446. ACM Press, New York (2004)

19. Emiris, I., Tsigaridas, E.: Computing with real algebraic numbers of small degree. In: Albers, S., Radzik, T. (eds.) ESA 2004. LNCS, vol. 3221, pp. 652–663. Springer, Heidelberg (2004)

20. Emiris, I., Tsigaridas, E.: Real solving of bivariate polynomial systems. In: Ganzha, V., Mayr, E., Vorozhtsov, E. (eds.) Proc. Computer Algebra in Scientific Computing (CASC), pp. 150–161. Springer, Heidelberg (2005)

21. Emiris, I., Tsigaridas, E., Tzoumas, G.: The predicates for the Voronoi diagram of ellipses. In: Proc. 22th Annual ACM Symp. on Computational Geometry, Sedona, USA, pp. 227–236 (2006)
22. Emiris, I., Tsigaridas, E.P.: Computations with one and two algebraic numbers. Technical report, ArXiv (Dec 2005)
23. Geddes, K., Czapor, S., Labahn, G.: Algorithms of Computer Algebra. Kluwer Academic Publishers, Boston (1992)
24. González-Vega, L., Lombardi, H., Recio, T., Roy, M.-F.: Sturm-Habicht Sequence. In: ISSAC, pp. 136–146 (1989)
25. Guibas, L., Karavelas, M., Russel, D.: A computational framework for handling motion. In: Proc. 6th Workshop Algor. Engin. & Experim. (ALENEX), January 2004, pp. 129–141 (2004)
26. Heindel, L.E.: Integer arithmetic algorithms for polynomial real zero determination. Journal of the Association for Computing Machinery 18(4), 533–548 (1971)
27. Johnson, J.: Algorithms for polynomial real root isolation. In: Caviness, B., Johnson, J. (eds.) Quantifier elimination and cylindrical algebraic decomposition, pp. 269–299. Springer, Heidelberg (1998)
28. Karamcheti, V., Li, C., Pechtchanski, I., Yap, C.: A CORE library for robust numeric and geometric computation. In: 15th ACM Symp. on Computational Geometry (1999)
29. Krandick, W.: Isolierung reeller Nullstellen von Polynomen. In: Herzberger, J. (ed.) Wissenschaftliches Rechnen, pp. 105–154. Akademie-Verlag, Berlin (1995)
30. Krandick, W., Mehlhorn, K.: New bounds for the Descartes method. JSC 41(1), 49–66 (2006)
31. Lane, J.M., Riesenfeld, R.F.: Bounds on a polynomial. BIT 21, 112–117 (1981)
32. Lickteig, T., Roy, M.-F.: Sylvester-Habicht Sequences and Fast Cauchy Index Computation. J. Symb. Comput. 31(3), 315–341 (2001)
33. Lombardi, H., Roy, M.-F., Safey El Din, M.: New Structure Theorem for Subresultants. J. Symb. Comput. 29(4-5), 663–689 (2000)
34. Mignotte, M.: Mathematics for Computer Algebra. Springer, Heidelberg (1992)
35. Mignotte, M.: On the Distance Between the Roots of a Polynomial. Appl. Algebra Eng. Commun. Comput. 6(6), 327–332 (1995)
36. Mourrain, B., Pavone, J.P., Trébuchet, P., Tsigaridas, E.: SYNAPS, a library for symbolic-numeric computation. In: 8th Int. Symposium on Effective Methods in Algebraic Geometry, MEGA, Sardinia, Italy, May 2005. Software presentation (2005)
37. Mourrain, B., Rouillier, F., Roy, M.-F.: Bernstein's basis and real root isolation. Mathematical Sciences Research Institute Publications, pp. 459–478. Cambridge University Press, Cambridge (2005)
38. Mourrain, B., Técourt, J., Teillaud, M.: On the computation of an arrangement of quadrics in 3d. Comput. Geom. 30(2), 145–164 (2005)
39. Mourrain, B., Vrahatis, M., Yakoubsohn, J.: On the complexity of isolating real roots and computing with certainty the topological degree. J. Complexity 18(2) (2002)
40. Pan, V.: Univariate polynomials: Nearly optimal algorithms for numerical factorization and rootfinding. J. Symbolic Computation 33(5), 701–733 (2002)
41. Reischert, D.: Asymptotically fast computation of subresultants. In: ISSAC, pp. 233–240 (1997)
42. Rioboo, R.: Towards faster real algebraic numbers. In: Proc. ACM Intern. Symp. on Symbolic & Algebraic Comput, Lille, France, pp. 221–228 (2002)
43. Rouillier, F., Zimmermann, Z.: Efficient isolation of polynomial's real roots. J. of Computational and Applied Mathematics 162(1), 33–50 (2004)

44. Roy, M.-F., Szpirglas, A.: Complexity of the Computation on Real Algebraic Numbers. J. Symb. Comput. 10(1), 39–52 (1990)
45. Schönhage, A.: The fundamental theorem of algebra in terms of computational complexity. Univ. of Tübingen, Germany (manuscript, 1982)
46. Sharma, V., Yap, C.: Sharp Amortized Bounds for Descartes and de Casteljau's Methods for Real Root Isolation (October 2005) (unpublished manuscript)
47. Tsigaridas, E.P., Emiris, I.Z.: Univariate polynomial real root isolation: Continued fractions revisited. In: Azar, Y., Erlebach, T. (eds.) ESA 2006. LNCS, vol. 4168, pp. 817–828. Springer, Heidelberg (2006)
48. von zur Gathen, J., Gerhard, J.: Fast Algorithms for Taylor Shifts and Certain Difference Equations. In: ISSAC, pp. 40–47 (1997)
49. von zur Gathen, J., Gerhard, J.: Modern Computer Algebra, 2nd edn. Cambridge Univ. Press, Cambridge (2003)
50. von zur Gathen, J., Lücking, T.: Subresultants revisited. Theor. Comput. Sci. 1-3(297), 199–239 (2003)
51. Yap, C.: Fundamental Problems of Algorithmic Algebra. Oxford University Press, New York (2000)

Verified Methods in Stochastic Traffic Modelling

Sebastian Kempken and Wolfram Luther

Universität Duisburg-Essen,
47048 Duisburg, Germany
{kempken, luther}@informatik.uni-duisburg.de,
http://www.informatik.uni-duisburg.de

Abstract. In this paper, we present two validated methods from different application fields in stochastic traffic modelling. First, we show how the autocorrelation of a semi-Markov arrival process can be described as a sum of exponential terms using validated numerics. Next, we use interval arithmetic as a reliable method to analyse the transient states of simple GI/G/1 queueing systems and compute the time required for the system to reach the equilibrium.

1 Introduction

The classical approach in queueing and service systems is to consider random variables for the interarrival times of events corresponding to the arrivals of packets, flows, connections or other units relevant for network elements. Two basic characteristics of the stochastic behavior of traffic are the distribution function of considered random variables and the autocorrelation of the process. To model the distribution function of arriving data as well as the autocorrelation function, semi-Markov processes (SMPs) are well suited [1].

Special emphasis is given to the computation of workload distributions of service systems and the modelling of the autocorrelation function of a $SMP(m)$ with m states as a superposition of $m - 1$ geometrical terms including complex coefficients. We use traffic data to recover the coefficients of the exponential sums. Interval arithmetic is applied to validate the parameter estimation and to guarantee the results of the analysis. Our goal is to apply an interval version of Prony's algorithm and to analyse its relation to other approaches in order to solve the parameter estimation problem.

Recently, we have proposed an algorithm to compute the verified stationary workload distributions of GI/GI/1 and SMP/GI/1 service systems accurately using factorisation approaches [2]. Therefore, another interesting task is to study the transient behavior of the queue and to compute the time required for the system to reach the equilibrium. In order to handle rounding errors, we use interval arithmetic to provide reliable results.

P. Hertling et al. (Eds.): Real Number Algorithms, LNCS 5045, pp. 83–101, 2008.
© Springer-Verlag Berlin Heidelberg 2008

2 Verified Parameter Estimation in Stochastic Traffic Modelling

2.1 Stochastic Traffic Modelling

The nature of traffic in telecommunication networks is unpredictable from a variety of viewpoints. On the one hand, service providers do not exactly know the transmission time and volume when transmission demands are created by the users and applications. On the other hand, sender and receiver cannot predict the amount of network resources they will require at the start of their communication. Randomly changing workload, routes and system parameters are relevant under normal operating conditions, failure events are unpredictable as well.

Therefore, stochastic traffic models represent an appropriate basis for describing the traffic flow over time. The usual stochastic traffic models in telecommunication consider random variables for

- the interarrival times of events corresponding to such things as arrivals of packets, flows, connections or other units relevant to network elements like the classical approach in queueing and service systems,
- the counting function of the number of arrivals in predefined intervals, for instance in time-slotted multiplexer models.

The two basic characteristics of the stochastic behavior of traffic are the distribution function of the random variables considered and, especially for workload analysis, the autocorrelation of the process.

The distribution of the amount of arriving data for a transmission line is essential to estimate overload situations leading to data loss. Autocorrelation is a measure for the dependency of values observed in a process between two different points in time as a function of their time distance.

Open queueing networks provide a typical approach for the analysis of telecommunication systems being decomposed into nodes representing switches and routers and edges as transmission links (cf. Figure 1). The main items of this approach are

- traffic flows on the edges of a network topology being characterized by stochastic processes,
- server systems for switching and forwarding traffic with the help of buffers to store data in phases of temporary overload and
- sources and destinations, which generate and terminate traffic flows in the network.

In order to model the arrival and service behavior at a network node in terms of arrival distribution and autocorrelation, we introduce a discrete-time homogeneous semi-Markov process as a special case of a general semi-Markov process (cf. [3]). A family of random variables $\{(R_t, \sigma_t) | t \in \mathbb{N}\}$ denotes such a process if

$$Pr\left(R_{t+1} = a, \sigma_{t+1} = j | R_k = a_k, \sigma_k = i_k, 1 \leq k \leq t\right) =$$

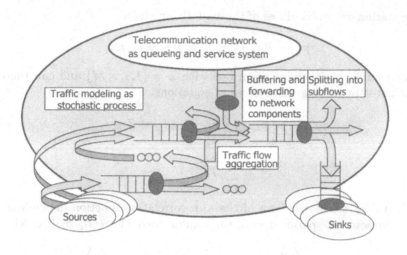

Fig. 1. A network with sources, sinks, buffering and forwarding components

$$Pr\left(R_{t+1} = a, \sigma_{t+1} = j | \sigma_t = i_t\right)$$

holds for all $t \in \mathbb{N}$ and $\{\sigma_t\}$ is a Markov chain. R_t denotes the t-th sojourn time. In the following, we restrict ourselves to the case of a finite Markov chain. If the underlying chain $\{\sigma_t\}$ has M states, the semi-Markov process is called $SMP(M)$.

To define a semi-Markov process completely, the transition matrix $P :=$ (p_{ij}) of the underlying chain $\{\sigma_t\}$ with M^2 distribution functions $f_{ij}(t)$, $i, j = 1, \ldots, M$ must be known.

A simplification of the transition is achieved using a special case of a semi-Markov process ($SSMP$): We have

$$P\{R_{t+1} = a, \sigma_{t+1} = j | R_k = a_k, \sigma_k = i_k, 1 \leq k \leq t\} =$$
$$P\{R_{t+1} = a, \sigma_{t+1} = j | \sigma_t = i_t\} =$$
$$p_{i_t j} P\{R_{t+1} = a | \sigma_t = i_t\}$$
$$f_{i_t j}(t) = p_{i_t j} f_{i_t}(t)$$

Please note that every SMP with M states can always be described as an $SSMP$ with M^2 states. To do so, we identify each transition from state i to state j of the SMP with a state (i, j) of the $SSMP$ (cf. [4]).

The main advantage of this notation is that a closed formula for the auto-correlation of the process can now be given. We assume the underlying Markov chain to be aperiodic and irreducible and introduce the $n-$step transition matrix $P^{(n)} = \underbrace{P \circ \ldots \circ P}_{n \text{ times}}$ with

$$p_{ij}^{(n)}(t) = P\left(\sigma_{t+n} = j | \sigma_t = i\right), \forall n \in \mathbb{N}.$$

The stationary probabilities of the underlying chain

$$p_j := \lim_{n \to \infty} P\{\sigma_{t+n} = j | \sigma_t = i\} \forall i, t$$

exist and fulfill the following equations with $Z = \{1, \ldots, M\}$ and can therefore be calculated by solving this system of equations:

$$\sum_{k \in Z} p_k p_{k,j} = p_j \text{ for all } j \in Z \tag{1}$$

$$\sum_{j \in Z} p_j = 1 \tag{2}$$

Then we have a closed form of the autocorrelation function depending only on the characteristic parameters of the random variables $\{(R_t, \sigma_t) | t \in \mathbb{N}\}$

$$A_R(n) = \left(\sum_{i=1}^{M} \sum_{j=1}^{M} p_i E_i(R) p_{ij}^{(n)} E_j(R) - E^2(R) \right) / \sigma_R^2$$

$$A_R(0) = 1 \tag{3}$$

The $E_i(R), i = 1, \ldots, M$, here denote the state dependent expectation values of the random process R, $E(R) = \sum_{i=1}^{M} p_i E_i, \sigma_R^2 = \sum_{i=1}^{M} p_i(E_i - E(R))^2$ the mean value and variance of the whole process R.

This formula contrasts with the standard estimation of the autocorrelation of sample data R_i with time-averaging of a given time interval $i \in [1, \ldots, N]$

$$A_R(n) = \frac{1}{\sigma_R^2 (N - n)} \sum_{j=1}^{N-n} (R_j - E(R)) (R_{j+n} - E(R)) \tag{4}$$

2.2 Parameter Estimation for the Autocorrelation

Parameter estimation via eigenvalues of the transition matrix. If the transition matrix is diagonalizable and the stochastic process $\{\sigma_j\}$ is aperiodic and irreducible, the autocorrelation function of an $SSMP(M)$ is also an exponential sum

$$A_R(n) = \sum_{j=2}^{M} \alpha_j \lambda_j^n \tag{5}$$

This can be proven using the fact that the transition matrix P has M eigenvalues λ_j with $\lambda_1 = 1$, and left eigenvector $\mathbf{p}_1 = (p_1, p_2, \ldots, p_M)$ and $|\lambda_j| < 1$ for $j = 2, \ldots, M$.

Proof. Equation 1 holds for the stationary probabilities, hence $\mathbf{p}_1$ is left eigenvector for the eigenvalue $\lambda_1 = 1$.

Let λ denote an eigenvalue of P with corresponding right eigenvector $x = (x_1, \ldots, x_M)^T$, $x \neq 0$. If i is chosen in such a way that

$$|x_i| = \max_{1 \leq j \leq M} |x_j|$$

holds, then the following inequation holds in conjunction with $\lambda x_i = \sum_{j=1}^{M} p_{ij} x_j$:

$$|\lambda| = \left| \sum_{j=1}^{M} p_{ij} \frac{x_j}{x_i} \right| \leq \sum_{j=1}^{M} p_{ij} \left| \frac{x_j}{x_i} \right| \leq \sum_{j=1}^{M} p_{ij}.$$

As P is a stochastic matrix, it follows that

$$|\lambda| \leq \sum_{j=1}^{M} p_{ij} = 1.$$

The strict inequality $|\lambda| < 1$ is obtained by the Perron-Frobenius theorem for primitive matrices [5, theorem 1.1], as the acyclic irreducible matrix P is also primitive [5, theorem 1.4].

The matrix P is assumed to be diagonalizable. Therefore, a matrix Q with $QP = \mathrm{diag}(\lambda_1, \ldots, \lambda_M)Q$ can be given:

$$\underbrace{\begin{pmatrix} p_1 & p_2 & \cdots & p_M \\ q_{2,1} & q_{2,2} & \cdots & q_{2,M} \\ \vdots & \vdots & \vdots & \vdots \\ q_{M,1} & q_{M,2} & \cdots & q_{M,M} \end{pmatrix}}_{Q} P = \underbrace{\begin{pmatrix} 1 & 0 & \cdots & 0 \\ 0 & \lambda_2 & \cdots & 0 \\ \vdots & \vdots & \vdots & \vdots \\ 0 & 0 & \cdots & \lambda_M \end{pmatrix}}_{\Lambda} \begin{pmatrix} p_1 & p_2 & \cdots & p_M \\ q_{2,1} & q_{2,2} & \cdots & q_{2,M} \\ \vdots & \vdots & \vdots & \vdots \\ q_{M,1} & q_{M,2} & \cdots & q_{M,M} \end{pmatrix}$$

It can be shown [6] that with $Q^{-1} = (\mathbf{q}_1', \ldots, \mathbf{q}_M')$, $\mathbf{q}_1' = (1, \ldots, 1)^T$:

$$\mathcal{A}_R(n)Var(R) + E^2(R) = (p_1 E_1, \ldots, p_M E_M) \underbrace{Q^{-1} \Lambda^n Q}_{P^n} (E_1, \ldots, E_M)^T$$

$$= \left(\sum_{i=1}^{M} p_i E_i \right)^2 + \lambda_2^n \sum_{i=1}^{M} p_i E_i q_{i,2}' \sum_{j=1}^{M} E_j q_{2,j} + \ldots + \lambda_M^n \sum_{i=1}^{M} p_i E_i q_{i,M}' \sum_{j=1}^{M} E_j q_{M,j}$$

Using $\sum_{i=1}^{M} p_i E_i = E(R)$, we get

$$\mathcal{A}_R(n) = \lambda_2^n \underbrace{\frac{\sum_{i=1}^{M} p_i E_i q_{i,2}' \sum_{j=1}^{M} E_j q_{2,j}}{Var(R)}}_{\alpha_2} + \ldots + \lambda_M^n \underbrace{\frac{\sum_{i=1}^{M} p_i E_i q_{i,M}' \sum_{j=1}^{M} E_j q_{M,j}}{Var(R)}}_{\alpha_M}.$$

$$(6)$$

Therefore, the autocorrelation has the form described in Equation 5.

Using the Cardano formulae, the eigenvalues can be computed explicitly in the case $M \leq 5$. For higher values of M the values can be found via numeric eigenvalue problem techniques.

Parameter estimation via Prony approximation. Another way to find a description for the autocorrelation function in the form of an exponential sum is via a polynomial equation solver using Prony's method (cf. [7]), which dates back to the 18th century [8]:

For arbitrary real constants $\alpha_1, \ldots, \alpha_m$ and distinct constants $\beta_1, \ldots, \beta_m$ and equidistant points $j = 1, \ldots, N$ and values

$$\mu_j = \alpha_1 \exp(\beta_1 j) + \ldots + \alpha_m \exp(\beta_m j)$$

there exist constants $g_0, \ldots, g_m$, such that

$$\begin{pmatrix} \mu_1 & \cdots & \mu_{m+1} \\ \vdots & \vdots & \vdots \\ \mu_{N-m} & \cdots & \mu_N \end{pmatrix} \begin{pmatrix} g_0 \\ \vdots \\ g_m \end{pmatrix} = \begin{pmatrix} 0 \\ \vdots \\ 0 \end{pmatrix} \tag{7}$$

and a further condition on g_j, e.g. $g_m = 1$ can be imposed. To recover the $\alpha_k, \beta_k, k = 1, \ldots, m$ from given $\mu_j, j = 1, \ldots, N$ ($N \geq 2m$), we first solve the equation system above to determine $g_0, \ldots, g_m$. As the roots of the polynomial equation

$$g_0 + g_1 z + \ldots + g_m z^m = 0$$

are $\exp(\beta_1), \ldots, \exp(\beta_m)$, the values β_k can be calculated. The coefficients α_k are obtained by solving a linear equation system with the Vandermonde matrix X:

$$\mu = X\alpha$$

with $\mu = (\mu_1, \ldots, \mu_n)^T$ and $\alpha = (\alpha_1, \ldots, \alpha_m)^T$ and

$$X = \begin{pmatrix} \exp(\beta_1) & \cdots & \exp(\beta_m) \\ \vdots & \vdots & \vdots \\ \exp(\beta_1 N) & \cdots & \exp(\beta_m N) \end{pmatrix}.$$

In our case, we compute at least $2(M-1)$ values of the autocorrelation function using formula 3 for the description of a $SSMP(M)$. The coefficients α_l and the values β_l, $j = 1, \ldots, M - 1$, can be calculated using the method described. In this way, we obtain a realization for the autocorrelation as a weighted sum of $M - 1$ exponential terms.

Estimation of parameter intervals. The presented Prony approach can be applied to estimate the autocorrelation function of a given SSMP as well as to approximate the autocorrelation function of a given data set. It has been noticed that the original Prony method is very sensitive to noise in the given data. Some improvements have been suggested, for example by Osborne and Smyth [9].

As an enhancement to the presented Prony approach, we considered the question of how to expand the parameter intervals to include short-term fluctuations in the approximation of a given data series. We assume a given autocorrelation function $a_0(x), x = 0, \ldots, N$ computed up to N from a given data set using equation 4. The idea is to find m interval parameters $[\alpha_j] = [\underline{\alpha_j}; \overline{\alpha_j}]$ and m parameters $[\lambda_j] = [\underline{\lambda_j}; \overline{\lambda_j}]$ so that

$$a_0(x) \in a(x) = \sum_{j=1}^{m} [\alpha_j][\lambda_j]^x, x = 0, \ldots, N. \tag{8}$$

Even harsher restrictions are given by

$$a_0(x) \leq \sum_{j=1}^{m} \overline{\alpha_j} \, \overline{\lambda_j}^x, a_0(x) \geq \sum_{j=1}^{m} \underline{\alpha_j} \, \underline{\lambda_j}^x. \tag{9}$$

As further criteria, the enclosure should be as tight as possible. In order to find these parameter intervals, we start with an ordinary Prony approximation with a chosen m of the given function $a_0(x)$ and yield the parameters α_j and $\lambda_j = exp(\beta_j)$. For the steps to come, it is irrelevant whether the traditional Prony method or the modifications of Osborne and Smyth have been used.

Our idea now is to enclose both types of parameters in tight intervals and to keep the original parameters $[\lambda_j]$ while expanding the intervals of $[\alpha_j]$. We do so by finding the upper and lower bounds of $[\alpha_j]$ separately using linear optimization techniques.

The autocorrelation function is limited to a codomain of $[-1; 1]$. Therefore, all coefficients λ_j that are found by approximation must be inside the unit disc. Otherwise, any resulting exponential sum would yield values outside the codomain of the autocorrelation function.

The interval approximation we are looking for must also be limited to this codomain. Hence, we introduce another constraint:

$$a(0) = \sum_{j=1}^{m} [\alpha_j] \subseteq [-1; 1]$$

Because all $[\lambda_j]$ are within the unit disc, subsequent function values are also included in this interval.

We define the following linear objective function, which is to be minimized, as another parameter of the optimization process in order to find the upper bound:

$$u(\overline{\alpha_1}, \ldots, \overline{\alpha_m}) := \sum_{j=1}^{m} mid([\lambda_j])^2 \cdot \overline{\alpha_j}.$$

λ_i denotes the midpoints of the intervals $[\lambda_j]$, which are considered constant in the optimization process. For the calculation of the lower bound, we use

$$l(\underline{\alpha_1}, \ldots, \underline{\alpha_m}) := - \sum_{j=1}^{m} mid([\lambda_j])^2 \cdot \underline{\alpha_j}$$

accordingly. The lower bounds for the optimization process are set to -1, as upper bounds we choose the computed values $\overline{\alpha_j}$. Hence,

$$\underline{\alpha_j} \in [-1; \overline{\alpha_j}] \Rightarrow \underline{\alpha_j} \le \overline{\alpha_j}.$$

Negative parameters $[\lambda_j]$ require further consideration: If the coefficients $\overline{\alpha_j}$ and $\underline{\alpha_j}$ are approximately 0 for these negative parameters $[\lambda_j]$, which is the case for a variety of video traces analysed in our experiments, the corresponding terms may be neglected. This is because the minimal value of the objective function $u(\dots)$ is achieved for the lower bound of the optimization process $\overline{\alpha_j} = 0$, if this is a valid solution. For the lower bound of the enclosure, this applies accordingly.

In the case described, the lower bound of the enclosure of the given auto-correlation function is defined by the lower bound of the parameter intervals, and upper bounds are given accordingly. Therefore, restriction 9 holds. If the coefficients $[\alpha_j]$ are not negligible, only relation 8 is satisfied.

Some numeric results of this approach are given in the following section. However, the approach presented is not always feasible: If, for instance,

$$a_0(x) > \sum_{j=0}^{m} \overline{\lambda_j}^x$$

for any $x \in \{0, \dots, N\}$, it is not possible to compute upper bounds for the intervals $[\alpha_j]$ that satisfy the mentioned constraints. For the lower bound, similar restrictions apply accordingly.

Discussion. Whereas the computation of Equation 3 has $O(nM^3)$-complexity, the evaluation of Formula 5 is possible in linear time. Thus, it is much more efficient to use the latter equation to calculate $\mathcal{A}_R(n)$ for larger n as the additional effort for the determination of the exponential parameters can be neglected.

Another advantage of the exponential sum form is numeric stability. The computation of Equation 3 requires an n-fold multiplication of the matrix P. As the first eigenvalue λ_1 equals one and the other eigenvalues have an absolute value smaller than one, the multiplications may lead to extinction of the eigenvalues other than λ_1, especially for larger n.

We have implemented the verified parameter estimation using the eigenvalue approach in INTLAB [10]. We use the method by Rump [11] included in this library to determine verified enclosures for the eigenvalues $\lambda_2, \dots, \lambda_M$ of a given transition matrix P. As the transition probabilities are computed by division from given integer data, we start with tight enclosures and obtain verified eigenvalue intervals. Using interval arithmetic also yields tight enclosures for the coefficients of the exponential sum. An example is given in the following section.

The main advantage of the interval parameter estimation is that the parameters $[\lambda_j]$ are kept very tight, whereas the coefficients $[\alpha_j]$ are expanded. As this notation corresponds to the autocorrelation of a SSMP, we hope to model given data traces with a SSMP by solving the inverse eigenvalue problem to compute a stochastic transition matrix that has eigenvalues within $[\lambda_j]$. The intervals $[\alpha_j]$

are furthermore dependent on the mean values of the particular states of the SSMP. By choosing mean values that result in coefficients within the parameter intervals $[\alpha_j]$, it is guaranteed that the autocorrelation function of the resulting SSMP is also within the bounds derived from the given data trace.

2.3 Analysis of Video Data

As an example, we model video data as a semi-Markov process and calculate the resulting autocorrelation of both the model and the actual data in short- and long-term respectively. The same approach has been taken by Rose to analyse MPEG-1 data [12]. We consider several MPEG-4 encoded video traces with 25 frames per second and duration between one and two hours. The trace data is provided by Fitzek and Reisslein [13].

Let F_j denotes the frame size at time j. The frame sizes typically vary between a few hundred and 20 thousand bytes. For the example clip presented here, which is taken from *Silence of the lambs*, we find $E(F) = 2876.3$ as mean value and standard deviation $\sigma_F = 2291.0$ for the frame sizes.

The encoding process produces sequences of frames compressed via different methods (so-called I-, P- and B-Frames). Thus, the MPEG4 standard provides pictures of three different types within a well-defined sequence leading to corresponding groups of pictures (GoP) - in our case, size 12. The periodic structure of the GoP, however, results in an important variation of the autocorrelation function, as can be seen in Figure 2. This can be avoided if the frame sizes are

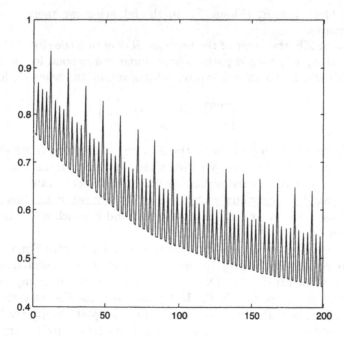

Fig. 2. Autocorrelation of frames, *Silence...* sequence

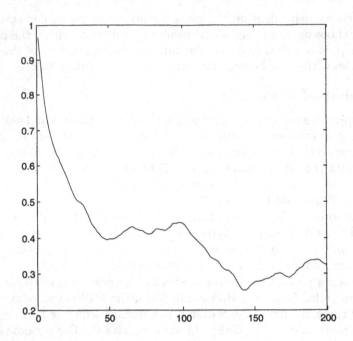

Fig. 3. Autocorrelation of groups of pictures, *Silence...* sequence

combined to GoP sizes R_j (Figure 3). In the following, we therefore consider groups of pictures.

To model a SSMP, the range of the function R denoting the size of the GoP is divided into M equally ranged parts. These ranges correspond to the M states of the $SSMP(M)$. The number of states M is given by the heuristic formula

$$M = \left\lfloor \frac{max_{j \in \{1,...,j_{max}\}}(R_j)}{\sigma_R} \right\rfloor \tag{10}$$

Each frame is then assigned to one of the M ranges according to its size which gives the sequence of the states. We count the number of transitions between the states and determine the transition matrix P. The steady state probabilities, however, can not be approximated by counting the given data. Instead, they have to be determined by solving the equations 1 and 2, which would not be met exactly otherwise.

For our sample clip, we chose $M = 8$ states in accordance with Equation 10. As the values are results of a division, the numbers cannot be represented accurately using floating point arithmetic. The lower bounds of the resulting enclosures, which are exact up to one unit in the last place (ulp), for the transition matrix and the stationary probabilities are given in Tables 1 and 2 respectively.

Next, we determine the eigenvalues for the transition matrix using Rump's algorithm [11]. The α-coefficients for the exponential sum can be calculated

Table 1. Transition matrix for SSMP(8) for *Silence...* clip

Columns 1 through 4			
0.94176413255360. . .	0.05604288499025. . .	0.00170565302144. . .	0.00048732943469. . .
0.09881255301102. . .	0.85156912637828. . .	0.04834605597964. . .	0.00084817642069. . .
0.00859598853868. . .	0.16762177650429. . .	0.76790830945558. . .	0.05300859598853. . .
0.00520833333333. . .	0.01041666666666. . .	0.20833333333333. . .	0.64062500000000. . .
0.00000000000000. . .	0.01149425287356. . .	0.01149425287356. . .	0.29885057471264. . .
0.00000000000000. . .	0.00000000000000. . .	0.00000000000000. . .	0.04166666666666. . .
0.00000000000000. . .	0.00000000000000. . .	0.00000000000000. . .	0.00000000000000. . .
0.00000000000000. . .	0.00000000000000. . .	0.00000000000000. . .	0.00000000000000. . .
Columns 5 through 8			
0.00000000000000. . .	0.00000000000000. . .	0.00000000000000. . .	0.00000000000000. . .
0.00042408821034. . .	0.00000000000000. . .	0.00000000000000. . .	0.00000000000000. . .
0.00143266475644. . .	0.00143266475644. . .	0.00000000000000. . .	0.00000000000000. . .
0.13541666666666. . .	0.00000000000000. . .	0.00000000000000. . .	0.00000000000000. . .
0.64367816091954. . .	0.03448275862069. . .	0.00000000000000. . .	0.00000000000000. . .
0.08333333333333. . .	0.66666666666666. . .	0.16666666666666. . .	0.04166666666666. . .
0.05555555555555. . .	0.22222222222222. . .	0.61111111111111. . .	0.11111111111111. . .
0.00000000000000. . .	0.00000000000000. . .	0.16666666666666. . .	0.83333333333333. . .

Table 2. Stationary probabilities for SSMP(8) for *Silence...* clip

Columns 1 through 4			
0.54924240609103. . .	0.31435009675894. . .	0.09240540377074. . .	0.02487486054065. . .
Columns 5 through 8			
0.01129776948681. . .	0.00313178534072. . .	0.00234883900554. . .	0.00234883900554. . .

according to equation 6 from the parameters of the SSMP. The resulting enclosures for the exponential sum parameters are given in Table 3.

Because the values for $[\lambda_j]$ are complex-valued, we may also yield complex results for the evaluation of the exponential sum. The small imaginary values for parameters $[\alpha_j], [\lambda_j], j = 2, \ldots, 6$ result from the usage of interval input data with corresponding diameter. However, if the coefficent intervals $[\alpha_j]$ were exact point values, the imaginary parts would be erased. But due to the expansion of the coefficients into intervals, the imaginary interval values of the resulting exponential sum are now symmetric around zero. For the approximation of a real-valued function, we therefore consider only the real part of the results.

The resulting traces of the autocorrelation function of both the data and the SSMP(8) model are displayed in Figure 4. As the interval values remain very tight, we draw only their midpoints.

We also include an approximation of the autocorrelation function of the model as an exponential sum, which was calculated using a floating-point implementation of Prony's method from the first $2(M-1) = 14$ values of the autocorrelation

Table 3. Enclosures for exponential sum parameters

j	α_j	λ_j
2	$0.578460644145^{53}_{44} \pm 0.00000000000004i$	$0.9568138718639^{2}_{1} \pm 0.00000000000001i$
3	$0.416090833405^{33}_{23} \pm 0.00000000000005i$	$0.9037083411072^{1}_{0} \pm 0.00000000000001i$
4	$0.00383187037^{703}_{695} \pm 0.00000000000004i$	$0.8257655866834^{8}_{7} \pm 0.00000000000001i$
5	$0.001335877538^{85}_{77} \pm 0.00000000000004i$	$0.7347175180457^{5}_{4} \pm 0.00000000000001i$
6	$0.0000893728940^{17}_{08} \pm 0.00000000000004i$	$0.6874134301784^{2}_{1} \pm 0.00000000000001i$
7	$0.0000957008197^{71}_{60} - 0.000098030122^{73}_{84}i$	$0.4241185462696^{9}_{8} - 0.005840800429^{9}_{2}i$
8	$0.0000957008197^{71}_{60} + 0.000098030122^{84}_{73}i$	$0.4241185462696^{9}_{8} + 0.005840800429^{2}_{1}i$

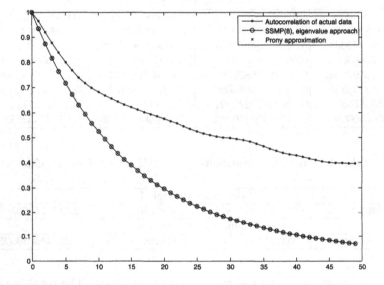

Fig. 4. Autocorrelation of given data and SSMP model

determined according to equation 3. An interval-based version of this approach is part of our current research (cf. [14]).

As we can see, the SSMP approach seems adequate for short-term modelling of actual video data, but the autocorrelation is still significantly different in the long term. A slight improvement, however, can be achieved by increasing the number of states of the SSMP [6].

Another result of the parameter estimation is that some of the coefficients α_j are almost zero, that is, negligible. This indicates that a more accurate modelling of a given video trace by semi-Markov processes may be achieved with a smaller number of states if the other parameters like mean values or the transition matrix are chosen more carefully.

Therefore, it is an interesting question how the parameters of a SSMP model can be modified to increase its quality. For example, how can the states of the

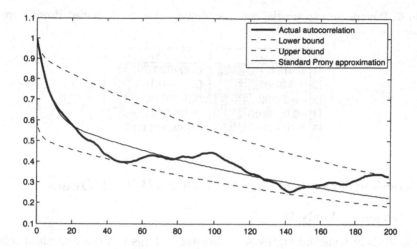

Fig. 5. Autocorrelation function bounded by exponential sums

Table 4. Bounds for exponential sum parameters - enclosure of original data

j	α_j	λ_j
1	$0.50371451646391, 0.90938513991036$	$0.9949220751_{16295}^{46157}$
2	$0.02327227858326, 0.02327227858327$	$0.8356161616_{02726}^{51959}$
3	$0.06734258150621, 0.06734258150623$	$0.5959113221_{195195}^{216477}$
4	$-0.00000000000004, 0.00000000000001$	$-0.3169324189_{4284}^{3635}$
5	$-0.00000000000009, 0.00000000000016$	$0.1324512051_{4122}^{5922}$

$SSMP(M)$ be modified in such a way that the eigenvalues of the transition matrix P perfectly match with the parameters obtained via a Prony approximation of the original data (Inverse eigenvalue problem, cf. [15])?

An example of the estimation of parameter intervals to include the autocorrelation of a given data series is given in Figure 5. We chose a number of 5 terms for the enclosing exponential sums. The coefficients we have determined using our approach are listed in Table 4. Please note that the indexing here begins at 1, as the parameters are not derived by the eigenvalue approach but by a Prony approximation. The values $[\lambda_i]$ are determined by a verified root-finding algorithm according to Prony's method, which is feasible in this case because the original floating-point autocorrelation data is enclosed in tight intervals. The parameters $[\alpha_i]$ are determined using the approach described in this paper.

The coefficients $[\alpha_j]$ of a standard Prony approximation carried out in interval arithmetic are given in Table 5 and the resulting exponential sum is also depicted in Figure 5 for comparison.

We see that in the given example the actual autocorrelation of a given data set can be enclosed in tight bounds given by exponential sums. Also, the coefficient $[\alpha_j]$ for the negative $[\lambda_j]$ is a tight enclosure of zero.

Table 5. Bounds for exponential sum parameters - standard verified Prony approximation

j	α_j	λ_j
1	$0.6154716_{0350666}^{2039956}$	$0.99492207_{16295}^{46157}$
2	$0.5494448_{89171306}^{98569680}$	$0.83561616_{02726}^{51959}$
3	$-0.2150237_{90372194}^{74580986}$	$0.59591132_{195195}^{216477}$
4	$-0.0289985_{9626237}^{5148103}$	$-0.3169324189_{4284}^{3635}$
5	$0.079359_{50683209}^{63230209}$	$0.1324512051_{4122}^{5922}$

3   Verified Transient Analysis of a GI/G/1 Queue

3.1   Transient Analysis

The analysis of transient states of stochastic systems is a basic method related to both steady state analysis and simulation. The first is included in the transient analysis as its eventual result, but the computation effort is often much greater than for direct steady state solutions. Compared to simulation, transient analysis has the advantage of providing complete distribution functions of system states, whereas simulation yields results subject to statistical deviations within confidence levels.

Breuer [16] provides a means for the transient analysis of queue sizes of GI/G/1 queues, but one of the main challenges in transient analysis is the effective limitation of the state space considered. Haßlinger and Kempken [17] describe a way to analyse the queue size in a compact state space. The transient workload of a queueing system can be analysed in a similar way. We consider the case of a discrete time GI/G/1 queue.

Let $\{A_n | n \in \mathbb{N}\}$ be a family of independent and identically distributed (i.i.d.) random variables, where A_n denotes the n-th interarrival time with

$$a_k := Pr(A_n = k) \text{ (independent of } n).$$

Correspondingly, let $\{S_n | n \in \mathbb{N}\}$ be a family of i.i.d. random variables, where S_n denotes the n-th service time with

$$s_k := Pr(S_n = k) \text{ (again independent of } n).$$

We assume the arrival function is limited by g and the service distribution by h respectively. Hence, a difference distribution U_n is given by $U_n = S_n - A_n$ and $u(k) = Pr(U_n = S_n - A_n = k)$ with range $-g \leq U_n \leq h$. The mean service time is assumed to be smaller than the mean interarrival time as a stability condition for a system with unlimited queue size. The service discipline is non-preemptive and the order of service is independent of the required service time (for example, first come, first served). Then the workload W_n immediately after the arrival of the n-th customer can be computed according to Lindley's equation [18]:

$$W_{n+1} = max(W_n + U_n, 0), n \in \mathbb{N}_0 \tag{11}$$

Using our given distributions, the workload distribution can be calculated iteratively by convolutions:

$$Pr(W_{n+1} = k) = \sum_{j=-h}^{g} Pr(W_n = k + j)Pr(U_n = -j), k \in \mathbb{N}, n \in \mathbb{N}_0$$

$$Pr(W_{n+1} = 0) = \sum_{k=-g}^{0} \sum_{j=-h}^{g} Pr(W_n = k + j)Pr(U_n = -j), n \in \mathbb{N}_0$$

$$Pr(W_n = -k) = 0, n \in \mathbb{N}_0, k \in \mathbb{N} \tag{12}$$

The workload cannot be negative; therefore the probabilities for the corresponding states have to be added to $Pr(W_{n+1} = 0)$. If we assume $E(A) > E(S)$, that is $E(U) < 0$, as a stability condition, the workload distribution converges to a steady state for all starting distributions. This has been shown by Lindley [18].

Therefore, for given ϵ and k with

$$w_n(k) = Pr(W_n = k), w(k) = \lim_{n \to \infty} w_n(k),$$

a point n can be given such that $|w(k) - w_n(k)| \leq \epsilon$.

In terms of interval arithmetic this implies that for arbitrary precision (limited by the implementation) there is a point n at which the intervals of the stationary and transient distributions intersect, which means that the difference between the actual probabilities and the stationary ones is smaller than the inaccuracy of the computation. In the following section, we compute the point n for two examples for standard IEEE intervals. Additionally, we are able to perform a verified "worst case" analysis for the number of iterations required to satisfy the mentioned constraint for different values of ϵ, which are limited by the diameter of the intervals involved. To do so, we determine the maximum difference between the interval bounds of the current transient state $w_n(k)$ and the stationary distribution $w(k)$ calculated by Wiener-Hopf analysis (cf. [2] [19]).

3.2   Examples

Two examples will illustrate our approach. The first is rather simple, considering the limited support of the arrival and service distributions. Let the arrival and service distributions be given by $a(1) = a(5) = 0.5$ and $s(1) = s(4) = 0.5$ with mean $E(A) = 3$, $E(S) = 2.5$ and $E(U) = -0.5$ (cf. [17]). For our observations, we assume an empty queue at the beginning and start in the instant of the first arrival. This state is described by $Pr(W = 0) = 1$, $Pr(W = k) = 0, k \in \mathbb{N}$. We computed the steady state probabilities with the tool *InterVerdiKom*, which yields validated enclosures [2] [19]. The sequences $\{w_n(k)\}$ of the transient states are calculated using an implementation in MATLAB with the INTLAB extension.

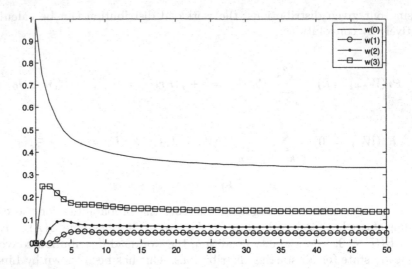

Fig. 6. Transient analysis of the workload for the first example

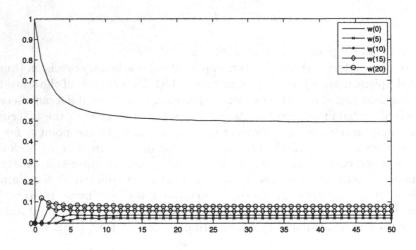

Fig. 7. Transient analysis of the workload for the Chaudhry example

The results of the transient analysis are shown in Figure 6, displaying the convergence of the empty system to the steady state probability distribution. The required number of iterations for different values of ϵ are given in Table 6. After 1135 iterations, the enclosures for the transient and steady state probabilities start to intersect.

The second example is taken from Chaudhry [20, example 4]. The arrival and service distributions are given by

Table 6. Required number of iterations for different ϵ

ϵ	Number of iterations (simple ex.)	Number of iterations (Chaudhry ex.)
10^{-1}	8	5
10^{-2}	45	21
10^{-3}	110	48
10^{-4}	191	80
10^{-5}	280	115
10^{-6}	375	151
10^{-7}	474	190
10^{-8}	575	229
10^{-9}	679	269
10^{-10}	783	309

$$Pr(A = 15) = 2/5 \; Pr(A = 30) = 3/5$$
$$Pr(S = 10) = 4/5 \; Pr(S = 50) = 1/5$$

As in the first example, we assume an empty system as the starting point. The results for this case are displayed in Figure 7 and Table 6 respectively. The intervals intersect after 444 iterations. In this special case, the probability of a specific workload i is only larger than zero if i is zero or a multiple of 5.

4   Conclusions and Further Work

We have presented two approaches on how interval arithmetic can be used to yield validated results in different application settings for stochastic traffic analysis.

It was shown how reliable parameters for the representation of the autocorrelation function of a SSMP as an exponential sum can be determined using eigenvalue decomposition. This features favorable numerical properties and yields reliable results for the parameters of this notation. As an alternative approach, these values can also be calculated using Prony's method.

The calculated coefficient intervals can be expanded in order to include short term oscillation of the autocorrelation evidenced in Figures 2 and 3. Presently we study alternative approaches to the root-finding in Prony's algorithm to implement an interval version of Prony's method.

We also focus on the verified computation in the field of transient analysis in a compact state space. Apart from yielding reliable and consistent results, an interval-based description of system states may allow a further reduction of the number of states considered, as, for instance, adjacent unlikely states (that is, very high workload) may be aggregated by extension of the describing intervals.

Acknowledgements

This research and software development for algorithms with result verification has been carried out by the authors in a recent project funded by the German

Research Council (DFG). We also thank the anonymous reviewers for their valuable comments which have improved this paper and its readability.

References

1. Fausten, D., Haßlinger, G.: Verified numerical analysis of the performance of switching systems in telecommunication. In: Alt, R., Frommer, A., Kearfott, R.B., Luther, W. (eds.) Dagstuhl Seminar 2003. LNCS, vol. 2991, pp. 209–228. Springer, Heidelberg (2004)
2. Traczinski, D., Luther, W., Haßlinger, G.: Polynomial factorization for servers with semi-Markovian workload: Performance and numerical aspects of a verified solution technique. Stochastic Models 21, 643–668 (2005)
3. Kleinrock, L.: Queueing Systems, vol. 1/2. Wiley, Chichester (1975)
4. Haßlinger, G.: Semi-Markovian modelling and performance analysis of variable rate traffic in ATM networks. Telecommunication Systems 7, 281–298 (1997)
5. Seneta, E.: Non-negative matrices and Markov chains. Springer, Heidelberg (1981)
6. Takes, P.: Modellierung von Datenverkehr mittels stochastischer Prozesse angewendet auf Video-Übertragungen in IP-basierten Netzen unter besonderer Berücksichtigung autokorrelierter und semi-Markov-prozesse. Master's thesis, Gerhard-Mercator-Universität Duisburg (2002)
7. Hildebrand, F.B.: Introduction to numerical analysis. McGraw-Hill Book Co., New York (1974)
8. de Prony, G.: Essai experimental et analytique sur les lois de la dilatabilité des fluides élastiques et sur celles de la force expansive de la vapeur de l'eau et de la vapeur de l'alkool, à différentes températures. Journal de l'Ecole Polytechnique 1(2), 24–76 (1975)
9. Osborne, M., Smyth, G.: A modified Prony algorithm for fitting sums of exponential functions. SIAM Journal of Scientific Computing 16, 119–138 (1995)
10. Rump, S.: INTLAB-Interval laboratory. Developments in Reliable Computing, 77–104 (1999)
11. Rump, S.: Computational error bounds for multiple or nearly multiple eigenvalues. Linear Algebra and its Applications 324, 209–226 (2001)
12. Rose, O.: Simple and efficient models for variable bit rate MPEG video traffic. Performance Evaluation 30(69-85) (1997)
13. Fitzek, F.H., Reisslein, M.: MPEG-4 and H.263 video traces for network performance evaluation (extended version). Technical Report TKN-00-06, TU Berlin, Dept. of Electrical Engineering, Telecommunication Networks Group (2000)
14. Garloff, J., Granvilliers, L., Smith, A.P.: Accelerating consistency techniques and Prony's method for reliable parameter estimation of exponential sums. In: Jermann, C., Neumaier, A., Sam, D. (eds.) COCOS 2003. LNCS, vol. 3478, pp. 31–45. Springer, Heidelberg (2005)
15. Chu, M.: Inverse eigenvalue problems. SIAM Review 40(1), 1–39 (1998)
16. Breuer, L.: Numerical results for the transient distribution of the GI/G/1 queue in discrete time. In: Proceedings of the 13th GI/ITG Conference on Measuring, Modelling and Evaluation of Computer and Communication Systems (MMB), pp. 209–218 (2006)
17. Haßlinger, G., Kempken, S.: Transient analysis of a single server system in a compact state space. In: Proceedings of the 13th International Conference on Analytical and Stochastic Modelling Techniques and Applications, pp. 91–96 (2006)

18. Lindley, D.: The theory of queues with a single server. In: Proc. Cambridge Philos. Soc., vol. 48, pp. 277–289 (1952)
19. Kempken, S., Luther, W., Traczinski, D.: Reliable computation of workload distributions using semi-Markov processes. In: Proceedings of the 13th International Conference on Analytical and Stochastic Modelling Techniques and Applications, pp. 111–117 (2006)
20. Chaudhry, M.: Alternative numerical solutions of stationary queueing-time distributions in discrete-time queues: GI/G/1. Journal of the Operational Research Society 44(10), 1035–1051 (1993)

Interval Arithmetic Using SSE-2

Branimir Lambov

BRICS[*], University of Aarhus
IT Parken, 8200 Aarhus N
Denmark
barnie@brics.dk

Abstract. We present an implementation of double precision interval arithmetic using the single-instruction-multiple-data SSE-2 instruction and register set extensions. The implementation is part of a package for exact real arithmetic, which defines the interval arithmetic variation that must be used: incorrect operations such as division by zero cause exceptions and loose evaluation of the operations is in effect. The SSE-2 extensions are suitable for the job, because they can be used to operate on a pair of double precision numbers and include separate rounding mode control and detection of the exceptional conditions. The paper describes the ideas we use to fit interval arithmetic to this set of instructions, shows a performance comparison with other freely available interval arithmetic packages, and discusses possible very simple hardware extensions that can significantly increase the performance of interval arithmetic.

1 Introduction

Numerical computations on a computer are plagued by the problem of roundoff error and its accumulation. Very often we trust results obtained using floating point computations, although we do not know anything about the quality of these results. Verifying their correctness may be a very difficult task, to solve which various mathematical or programming tools may be used.

In this paper we will describe a package which implements interval arithmetic, one of the methods that can be used as one of the steps towards solving the problem, in a very efficient way. The package itself is part of a bigger library for exact real computation, which can be used to solve the roundoff error problems completely by providing certified accuracy.

Interval arithmetic is a method of finding lower and upper bounds for the possible values of a result by performing a computation in a manner which preserves these bounds (for an introduction, see [4]). The IEEE-754 standard for floating point arithmetic [7] has useful features to aid fast interval arithmetic, namely the directed rounding modes that should be present with every IEEE-754 implementation. Unfortunately, in some processor architectures, notably Intel's x86, it is non-trivial to effectively use them, as switching the rounding mode for an operation requires significantly more time

---

[*] Basic Research in Computer Science (www.brics.dk), funded by the Danish National Research Foundation.

P. Hertling et al. (Eds.): Real Number Algorithms, LNCS 5045, pp. 102–113, 2008.

than the operation itself. Even when one takes into account the fact that one of the directed rounding modes can be emulated by operations on negated values rounded in the other direction, an interval arithmetic package has to be aware that users may mix interval with standard floating point arithmetic and would still require repeatedly switching the rounding modes.

Fortunately, the newer generations of the x86 architecture provide an additional set of registers with its own rounding control, the SSE-2 double-precision floating point registers [8]. They can coexist with the old x87-style floating point, which is still the register and instruction set used most widely. Thus, to serve all purposes, we can reserve the SSE-2 register and operation set for interval arithmetic and leave x87-style floating point for any standard floating point operations that the user may be performing.

The SSE-2 instruction set can also work on packed data, as every SSE-2 register can contain and operate on a pair of double-precision floating point numbers. Since an interval is in fact a pair of bounds, one SSE-2 register can be used to hold an interval, which nullifies the additional register pressure that interval arithmetic would normally exert.

The package described in this paper is part of the *RealLib* library for exact real number computations [5] and makes use of this register and instruction set to implement very fast machine precision interval arithmetic.

Often an interval arithmetic package would be expected to continue computations even if it reaches exceptional conditions, such as division by an interval that contains zero. This does not make sense for interval arithmetic within exact arithmetic, as in the latter one assumes that in any such case the bad interval will improve in subsequent reiterations at higher precision and ultimately one can reach a precision which avoids the exceptional condition. Thus, contrary to the modern trends of interval arithmetics, for us actually throwing an exception in such a case is the desirable action, so that computations can restart at higher precision as soon as possible.

Additionally, the package implements loose interval arithmetic, i.e. it ignores the portions of the argument of an operation that are outside its domain, e.g. the negative parts of the argument to a square root, meaning for example that $\sqrt{[-1,4]} = [0,2]$. This is the proper mode of operation to ensure that $\sqrt{0}$ is computable in exact real arithmetic.

2 Key Ideas

Normally, interval arithmetic based on floating point would use two rounding operations, Δ (rounding towards $+\infty$) and ∇ (rounding towards $-\infty$). By default IEEE floating point uses rounding to nearest, which is not useful for our purposes.

We already mentioned that switching the rounding mode has a detrimental effect on the performance of floating point operations, thus we would want to avoid all rounding mode switches. We will only do this once, at the beginning of a computation[1], setting the rounding mode to rounding towards $-\infty$. To compute lower bounds of the results, we will directly use the floating point operation. To compute upper bounds, we will

---

[1] This is accomplished by the construction of a special object that also takes care of restoring the previous rounding mode after the interval computation has completed.

make sure that the result of the floating point operation is negated, thus making use of the identity

$$\Delta(x) = -\nabla(-x).$$

Seeing operations in the form above, compilers are usually overzealous[2] to fold the pair of negations and destroy the effect we want to achieve. To avoid this, at the same time keeping down the number of required operations, we make sure that we always keep the high bound of the interval negated, i.e. our representation of the interval $x = [\underline{x}, \overline{x}]$ is the pair $\langle \underline{x}, -\overline{x} \rangle$. (in the rest of this chapter we will assume every interval is represented in this fashion and will simply write x to mean $[\underline{x}, \overline{x}]$ and $\langle \underline{x}, -\overline{x} \rangle$)

Three observations can be made directly from this:

- in this setting, the sum of x and y is evaluated by $\langle \nabla(\underline{x} + \underline{y}), -\nabla(-\overline{x} - \overline{y}) \rangle$ which is achieved by a single instruction, _mm_add_pd.
- changing the sign of an interval x is achieved by simply swapping the two bounds, i.e. $\langle -\overline{x}, \underline{x} \rangle$, achieved by a single instruction, _mm_shuffle_pd,
- joining two intervals (i.e. finding an interval containing all numbers in both, or finding the minimum of the lower bounds and the maximum of the higher bounds) is performed as $\langle \min(\underline{x}, \underline{y}), -\min((-\overline{x}), (-\overline{y})) \rangle$ in a single instruction, _mm_min_pd.

The latter is used extensively in the computation of multiplication, division and other operations.

3 Operations

In this section we will give short remarks on our implementation of the basic operations on intervals. The operations include the arithmetic operators, including the special cases $-x$, $\frac{1}{x}$, and x^2, absolute value and square root.

All the operations give tight bounds (i.e. the best possible enclosures after rounding).

3.1 Addition

Definition:

$$x + y = [\underline{x} + \underline{y}, \overline{x} + \overline{y}] \subseteq \langle \nabla(\underline{x} + \underline{y}), -\nabla((-\overline{x}) + (-\overline{y})) \rangle$$

Addition is implemented as a single _mm_add_pd instruction. The negated sign of the higher bound ensures the proper direction of the rounding.

3.2 Sign Change

Definition:

$$-x = [-\overline{x}, -\underline{x}] = \langle -\overline{x}, \underline{x} \rangle$$

This is a single swap of the two values, implemented as a _mm_shuffle_pd instruction. No rounding is performed here.

---

[2] The two negations have no effect on the rounding-to-nearest mode which is normally in place in C/C++ code, and on which many standard functions rely, thus this optimization is perfectly legal. Only our specific (non-standard) use of floating point makes it unwanted.

3.3 Subtraction

Definition:

$$x - y = [\underline{x} - \overline{y}, \overline{x} - \underline{y}] \subseteq \langle \nabla(\underline{x} + (-\overline{y})), -\nabla((-\overline{x}) + \underline{y}) \rangle$$

Subtraction is implemented as $x + (-y)$, which corresponds to two processor instructions. This is the best that can be achieved with packed SSE-2 instructions, because the formula requires a combination of the high bound of one of the arguments with the low bound of the other.

3.4 Multiplication

Definition:

$$xy = [\min(\underline{x}\underline{y}, \underline{x}\overline{y}, \overline{x}\underline{y}, \overline{x}\overline{y}), \max(\underline{x}\underline{y}, \underline{x}\overline{y}, \overline{x}\underline{y}, \overline{x}\overline{y})] \qquad (1)$$

Unfortunately, the rounding steps are inseparable parts of the operations, hence the equation above requires 8 multiplications. Using the fact that $\Delta(\nabla(r) + \epsilon) \geq \Delta(r)$ (for ϵ being the smallest representable positive number), one can do with 4 multiplications at the expense of some accuracy.

A widely used alternative examines the signs of both factors to choose one of 9 possible execution paths of different complexity. One of the paths involves taking the minimum and maximum of the result of two pairs of multiplications. Implementing this idea requires branching, which is one of the main performance problems in current processor architectures.

In our implementation we chose an approach where we perform the same amount of work as in the worst case scenario of the case distinction, but we avoid the branching by using boolean arithmetic to select one of the arguments of each multiplication based on the signs of $\underline{x}$ and $\overline{x}$. More specifically, we use these observations:

$$xy = \begin{cases} [\min(\underline{x}\underline{y}, \overline{x}\underline{y}), \max(\overline{x}\overline{y}, \underline{x}\overline{y})], & \text{if } 0 \leq \underline{x} \leq \overline{x} \\ [\min(\underline{x}\overline{y}, \overline{x}\underline{y}), \max(\overline{x}\overline{y}, \underline{x}\underline{y})], & \text{if } \underline{x} < 0 \leq \overline{x} \\ [\min(\underline{x}\overline{y}, \overline{x}\overline{y}), \max(\overline{x}\underline{y}, \underline{x}\underline{y})], & \text{if } \underline{x} \leq \overline{x} < 0 \end{cases} \qquad (2)$$

to conclude that the formula

$$xy \subseteq \langle \min(\nabla(a\underline{x}), \nabla(b(-\overline{x}))), -\min(\nabla(c(-\overline{x})), \nabla(d\underline{x})) \rangle,$$

where

$$a = \begin{cases} \underline{y} & \text{if } 0 \leq \underline{x} \\ -(-\overline{y}) & \text{otherwise} \end{cases}$$

$$b = \begin{cases} -\underline{y} & \text{if } (-\overline{x}) \leq 0 \\ (-\overline{y}) & \text{otherwise} \end{cases}$$

$$c = \begin{cases} -(-\overline{y}) & \text{if } (-\overline{x}) \leq 0 \\ \underline{y} & \text{otherwise} \end{cases}$$

$$d = \begin{cases} (-\overline{y}) & \text{if } 0 \leq \underline{x} \\ -\underline{y} & \text{otherwise} \end{cases}$$

computes the rounded results of the multiplication formula in (1). It uses more instructions than the direct implementation with 8 multiplications, but achieves better performance.

3.5 Multiplication by a Positive Number

When one of the numbers is known to be positive (e.g. a known constant), one can use one of the cases in (2) directly:

$$xy \stackrel{x \geq 0}{=} [\min(\underline{x}\underline{y}, \overline{x}\underline{y}), \max(\overline{x}\overline{y}, \underline{x}\overline{y})]$$

This is significantly faster than the general case multiplication, involving only 5 instructions (4 for constants).

3.6 Multiplication of Two Positive Numbers

If both multiples are known to be positive, multiplication can be achieved by simply changing the sign of the higher bound of one of the arguments followed by _mm_mul_pd. If one of the numbers is a constant, one can prepare it in a suitable way to avoid the sign change and implement the multiplication as a single instruction.

3.7 Division

Definition:

$$\frac{x}{y} = \left[\min\left(\frac{\underline{x}}{\underline{y}}, \frac{\underline{x}}{\overline{y}}, \frac{\overline{x}}{\underline{y}}, \frac{\overline{x}}{\overline{y}}\right), \max\left(\frac{\underline{x}}{\underline{y}}, \frac{\underline{x}}{\overline{y}}, \frac{\overline{x}}{\underline{y}}, \frac{\overline{x}}{\overline{y}}\right)\right],$$

undefined if $0 \in y$.

Again, this computation would require 8 divisions. Unfortunately, division is a rather slow operation, that is why we would prefer to use as few divisions as possible. One way to do this is to use $\frac{x}{y} = x\frac{1}{y}$, using the definition of reciprocal below, which uses only two divisions but quite a few other operations.

A more efficient (as well as more precise) approach turns out to be the use of case distinction similar to (2). By examining the divisor, we end up with fewer possible cases and easy recognition of the exceptional cases. More specifically, the operation becomes:

$$\frac{x}{y} = \begin{cases} \left[\min\left(\frac{\underline{x}}{\underline{y}}, \frac{\underline{x}}{\overline{y}}\right), \max\left(\frac{\overline{x}}{\underline{y}}, \frac{\overline{x}}{\overline{y}}\right)\right], & \text{if } 0 < \underline{y} \leq \overline{y} \\ \text{exception}, & \text{if } \underline{y} \leq 0 \leq \overline{y} \\ \left[\min\left(\frac{\overline{x}}{\underline{y}}, \frac{\overline{x}}{\overline{y}}\right), \max\left(\frac{\underline{x}}{\underline{y}}, \frac{\underline{x}}{\overline{y}}\right)\right], & \text{if } \underline{y} \leq \overline{y} < 0 \end{cases} \quad (3)$$

The final formula we use is

$$\frac{x}{y} \subseteq \left\langle \min\left(\nabla\left(\frac{a}{\underline{y}}\right), \nabla\left(\frac{-a}{(-\overline{y})}\right)\right), -\min\left(\nabla\left(\frac{-b}{(-\overline{y})}\right), \nabla\left(\frac{b}{\underline{y}}\right)\right)\right\rangle,$$

where

$$a = \begin{cases} \underline{x} & \text{if } (-\overline{y}) \leq 0 \\ -(-\overline{x}) & \text{otherwise} \end{cases}$$

$$b = \begin{cases} (-\overline{x}) & \text{if } 0 \leq \underline{y} \\ -\underline{x} & \text{otherwise} \end{cases}$$

with an additional check to throw an exception if $\underline{y} \leq 0 \leq \overline{y}$.

3.8 Reciprocal

Definition:

$$\frac{1}{x} = \left[\frac{1}{\overline{x}}, \frac{1}{\underline{x}}\right] \subseteq \left\langle \nabla\left(\frac{-1}{(-\overline{x})}\right), \nabla\left(\frac{-1}{\underline{x}}\right)\right\rangle,$$

undefined if $0 \in x$.

This is implemented as a check if the argument contains zero, followed by division of -1 by the argument and swapping the two components.

3.9 Absolute Value

Definition:

$$|x| = [\max(\underline{x}, -\overline{x}, 0), \max(-\underline{x}, \overline{x})] = \langle\max(0, \underline{x}, (-\overline{x}))), -\min(\underline{x}, (-\overline{x}))\rangle.$$

3.10 Square

Implemented as $x^2 = |x||x|$, using multiplication of positive numbers.

3.11 Square Root

Definition:

$$\sqrt{x} = \left[\sqrt{\underline{x}}, \sqrt{\overline{x}}\right]$$

only defined if $0 \le \underline{x}$.

Since the rounding is an integral part of the square root operation, and in this case we cannot achieve a negated result, we need to use another method to ensure rounding in the correct direction. We use the fact already mentioned in the subsection on multiplication, $\Delta(r) \le -\nabla(-\epsilon - \nabla(r))$.

The formula we use is:

$$\sqrt{x} \subseteq \begin{cases} \left\langle \nabla\left(\sqrt{\underline{x}}\right), -\nabla\left(\sqrt{-(-\overline{x})}\right)\right\rangle, & \text{if } \nabla\left(\nabla\left(\sqrt{-(-\overline{x})}\right)\right)^2 = -(-\overline{x}) \\ \left\langle \nabla\left(\sqrt{\underline{x}}\right), \nabla\left(\nabla\left(-\epsilon - \sqrt{-(-\overline{x})}\right)\right)\right\rangle, & \text{otherwise} \end{cases}$$

(where ϵ is the smallest representable positive number).

The condition for making the first choice in this formula is only satisfied if the result of $\sqrt{-(-\overline{x})}$ is exactly representable, in which case $\nabla\left(\sqrt{-(-\overline{x})}\right) = \Delta\left(\sqrt{-(-\overline{x})}\right)$. Otherwise the second choice adjusts the high bound to the next representable number.

Note that if we don't require tight bounds, using only the second choice in the equation above is sufficient to implement interval square root.

If the argument is entirely negative, the implementation will raise an exception. If it contains a negative part, the implementation will crop it to only its non-negative part, to allow that computations such as $\sqrt{0}$ can be carried out in exact real arithmetic.

4 Performance

We compare the performance of this implementation to the performance of two other packages for interval arithmetic freely available on the internet: the interval part of the *Boost* project (version 1.33.0, [6]) and the library *filib++* (version 2.0, [2]). For the latter, we tried the macro version as well as two of the available rounding policies, *multiplicative* and *native_onesided_global*, the latter corresponding most closely to our method of rounding.

The results of the benchmark are summarized in the following table, showing the ratio between the performance of the respective library and double precision floating point:

Table 1. Performance comparison (Pentium-M 1.8GHz, Windows XP + Cygwin, GCC 3.4.4)

operation	filib++, macro	filib++, onesided	filib++, mult.	Boost	RealLib
+	6.12	2.45	6.22	9.90	1.05
*	6.78	3.57	6.97	124.27	6.35
/	12.33	3.63	9.24	8.62	3.72
$\sqrt{\cdot}$	27.06	62.23	61.79	15.77	1.73
$\lvert \cdot \rvert$	30.71	23.76	23.76	2.03	3.16
$\sum_{i=1}^{1000000} \frac{1}{i}$	3.67	1.74	2.33	4.90	1.70
dot product	12.11	6.28	13.19	148.86	3.72

RealLib is faster almost everywhere, with the notable exception of multiplication in *filib++*'s *native_onesided_global* mode. In this case *filib++* uses a case distinction, which in our test only reaches the shortest of the 9 possible paths, giving only the best case performance of *filib++*'s multiplication code. In contrast, our implementation uses only one execution path for all multiplications, thus the ratio given in the table is both best and worst case performance. In the dot product operation the reader can see the effect that varying signs of the argument has on the performance of multiplication. The time spent in evaluating a dot product is dominated by the time spent in multiplications, but since our implementation has constant performance while *filib++*'s efficiency deteriorates, our SSE-2 code turns out to be significantly faster.

5 Intel's SSE-3

The latest multimedia extension set introduced by Intel, the SSE-3 [10], aimed at improving complex number computations, does not provide any benefit for interval computations. Intel chose to improve complex multiplications and divisions by introducing the instruction _mm_addsub_pd, which combines two packed registers by adding one of the two components and subtracting the other [9]. Unfortunately, the use of this instruction in complex multiplications leads to incorrect results if a directed rounding mode is in effect, because the multiplication that precedes the subtraction is rounded in the wrong direction.

A better handling of complex multiplications would have been the introduction of a multiplication instruction "*mulpn*" (for multiply positive negative) that changes the sign of one of the components of one of the arguments. This would require the same effort that the instruction _mm_addsub_pd required, but would have the correct behavior in directed rounding modes, i.e. complex multiplication code using *mulpn* would yield upper bounds for the result of the multiplication if rounding towards $+\infty$ is in effect, and lower bounds in the case of rounding towards $-\infty$.

Unlike _mm_addsub_pd, a *mulpn* instruction would have been useful and advantageous for interval arithmetic. Multiplication of two positive numbers could be implemented as a single *mulpn*, which would also speed up the implementations of transcendental interval functions.

6   Suggestions for a Hardware Implementation

We hope that the presentation until this point has convinced the reader that the use of the storage $\langle \underline{x}, -\overline{x} \rangle$ for intervals in SSE-2 registers is clearly superior to the traditional method of storing intervals as simply the pair of the two bounds. This mode of storage avoids the need for special rounding modes in a hardware implementation, and even turns some existing instructions into meaningful interval operations.

We propose this storage to be adopted as the preferred storage format for intervals in hardware implementations.

To further speed up computations on intervals, we propose the introduction of a special selection instruction we call *ivchoice* (for interval choice) that can be used to prepare the arguments for multiplication and division. The action of this instruction should correspond to the following function:

$$
\textit{ivchoice}\left(\langle a_0, a_1 \rangle, \langle b_0, b_1 \rangle\right) = \begin{cases} \langle a_1, a_1 \rangle, & \text{if } b_0 \geq +0 \wedge b_1 \geq +0 \\ \langle a_1, -a_0 \rangle, & \text{if } b_0 \geq +0 \wedge b_1 \leq -0 \\ \langle -a_0, a_1 \rangle, & \text{if } b_0 \leq -0 \wedge b_1 \geq +0 \\ \langle -a_0, -a_0 \rangle, & \text{if } b_0 \leq -0 \wedge b_1 \leq -0 \end{cases} .
$$

This can also be given as the following *C++* function:

```
__m128d ivchoice(__m128d a, __m128d b)
{
    a = _mm_xor_pd(a, _mm_set_pd(0.0, -0.0));
    a = _mm_shuffle_pd(a, a, _mm_movemask_pd(b));
    return a;
}
```

This code does not compile, because _mm_shuffle_pd cannot be performed based on a non-const integer. A software implementation of the above requires a switch statement, which slows the execution considerably, especially in cases where the signs of the multiples cannot be predicted.

If such an instruction is available, the multiplication code becomes:

```
__m128d IntervalMul(__m128d x, __m128d y)
{
    __m128d a, b;
    a = _mm_shuffle_pd(x, x, 1);          // 1
    b = _mm_shuffle_pd(y, y, 1);          // 2
    y = ivchoice(y, a);                   // 3
    b = ivchoice(b, x);                   // 4
    y = _mm_mul_pd(y, a);                 // 5
    b = _mm_mul_pd(b, x);                 // 6
    y = _mm_min_pd(b, y);                 // 7
    return y;
}
```

The following is a pseudocode translation of the function for the readers that are not familiar with the SSE-2 instructions:

1. Swap the two components of x and store the result in a.
2. Swap the two components of y and store the result in b.
3. Apply *ivchoice* to y and a and store the result in y.
4. Apply *ivchoice* to b and x and store the result in b.
5. Multiply y and a componentwise and store the result in y.
6. Multiply b and x componentwise and store the result in b.
7. Return the componentwise minimum of b and y.

If the latency of the proposed instruction can be the same as the latency of the instruction _mm_shuffle_pd, this sequence of instructions will run about 30% faster than the current implementation.

Moreover, since the multiplications above only use the results of *ivchoice* with the same second argument, it is even possible to fuse *ivchoice* with the multiplication that is applied to the result:

```
__m128d ivmul(__m128d a, __m128d b)
{
    a = _mm_xor_pd(a, _mm_set_pd(0.0, -0.0));
    a = _mm_shuffle_pd(a, a, _mm_movemask_pd(b));
    a = _mm_mul_pd(a, b);
    return a;
}
```

Correspondingly, the *IntervalMul* function will in this case change to:

```
__m128d IntervalMul(__m128d x, __m128d y)
{
    __m128d a, b;
    a = _mm_shuffle_pd(x, x, 1);
    b = _mm_shuffle_pd(y, y, 1);
    y = ivmul(y, a);
    b = ivmul(b, x);
    y = _mm_min_pd(b, y);
    return y;
}
```

The extent to which such fusion can be beneficial depends on the actual hardware implementation. If the latency of *ivchoice* can be folded completely (which seems possible) or partially, interval multiplication using the fused *ivmul* could reach a latency close to the latency of two dependant double precision multiplications.

Apart from an additional test if the divisor contains zero and the use of _mm_div_pd instead of _mm_mul_pd, the division code is identical to the multiplication one:

```
__m128d IntervalDiv(__m128d y, __m128d x)
{
    __m128d a, b;
    if (_mm_movemask_pd(x)==3)
        throw exception;
    a = _mm_shuffle_pd(x, x, 1);
    b = _mm_shuffle_pd(y, y, 1);
    y = ivchoice(y, a);
    b = ivchoice(b, x);
    y = _mm_div_pd(y, a);
    b = _mm_div_pd(b, x);
    y = _mm_min_pd(y, b);
    return y;
}
```

Fused *ivchoice* and division ("*ivdiv*") is also possible, and the changes to the division code are exactly as above.

Of course, one would prefer to have a complete hardware implementation of interval arithmetic that provides instructions for the four basic operations on intervals. In our mode of operation addition already has a hardware implementation as a single instruction. Subtraction would require a fusion of swapping and addition ("*ivsub*") which should be easy to accomplish in hardware without extra latency compared to addition.

On the other hand, multiplication and division seem too complex to be directly implemented. A pure hardware implementation of multiplication may be able to choose execution paths without the delays associated with incorrect branch predictions, thus probably the preferable hardware design would examine the signs of the four components to choose one of 9 possible combinations and perform a single pair of multiplications in 8 of the possible cases. In the 9'th case, however, the operation would require the same amount of work as the function *IntervalMul* above.

Since the worst-case latency would be the same as the algorithm above, the latter should not be ignored as a possible basis for a pure hardware implementation of interval multiplication.

To conclude, we suggest that hardware assistance for interval computations should be provided as the adoption of the $\langle \underline{x}, -\overline{x} \rangle$ storage format and the introduction of the instructions of one of the following three levels:

> basic *mulpn, ivsub, ivchoice*
>
> advanced *mulpn, ivsub, ivmul, ivdiv*
>
> full *ivsub, IntervalMul, IntervalDiv*

The advanced level seems to be the best combination of feasibility and performance.

7 Related Work

In [1], von Gudenberg discusses the efficiency of implementations of interval arithmetic using the multimedia extensions Intel's SSE, AMD's 3DNow! and Motorola's AltiVec. The paper concludes that the use of multimedia extensions only leads to a very modest improvement in multiplication with Intel's SSE in comparison to standard floating point, and only due to the fact that four single-precision operations can be executed in parallel.

Unlike SSE, the double precision second version of the extensions, SSE-2, is a natural candidate for interval arithmetic because the packed registers hold two double precision values.

Von Gudenberg used a variety of rounding policies, the fastest of which is global onesided rounding, the method we use, but did not store one of the components negated in memory. Consequently, handling the negations required to perform rounding in the proper direction increases the number of instructions needed for every operation. If we were to use SSE-2 in a similar mode of operation, the required number of instructions for addition would be four instead of one, for sign change – two instead of one, for subtraction – five instead of two, and for multiplication of positive intervals – three instead of two.

Additionally, instead of 9-case branching on the signs of the 4 components, we prefer to use 4 multiplications with selected arguments (the selection is branch-free), which gives us stable performance that is not affected by branch mispredictions or longer latency execution paths, although with a worse best-case performance.

In [3], Kolla, Vodopivec and von Gudenberg discuss the possibility of hardware extensions supporting interval arithmetic similar to the multimedia extensions 3DNow!, via packed storage of single precision numbers in a double precision register. For addition and subtraction they require special instructions that round each component of the pair in the appropriate direction, and for multiplication they describe a case selection method that can easily be implemented and be very efficient for 8 of the 9 possible cases and requires a sequence of operations and longer latency for the (rare) 9'th case.

We are quite skeptical about the chances of such a complicated multiplication instruction ever being implemented in hardware. Instead, we give a much more modest proposal that can also lead to very good performance at the cost of little extra hardware. It also has the benefit that one of the operations, addition, already has a hardware implementation.

References

1. von Gudenberg, J.W.: Interval Arithmetic on Multimedia Architectures. Reliable Computing 8(4) (2002)
2. Hofschuster, W., Krämer, W., Lerch, M., Tischler G., von Gudenberg, J.W.: The Interval Library fi_lib++ 2.0 Design, Features and Sample Programs. Preprint 2001/4,Universität Wuppertal (2001),
   http://www.math.uni-wuppertal.de/wrswt/preprints/prep_01_4.pdf

3. Kolla, R., Vodopivec, A., von Gudenberg, J.W.: The IAX Architecture – Interval Arithmetic Extension. Universität Würzburg, Institut für Informatik, Techn. Report TR225 (1999), http://www2.informatik.uni-wuerzburg.de/mitarbeiter/wvg/Public/iax.ps.gz
4. Kearfott, R.B.: Interval Computations: Introduction, Uses, and Resources. Euromath Bulletin 2(1), 95–112 (1996)
5. Lambov, B.: RealLib: An Efficient Implementation Exact Real Arithmetic. Mathematical Structures in Computer Science (to appear), http://www.brics.dk/~barnie/RealLib/
6. Boost Interval Arithmetic Library, http://www.boost.org/libs/numeric/interval/doc/interval.htm
7. IEEE Standards Committee 754, IEEE Standard for Binary Floating-Point Arithmetic, ANSI/IEEE Standard 754-1985. Institute of Electrical and Electronics Engineers, New York (1985); reprinted in SIGPLAN Notices, 22(2), 9–25 (1987)
8. Intel Corp. IA-32 Intel Architecture Software Developer's Manual, Volumes 1-3, http://developer.intel.com/design/pentium4/manuals/index_new.htm
9. Intel Corp. Using SSE3 Technology in Algorithms with Complex Arithmetic, http://www.intel.com/cd/ids/developer/asmo-na/eng/dc/pentium4/optimization/66717.htm
10. Intel Corp. Next Generation Intel Processor: Software Developers Guide, http://www.intel.com/cd/ids/developer/asmo-na/eng/dc/pentium4/optimization/66756.htm

Worst Cases for the Exponential Function in the IEEE 754r decimal64 Format

Vincent Lefèvre[1], Damien Stehlé[2,*], and Paul Zimmermann[3]

[1] INRIA/ÉNS Lyon/Université de Lyon/LIP,
46 allée d'Italie, F-69364 Lyon Cedex 07, France
Vincent.Lefevre@inria.fr
[2] CNRS/ÉNS Lyon/Université de Lyon/LIP/INRIA Arenaire,
46 allée d'Italie, F-69364 Lyon Cedex 07, France
damien.stehle@gmail.com
[3] LORIA/INRIA Lorraine, Bâtiment A, Technopôle de Nancy-Brabois,
615 rue du jardin botanique, F-54602 Villers-lès-Nancy Cedex, France
Paul.Zimmermann@loria.fr

Abstract. We searched for the worst cases for correct rounding of the exponential function in the IEEE 754r decimal64 format, and computed all the bad cases whose distance from a breakpoint (for all rounding modes) is less than 10^{-15} ulp, and we give the worst ones. In particular, the worst case for $|x| \geq 3 \times 10^{-11}$ is $\exp(9.407822313572878 \times 10^{-2}) = 1.098645682066338\,5\,0000000000000000\,278\ldots$. This work can be extended to other elementary functions in the decimal64 format and allows the design of reasonably fast routines that will evaluate these functions with correct rounding, at least in some domains.

1 Introduction

Most computers nowadays support the IEEE 754-1985 standard for binary floating-point arithmetic [1], which requires that all four arithmetic operations $(+, -, \times, \div)$ and the square root are *correctly rounded*. However radix 10 is more suited to some applications, such as financial and commercial ones, and there have been propositions to normalize it as well and also design hardware implementations. The IEEE 854-1987 standard for radix-independent floating-point arithmetic [2] has been a first step in this direction, but this standard just gives some constraints on the value sets and is not even specific to radix 10. The article [3] describes a first specification of a decimal floating-point arithmetic; it has been improved and the specification included in the current working draft of the revision of the IEEE 754 standard (754r) is described in [4].

One also seeks to extend the IEEE 754 standard to elementary functions, such as the exponential, logarithm and trigonometric functions, by requiring correct rounding on these functions too. Unfortunately fulfilling this requirement is much more complicated than with the basic operations. Indeed, while efficient algorithms to guarantee the correct rounding are known for the basic operations, the

---

* Hosted and partially funded by the MAGMA group (University of Sydney) during the completion of this work.

P. Hertling et al. (Eds.): Real Number Algorithms, LNCS 5045, pp. 114–126, 2008.

only known way to evaluate $f(x)$, where f is an elementary function and x is a machine number[1], is to compute an approximation to $f(x)$ without any useful knowledge except an error bound; and the exact result $f(x)$ may be very close to a machine number or to the middle of two consecutive machine numbers (which are the discontinuity points of the rounding functions), in which case correct rounding can be guaranteed only if the error on the approximation is small enough. This problem is known as the *Table Maker's Dilemma* (TMD). Some cases can be decided easily, but the only known way to obtain a bound on the acceptable error for any input value is to perform an exhaustive search (with a 64-bit format, as considered below, there are at most 2^{64} possible input values). The arguments x for which the values $f(x)$ are the hardest to round are called *worst cases*.

Systematic work on the TMD in radix 2 was first done by Lefèvre and Muller [5], who published worst cases for many elementary functions in double precision, over the full IEEE 754 range for some functions. And correct rounding requirements for some functions in some domains have been added to the 754r working draft. Improved algorithms to deal with higher precisions are given in [6], and in the present paper, the practical feasibility of the method for decimal formats is demonstrated. Indeed the worst cases depend on the representation (radix and precision) and the mathematical function.

Section 2 describes the decimal formats, how worst cases are expressed and briefly recalls the algorithms (in the decimal context) to search for these worst cases. Section 3 gives all worst cases of the exponential function in the 64-bit decimal format. These results allow us to give Theorem 1.

Theorem 1. *In the IEEE 754r decimal64 format, among all the finite values $|x| \geq 3 \times 10^{-11}$ such that $\exp(x)$ does not yield an exception, the input x such that $\exp(x)$ is nearest from a breakpoint, both for rounding-to-nearest and directed rounding modes, is $9.407822313572878 \times 10^{-2}$, and for this input, the exact value of $\exp(x)$ is:*

$$\underbrace{1.098645682066338}_{16 \ digits}\underbrace{50000000000000000}_{17 \ digits}278\dots.$$

Among all finite values x such that $\exp(x)$ does not yield an exception and $\exp(x) \notin [1 - 10^{-16}/2, 1 + 10^{-15}/2]$, the input x such that $\exp(x)$ is nearest from a breakpoint is $9.999999999999995 \times 10^{-16}$, and for this input, the exact value of $\exp(x)$ is:

$$\underbrace{1.000000000000000}_{16 \ digits}\underbrace{99999\dots99999}_{30 \ digits}666\dots.$$

2   The Table Maker's Dilemma in Decimal

In this section, the decimal formats are described in Section 2.1, then the general form of worst cases is given, along with a few illustrating examples (Section 2.2). Finally, the algorithms to search for these worst cases are briefly recalled and applied to radix 10 (Section 2.3).

---

[1] A number that is exactly representable in the floating-point system.

2.1   The Decimal Formats

As specified by the IEEE 854 standard [2], a non-special[2] decimal floating-point number x in precision n has the form:

$$x = (-1)^s \, 10^E \, d_0.d_1d_2 \ldots d_{n-1}$$

where $s \in \{0, 1\}$, the exponent E is an integer between two given integers $E_{\min}$ and $E_{\max}$, and the mantissa $d_0.d_1d_2 \ldots d_{n-1}$ is a fixed-point number written in radix 10; i.e., for i between 0 and $n - 1$, one has: $0 \le d_i \le 9$.

As d_0 may be equal to 0, some numbers have several representations and the standard does not distinguish them. Without changing the value set, one can require that if $E \ne E_{\min}$, then $d_0 \ne 0$, and this will be done in the following for the sake of simplicity. A number such that $d_0 \ne 0$ is called a *normal number*, and a number such that $d_0 = 0$ (in which case $E = E_{\min}$) is called a *subnormal number*. In this way, the representation of a floating-point number is uniquely defined.

Below $\mathrm{ulp}(x)$ denotes the weight of the digit d_{n-1} in this unique representation; i.e., $\mathrm{ulp}(x) = 10^{E-n+1}$.

The document [4], based on the IEEE 854 standard, defines three decimal formats, whose parameters are given in Table 2.1: decimal32, decimal64 and decimal128, with an encoding on 32, 64 and 128 bits respectively. This specification has been included in the 754r working draft.

Table 1. The parameters of the three 754r decimal formats

Format	decimal32	decimal64	decimal128
Precision n (digits)	7	16	34
$E_{\min}$	−95	−383	−6143
$E_{\max}$	96	384	6144

2.2   The Bad and Worst Cases

Given a floating-point format, let us call a *breakpoint* a value where the rounding changes in one of the rounding modes, i.e., a discontinuity point of the rounding functions. A breakpoint is either a machine number (for the directed rounding modes) or the middle of two consecutive machine numbers (for the rounding-to-nearest mode).

For a given function f and a "small" positive number ε, a machine number x is a *bad case* when the distance between the exact value of $f(x)$ and the nearest breakpoint(s) is less than $\varepsilon \cdot \mathrm{ulp}(f(x))$. For instance, if f is the exponential function in the decimal64 format ($n = 16$ digits), then the machine numbers 0.5091077534282133 and 0.7906867968553504 are bad cases for $\varepsilon = 10^{-16}$, since

---

[2] The special floating-point numbers are not-a-number (NaN), the positive and negative infinities, and the positive and negative zeros.

for these values, $\exp(x)$ is close enough to the middle of two consecutive machine numbers:

$$\exp(0.5091077534282133) = 1.\underbrace{6638060072615095}_{\text{16 digits}}\underbrace{0000000000000000}_{\text{16 digits}}49\ldots$$

and

$$\exp(0.7906867968553504) = \underbrace{2.204910231771509}_{\text{16 digits}}\underbrace{4999999999999999}_{\text{16 digits}}16\ldots,$$

i.e., rounding $\exp(x)$ in the rounding-to-nearest mode requires to evaluate $\exp(x)$ in a precision significantly higher than the target precision. Similarly, with the following bad cases, $\exp(x)$ is very close to a machine number, so that rounding it in directed rounding modes also requires to evaluate it in a precision significantly higher than the target precision:

$$\exp(0.001548443067391468) = 1.\underbrace{0015496425243749}_{\text{16 digits}}\underbrace{9999999999999999}_{\text{16 digits}}26\ldots$$

and

$$\exp(0.2953379504777270) = \underbrace{1.343580345589067}_{\text{16 digits}}\underbrace{0000000000000000}_{\text{16 digits}}86\ldots.$$

2.3   Searching for Bad and Worst Cases

Searching for bad cases in decimal is very similar to the search in binary. First the domain of the tested function is selected: arguments that give an underflow or an overflow are not tested, and some other arguments do not need to be tested either when a simple reasoning can be carried out (see Section 3.1 as an example). And like in binary [7,8,9], probabilistic hypotheses allow us to guess that the smallest distance amongst all the arguments to be tested is of the order of 10^{-n} ulp (divided by the number of exponents E), so that we can choose $\varepsilon \sim 10^{-n}$ to get only a few bad cases[3]; i.e., we search for bad cases with at least n (or $n-1$) identical digits 0 or 9 (possibly except the first one, which may be respectively 5 or 4) after the n-digit mantissa.

In the decimal32 format, the number of arguments to be tested is small enough for a naive algorithm to be sufficient: for each argument x, one computes $f(x)$ in a higher precision to eliminate the values x for which the distance between $f(x)$ and the nearest breakpoint(s) is larger than $\varepsilon \cdot \mathrm{ulp}(f(x))$. Since finding

---

[3] This may not be true in some domains, for instance when the function can be approximated accurately by a simple degree-2 polynomial, such as $\exp(x) \simeq 1+x+x^2/2$ for x sufficiently close to 0; in this case, one can get bad cases which are much closer to breakpoints and more numerous than what can be estimated with the probabilistic hypotheses. This is not a problem in practice: A simple reasoning is usually sufficient instead of an exhaustive search in this domain.

bad cases is rather easy for the decimal32 format, this paper will not focus on this format; the reader may find some results for the exponential function at http://www.loria.fr/~zimmerma/wc/decimal32.html.

In the decimal64 format, the number of remaining arguments after reducing the domain is still very large (say, around 10^{17} to 10^{19}, depending on the function), and a naive algorithm would require several centuries of computations. Like in the binary double precision, one needs specific algorithms, and since the decimal arithmetic has the same important properties as the binary one (the machine numbers are in arithmetic progression except at exponent changes, the breakpoints have a similar form...), the same methods can be applied.

In radix 2, bad cases for precision n and any rounding mode are the same as bad cases for precision $n+1$ and directed rounding modes[4], so that the problem was restricted to directed rounding modes in [6]. This property is no longer true in radix 10, but the breakpoints are still in an arithmetic progression (except when the exponent changes, just like in radix 2), which is the only important property used by our algorithms. Indeed in each domain where the exponent of $f(x)$ does not change, one needs to search for the solutions of:

$$|f(x) \bmod (u/2)| < \varepsilon u,$$

where $u = \mathrm{ulp}(f(x))$, which is a constant in the considered domain.

To solve this problem, one splits the domain into subintervals, and in each subinterval, one approximates the function f by a polynomial P of small degree and scales/translates the input and output values to reduce the problem to the following (as in the binary case [6]):

Real Small Value Problem (Real SValP). Given positive integers M and T, and a polynomial P with real coefficients, find all integers $|t| < T$ such that:

$$|P(t) \bmod 1| < \frac{1}{M}. \tag{1}$$

The coefficients of the polynomial are computed using the MPFR library [10] in order to obtain guaranteed error bounds.

Then several fast algorithms can be used to solve the Real SValP. Lefèvre's algorithm needs degree-1 polynomial approximations; as these approximations are valid on very small intervals, one also needs a way to determine these approximations very quickly [11]. The Stehlé-Lefèvre-Zimmermann (SLZ) algorithm allows to have polynomials of higher degrees and has a smaller asymptotic complexity [6], but with a high constant factor. It is based on Coppersmith's technique to find the small roots of multivariate polynomials modulo an integer: informally, in our situation, we look for small roots of $P(x) + y$ modulo 1. Coppersmith's technique was first introduced in a cryptographic context [12], and heavily relies on the LLL algorithm for reducing Euclidean lattice bases [13]. Heuristically, LLL takes as input a basis derived from the multivariate polynomial and its

---

[4] Said otherwise, in radix 2, the breakpoints for precision n and all rounding modes are the machine numbers in precision $n+1$.

powers: this basis contains the information we are interested in (the roots of the initial polynomial), but in an inconvenient way (there is no known way to efficiently compute roots modulo an arbitrary integer). LLL outputs a basis made of shorter vectors. In particular, if all the various parameters are chosen adequately, the first output vectors will be short enough to ensure that the corresponding polynomials contain among their roots (over the integers, without the modulus) the roots of the initial polynomial.

In order to make the implementation of the SLZ algorithm as efficient as possible, it is crucial to use an efficient LLL code. For instance, one should avoid using the text-book LLL algorithm making use of a rational arithmetic. In the implementation of the SLZ algorithm, it is better to use variants of the LLL algorithm relying on floating-point arithmetic rather than rational arithmetic within the Gram-Schmidt computations (central in LLL).

In his PhD thesis [14], Stehlé describes three floating-point variants of LLL, respectively called "fast", "heuristic" and "proved". The corresponding codes are available at http://perso.ens-lyon.fr/damien.stehle. The proved variant implements the algorithm described in [15], whereas the other two can fail[5] but are usually more efficient.

Remark 1. The above methods may no longer work well for the smallest subnormals, due to the loss of precision for these numbers. For instance, a low-degree polynomial approximation may be valid on an interval that contains only very few machine numbers. Nevertheless these few values may be tested separately with a naive algorithm, if need be.

3 The Exponential Function

We now show the feasibility of our method on the exponential function, denoted exp, in the decimal64 format. This is just an example: a similar work can be carried out for other functions. After a simple analysis of the function (Section 3.1), we search for bad cases (Section 3.2).

3.1 Correctly Rounding the Exponential Function

Let us first recall the parameters of the decimal64 format, with a few more details. A non-special floating-point number x has the form:

$$x = (-1)^s \, 10^E \, d_0.d_1 d_2 \ldots d_{15}$$

where $s \in \{0, 1\}$ and $-383 \le E \le 384$. So, the largest finite machine number is $10^{385} - 10^{369}$, the smallest positive normal machine number is 10^{-383} and the smallest positive machine number is 10^{-398}.

Now let us briefly analyze the exponential function, assuming that the argument is a finite number, to eliminate the special cases. The exponential function

---

[5] In practice, when they fail, they loop forever; they may also return a badly-reduced basis. But in both situations, no bad case will be missed.

is mathematically defined on the whole domain of real numbers, so that the value will never be a NaN. It is increasing, with $\exp(x) \to +\infty$ when $x \to +\infty$, and $\exp(x) \to 0$ when $x \to -\infty$. And the mathematical properties of the exponential function are such that there will be an overflow when x is larger than some value and an underflow when x is smaller than some value. Moreover, $\exp(0) = 1$, meaning that for values of x close to 0, the rounding of $\exp(x)$ is determined only by the rounding mode and the sign of x.

So, there are four couples of consecutive machine numbers (a^-, a^+), (b^-, b^+), (c^-, c^+) and (d^-, d^+) that determine the following five intervals:

$$\underbrace{-\infty \ldots a^-}_{+0} \ \underbrace{a^+ \ldots b^-}_{\text{search}} \ \underbrace{b^+ \ldots c^-}_{1} \ \underbrace{c^+ \ldots d^-}_{\text{search}} \ \underbrace{d^+ \ldots +\infty}_{+\infty}$$

where in intervals 1, 3 and 5, the rounded values in the rounding-to-nearest mode are respectively $+0$, 1 and $+\infty$ (the rounded values in the directed rounding modes can also be determined, keeping the same interval bounds for the sake of simplicity), and in intervals 2 and 4, a search for bad cases is needed. These interval bounds are determined below.

An argument x generates an overflow when the *rounded* result obtained assuming an unbounded exponent range exceeds the largest finite machine number $10^{385} - 10^{369}$. One has:

$$\log(10^{385} - 10^{369}/2) = \underbrace{886.4952608027075}_{\text{16 digits}}882469 \ldots,$$

so that one gets an overflow if and only if $x \geq d^+$, with $d^+ = 886.4952608027076$ (x being a machine number).

Concerning a^-, one has:

$$\log(10^{-398}/2) = -\underbrace{917.1220141921901}_{\text{16 digits}}2 \ldots,$$

so that in any rounding mode, $\exp(x)$ is rounded to the same value for any $x \leq a^-$, with $a^- = -917.1220141921902$: It is rounded to 10^{-398} in the rounding to $+\infty$ mode, and $+0$ in the other rounding modes.

Concerning b^+ and c^-, one has:

$$\log(1 - 10^{-16}/2) = -\underbrace{5.000000000000000}_{\text{16 digits}}125 \ldots \times 10^{-17}$$

and

$$\log(1 + 10^{-15}/2) = \underbrace{4.999999999999998}_{\text{16 digits}}750 \ldots \times 10^{-16},$$

so that one chooses $b^+ = -5 \times 10^{-17}$ and $c^- = 4.999999999999998 \times 10^{-16}$.

Finally, in the other domains, that is for x in

$$[a^+, b^-] = [-917.1220141921901, -5.000000000000001 \times 10^{-17}]$$

and in

$$[c^+, d^-] = [4.999999999999999 \times 10^{-16}, 886.4952608027075],$$

a search for bad cases needs to be done to be able to round $\exp(x)$ correctly in any rounding mode.

Remark 2. When x is close enough to 0, one could use the approximation $\exp(x) \simeq 1 + x + x^2/2$ to find bad cases with much less computing time in this domain. But globally, one would gain very little since this is an easy domain (as the error on a polynomial approximation is very small compared to higher values of x, and the algorithms work much better).

3.2   Searching for Bad and Worst Cases of the Exponential Function

To search for bad cases, one first splits the tested domain into intervals in which both the argument x and the result $\exp(x)$ have a constant (possibly different) exponent. This has been done with a small Maple program.

Table 2. All worst cases of the decimal64 exponential function for $x \geq 10^{-9}$, whose distance from a breakpoint is less than 5×10^{-17} ulp. The notation d^k means that the digit d is repeated k times.

x	$\exp(x)$
$6.581539478341669 \times 10^{-9}$	$1.000000006581539\,5\,0^{15}\,177\ldots$
$2.662858264545929 \times 10^{-8}$	$1.000000026628583\,0\,0^{15}\,318\ldots$
$3.639588333766983 \times 10^{-8}$	$1.000000036395884\,0\,0^{15}\,240\ldots$
$6.036998017773271 \times 10^{-8}$	$1.000000060369982\,0\,0^{15}\,379\ldots$
$6.638670361402304 \times 10^{-7}$	$1.000000663867256\,4\,9^{15}\,569\ldots$
$9.366572213364879 \times 10^{-7}$	$1.000000936657659\,9\,9^{15}\,883\ldots$
$7.970613003079781 \times 10^{-6}$	$1.000007970644768\,5\,0^{15}\,362\ldots$
$3.089765552852523 \times 10^{-5}$	$1.000030898132866\,0\,0^{15}\,241\ldots$
$1.302531956641873 \times 10^{-4}$	$1.000130261678980\,0\,0^{16}\,798\ldots$
$2.241856702421245 \times 10^{-4}$	$1.000224210801727\,5\,0^{15}\,118\ldots$
$7.230293679121590 \times 10^{-4}$	$1.000723290816653\,4\,9^{16}\,127\ldots$
$5.259640428979129 \times 10^{-3}$	$1.005273496619909\,4\,9^{15}\,739\ldots$
$9.407822313572878 \times 10^{-2}$	$1.098645682066338\,5\,0^{16}\,278\ldots$
$1.267914924960933 \times 10^{-1}$	$1.135180299492843\,0\,0^{16}\,706\ldots$
$5.091077534282133 \times 10^{-1}$	$1.663806007261509\,5\,0^{15}\,492\ldots$
3.359104074009002	$28.76340944572687\,5\,0^{16}\,904\ldots$
19.10511686234796	$1.982653538414981\,9\,9^{15}\,735\ldots \times 10^8$
294.9551257293143	$1.251363586659789\,5\,0^{15}\,108\ldots \times 10^{128}$
587.9131381356093	$2.125356221825522\,4\,9^{15}\,594\ldots \times 10^{255}$

Table 3. All worst cases of the decimal64 exponential function for $x \leq -10^{-10}$, whose distance from a breakpoint is less than 5×10^{-17} ulp. The notation d^k means that the digit d is repeated k times.

x	$\exp(x)$
$- 2.090862502185853 \times 10^{-9}$	$0.9999999979091375\,0\,0^{15}\,371\ldots$
$- 3.803619857233762 \times 10^{-9}$	$0.9999999961963801\,4\,9^{15}\,841\ldots$
$- 7.170496225708008 \times 10^{-9}$	$0.9999999928295038\,0\,0^{15}\,252\ldots$
$- 9.362256793825926 \times 10^{-9}$	$0.9999999906377432\,4\,9^{15}\,580\ldots$
$- 4.024416580979643 \times 10^{-8}$	$0.9999999597558350\,0\,0^{15}\,308\ldots$
$- 6.306378165019860 \times 10^{-7}$	$0.9999993693623823\,5\,0^{15}\,301\ldots$
$- 7.720146779532548 \times 10^{-7}$	$0.9999992279856200\,4\,9^{15}\,612\ldots$
$- 9.753167969712726 \times 10^{-7}$	$0.9999990246836786\,4\,9^{16}\,120\ldots$
$- 5.911964024384330 \times 10^{-5}$	$0.9999408821072876\,5\,0^{15}\,384\ldots$
$- 8.232272117182855 \times 10^{-5}$	$0.9999176806672504\,0\,0^{15}\,312\ldots$
$- 8.232461306131942 \times 10^{-5}$	$0.9999176787755166\,4\,9^{15}\,555\ldots$
$- 8.496743395712491 \times 10^{-2}$	$0.9185421971989605\,4\,9^{15}\,843\ldots$
$- 9.250971335383380 \times 10^{-2}$	$0.9116403558361098\,9\,9^{15}\,563\ldots$
$- 9.337621398029658 \times 10^{-2}$	$0.9108507610382665\,0\,0^{15}\,400\ldots$
$- 9.341228128742237 \times 10^{-2}$	$0.9108179096965556\,4\,9^{16}\,587\ldots$
$- 9.998733949173545 \times 10^{-2}$	$0.9048488738100865\,0\,0^{15}\,330\ldots$
$- 1.452866822458144$	$0.2338987797314129\,0\,0^{15}\,413\ldots$
$- 5.085363904672046$	$6.186635335115975\,4\,9^{15}\,774\ldots \times 10^{-3}$
$- 5.815903811599861$	$2.979785944945804\,5\,0^{15}\,173\ldots \times 10^{-3}$
$- 11.93382527979436$	$6.564558652611456\,9\,9^{15}\,658\ldots \times 10^{-6}$
$- 46.84177248885496$	$4.538127418220535\,9\,9^{15}\,769\ldots \times 10^{-21}$
$- 84.88822783213444$	$1.359912838893469\,5\,0^{15}\,266\ldots \times 10^{-37}$
$- 495.9839910528425$	$3.952661043031169\,5\,0^{15}\,371\ldots \times 10^{-216}$
$- 524.2585830842744$	$2.076778963867845\,0\,0^{15}\,287\ldots \times 10^{-228}$

As said in [11] and [5], one could test the inverse function, i.e., the logarithm, instead of the exponential when x is small enough (say, $|x| < 1$). The reason is that there are fewer machine numbers to test in this domain for the inverse function. However this domain requires very little computation time compared to those with high values of x.

The search for bad cases was performed with BaCSeL[6], running on a few machines. The chosen parameters were: a working precision of 200 bits, $m = 14.6$ (the quality of the bad cases, i.e., $-\log_{10}(2\varepsilon)$, to get all bad cases for $\varepsilon = 10^{-15}$), $t = 5.5$ (a parameter that fixes the size of the sub-intervals), $d = 3$ (the degree

---

[6] Available on http://perso.ens-lyon.fr/damien.stehle

of the polynomials) and $\alpha = 2$ (a parameter for Coppersmith's technique). For values of x close enough to 0, the fast LLL variant fails, so that the proved variant is used in this domain.

Tables 2 and 3 present all the bad cases for $x \geq 10^{-9}$ and for $x \leq -10^{-10}$ respectively, whose distance from a breakpoint is less than 5×10^{-17} ulp.

For $-10^{-9} < x < 10^{-8}$ (and in particular for the smaller domain $-10^{-10} < x < 10^{-9}$), many bad cases have some patterns in their mantissa. For instance, one has the following bad cases with $\varepsilon = 3 \times 10^{-15}$ (look at the 8th, 9th and 10th digits):

$$3.897940992403028 \times 10^{-9},$$
$$4.230932991049603 \times 10^{-9},$$
$$4.291382990792016 \times 10^{-9},$$
$$4.581289989505891 \times 10^{-9}.$$

This comes from the fact that $\exp(x)$ can be approximated by $1 + x + x^2/2 + x^3/6$ in these domains, and even by $1 + x + x^2/2$ for smaller values of x. Tables 4 and 5 give some other bad cases for $c^+ \leq x < 10^{-9}$ and $-10^{-10} < x \leq b^-$ respectively.

The complete list of all worst cases which are at a distance less than 10^{-15} ulp from a breakpoint is available at http://www.loria.fr/~zimmerma/wc/decimal64.html.

Table 4. Some bad cases of the exponential function in the decimal64 format, for $c^+ = 4.999999999999999 \times 10^{-16} \leq x < 10^{-9}$. At most two bad cases (the worst ones) are given per exponent.

x	$\exp(x)$
$6.000119998199928 \times 10^{-10}$	$1.000000000600011\,9\,9^{16}\,567\ldots$
$5.999879998200072 \times 10^{-10}$	$1.000000000599988\,0\,0^{16}\,431\ldots$
$1.039999999994592 \times 10^{-11}$	$1.000000000010399\,9\,9^{17}\,625\ldots$
$1.019999999994798 \times 10^{-11}$	$1.000000000010199\,9\,9^{17}\,646\ldots$
$1.199999999999280 \times 10^{-12}$	$1.000000000001199\,9\,9^{20}\,423\ldots$
$1.099999999999395 \times 10^{-12}$	$1.000000000001099\,9\,9^{20}\,556\ldots$
$1.399999999999902 \times 10^{-13}$	$1.000000000000139\,9\,9^{23}\,085\ldots$
$1.199999999999928 \times 10^{-13}$	$1.000000000000119\,9\,9^{23}\,423\ldots$
$2.999999999999955 \times 10^{-14}$	$1.000000000000029\,9\,9^{25}\,099\ldots$
$1.999999999999980 \times 10^{-14}$	$1.000000000000019\,9\,9^{25}\,733\ldots$
$3.999999999999992 \times 10^{-15}$	$1.000000000000003\,9\,9^{27}\,786\ldots$
$1.999999999999998 \times 10^{-15}$	$1.000000000000001\,9\,9^{28}\,733\ldots$
$9.999999999999995 \times 10^{-16}$	$1.000000000000000\,9\,9^{29}\,666\ldots$

Table 5. Some bad cases of the exponential function in the decimal64 format, for $-10^{-10} < x \leq b^- = -5.000000000000001 \times 10^{-17}$. At most two bad cases (the worst ones) are given per exponent.

x	$\exp(x)$
$-1.020000000005202 \times 10^{-11}$	$0.9999999999898000\,0\,0^{16}\,353\ldots$
$-1.000000000005000 \times 10^{-11}$	$0.9999999999900000\,0\,0^{16}\,333\ldots$
$-1.100000000000605 \times 10^{-12}$	$0.9999999999989000\,0\,0^{19}\,443\ldots$
$-1.000000000000500 \times 10^{-12}$	$0.9999999999990000\,0\,0^{19}\,333\ldots$
$-1.200000000000072 \times 10^{-13}$	$0.9999999999998800\,0\,0^{22}\,575\ldots$
$-1.000000000000050 \times 10^{-13}$	$0.9999999999999000\,0\,0^{22}\,333\ldots$
$-2.000000000000020 \times 10^{-14}$	$0.9999999999999800\,0\,0^{24}\,266\ldots$
$-1.000000000000005 \times 10^{-14}$	$0.9999999999999900\,0\,0^{25}\,333\ldots$
$-4.000000000000008 \times 10^{-15}$	$0.9999999999999960\,0\,0^{26}\,213\ldots$
$-2.000000000000002 \times 10^{-15}$	$0.9999999999999980\,0\,0^{27}\,266\ldots$

4   Conclusion

Like in binary arithmetic, correct rounding can be guaranteed in decimal arithmetic at a reasonable cost if the upper bound on the necessary precision for the intermediate computations is determined. This requires exhaustive tests on the whole input domain. While some subdomains can easily be handled, a large number of input values need to be tested.

For the 754r decimal32 format, the tests can be carried out with naive algorithms. However, for the 754r decimal64 format, specific algorithms needed to be designed and implemented. The complete results for the exponential function have been given in this paper. The worst case for $|x| \geq 3 \times 10^{-11}$ (i.e., if we disregard very small values) is

$$\exp(9.407822313572878 \times 10^{-2})$$
$$= \underbrace{1.098645682066338}_{16 \text{ digits}}\underbrace{50000000000000000}_{17 \text{ digits}}278\ldots,$$

meaning that a faithful approximation to 34 digits, which corresponds to the decimal128 format, would be enough to guarantee correct rounding for the exponential in the decimal64 format in this domain. For the smaller values of x, the worst case is

$$\exp(9.999999999999995 \times 10^{-16}) = \underbrace{1.000000000000000}_{16 \text{ digits}}\underbrace{999\ldots999}_{30 \text{ digits}}666\ldots,$$

so that a faithful approximation to $\exp(x) - 1$, also known as expm1(x), in the decimal128 format would be enough to guarantee correct rounding for the exponential in the decimal64 format in this domain.

Ziv's strategy [16] can be used to evaluate the decimal64 exponential function; it consists in carrying out the computations in a small precision (e.g., 22 digits) first, and increasing the precision only in the very unlikely case where the correct rounding cannot be decided. The results presented in this paper can be used to implement Ziv's strategy in an efficient way and prove that the algorithm terminates within limited time and memory.

Other elementary functions could be tested as well, with the same algorithms. As a consequence, standards could recommend (or even require) correct rounding for these functions in these formats.

Acknowledgements

The writing of this paper was completed while the second author was visiting the University of Sydney, whose hospitality is gratefully acknowledged. In particular, part of the computations described in the present article was performed on the machines of the MAGMA team.

The computations were also partly performed on machines of the Laboratoire de l'Informatique du Parallélisme (at the École Normale Supérieure de Lyon, France).

The third author acknowledges the support from the Schloss Dagstuhl International Conference and Research Center for Computer Science, in particular the Dagstuhl Seminar 06021 *Reliable Implementation of Real Number Algorithms: Theory and Practice*, which stimulated the writing of this article.

The authors also thank the anonymous reviewers for their helpful comments.

References

1. IEEE: IEEE Standard for Binary Floating-Point Arithmetic, ANSI/IEEE Standard 754-1985. Institute of Electrical and Electronics Engineers, New York (1985)
2. IEEE: IEEE Standard for Radix-Independent Floating-Point Arithmetic, ANSI/IEEE Standard 854-1987. Institute of Electrical and Electronics Engineers, New York (1987)
3. Cowlishaw, M., Schwarz, E.M., Smith, R.M., Webb, C.F.: A decimal floating-point specification. In: Burgess, N., Ciminiera, L. (eds.) Proceedings of the 15th IEEE Symposium on Computer Arithmetic, Vail, Colorado, USA, pp. 147–154. IEEE Computer Society Press, Los Alamitos (2001)
4. Cowlishaw, M.: Decimal arithmetic encoding strawman 4d, draft version 0.96. Report, IBM UK Laboratories, Hursley, UK (2003)
5. Lefèvre, V., Muller, J.M.: Worst cases for correct rounding of the elementary functions in double precision. In: Burgess, N., Ciminiera, L. (eds.) Proceedings of the 15th IEEE Symposium on Computer Arithmetic, Vail, Colorado, pp. 111–118. IEEE Computer Society Press, Los Alamitos (2001)
6. Stehlé, D., Lefèvre, V., Zimmermann, P.: Searching worst cases of a one-variable function using lattice reduction. IEEE Transactions on Computers 54(3), 340–346 (2005)
7. Dunham, C.B.: Feasibility of "perfect" function evaluation. ACM Sigum Newsletter 25(4), 25–26 (1990)

8. Gal, S., Bachelis, B.: An accurate elementary mathematical library for the IEEE floating point standard. ACM Transactions on Mathematical Software 17(1), 26–45 (1991)
9. Muller, J.M.: Elementary Functions, Algorithms and Implementation. Birkhauser, Boston (1997)
10. Fousse, L., Hanrot, G., Lefèvre, V., Pélissier, P., Zimmermann, P.: MPFR: A multiple-precision binary floating-point library with correct rounding. Research report RR-5753, INRIA (2005)
11. Lefèvre, V.: Moyens arithmétiques pour un calcul fiable. PhD thesis, École Normale Supérieure de Lyon, Lyon, France (2000)
12. Coppersmith, D.: Small solutions to polynomial equations, and low exponent RSA vulnerabilities. Journal of Cryptology 10(4), 233–260 (1997)
13. Lenstra, A.K., Lenstra Jr., H.W., Lovász, L.: Factoring polynomials with rational coefficients. Mathematische Annalen 261, 513–534 (1982)
14. Stehlé, D.: Algorithmique de la réduction de réseaux et application à la recherche de pires cas pour l'arrondi de fonctions mathématiques. PhD thesis, Université Henri Poincaré – Nancy 1, Nancy, France (2005)
15. Nguyen, P., Stehlé, D.: Floating-point LLL revisited. In: Cramer, R.J.F. (ed.) EUROCRYPT 2005. LNCS, vol. 3494, pp. 215–233. Springer, Heidelberg (2005)
16. Ziv, A.: Fast evaluation of elementary mathematical functions with correctly rounded last bit. ACM Transactions on Mathematical Software 17(3), 410–423 (1991)

Robustness and Randomness

Dominique Michelucci[1], Jean Michel Moreau[2], and Sebti Foufou[1]

[1] LE2I UMR CNRS 5158, UFR Sciences, Université de Bourgogne,
BP 47870, 21078 Dijon Cedex, France
dmichel@u-bourgogne.fr, sfoufou@u-bourgogne.fr
[2] LIRIS UMR 5205, Nautibus, Université Claude Bernard Lyon 1,
46 Bd du 11 Nov. 1918, 69622, Villeurbanne, France
Jean-Michel.Moreau@liris.univ-lyon1.fr

Abstract. The study of robustness problems for computational geometry algorithms is a topic that has been subject to intensive research efforts from both computer science and mathematics communities. Robustness problems are caused by the lack of precision in computations involving floating-point instead of real numbers. This paper reviews methods dealing with robustness and inaccuracy problems. It discusses approaches based on exact arithmetic, interval arithmetic and probabilistic methods. The paper investigates the possibility to use randomness at certain levels of reasoning to make geometric constructions more robust.

1 Introduction

Mastering the robustness of computational geometry algorithms (algorithms intended to solve geometric computing problems such as surfaces intersections, shortest paths on surfaces, planification of trajectories, etc.) is a topic that has attracted big attention from both computer science and mathematics communities. Robustness problems are caused by the lack of precision in computations involving floating-point instead of real numbers, in that case robust implementation of geometric algorithms is highly nontrivial and the strange behaviors (crashes, infinite loops, inconsistent outputs, etc.) of these algorithms are due to their inaccurate computations. Although a lot of work has been done to solve this problem, it is still impossible to find a systematic, simple and fast method that eliminates the sources of all these robustness problems.

Robustness and non-robustness issues in geometric computations have important scientific and economic impact (barrier to full automation, programmers' productivity, failure of critical missions, etc.). This impact has motivated intensive research on the subject during the last twenty years, which generated a large literature and surveys [1,2,3,4]. We refer the reader to the excellent recent survey by C. Yap in the Handbook of Discrete and Computational Geometry [4]. In another good survey J. Keyser classified robustness problems in two main categories: problems due to precision and problems due to degeneracies [5]. He presented the issues involved with each of these classes and discussed some of

P. Hertling et al. (Eds.): Real Number Algorithms, LNCS 5045, pp. 127–148, 2008.

the solutions that have been proposed for dealing with them. In an earlier paper, D. Goldberg presented a tutorial on the aspects of floating-point that have a direct impact on designers of computer systems [1]. The paper begins with a background on floating-point representation and rounding error and continues with a discussion of the IEEE floating-point standard. Robustness problems in computer aided design and geometric modelling have been studied by C. Hoffmann in [3] where exact arithmetic, symbolic reasoning, and reliable calculations (interval arithmetic) was identified as possible strategies to address this problem. Sugihara and Iri introduced the topology-based approach which avoids failure by using floating-point arithmetic, but places higher priority on topological consistency than on numerical values [6]. So decisions by this approach ensure that the result is always coherent from the topology point of view. But, there is no guarantee that any other software that works with such a result will give coherent outputs from it. Topology-oriented implementations have been applied to a number of geometric problems such as Voronoi diagram computations and convex polyhedra intersections [7,8].

In this paper we survey some inaccuracy issues in computational geometry, we discuss some of the classical solutions that have been suggested in the last twenty years and emphasize on interval analysis and the probabilistic approach. We show how randomness may sometimes be used in order to help reduce the impact of inaccuracy in geometric computations. The probabilistic approach has received less attention than other robustness tackling methods, our intention here is to highlight the positive role probabilistic algorithms can play. The use of randomness is not intended to solve all robustness problems of geometric algorithms, but to provide probabilistic algorithms as an alternative for geometric computations. These algorithms are costly in time, but tolerant and can resist to inaccuracies. They operate using weak oracles that very often converge to the true decision, but can also say "I do not know" when the right decision is out of reach.

The paper is organized as follow: Section 2 describes some of the problems caused by inaccuracy. Section 3 reviews the classical solutions that were designed to help prevent these inaccuracy problems. Section 4 explores new methods, based on probabilistic approaches already used in various unrelated domains.

2   Consequences of Inaccuracy on Geometric Algorithms

The inaccuracy of floating-point arithmetic has dramatic consequences on geometric computations. Inaccuracy causes inconsistencies both in geometric programs and their data structures, so that geometric programs may crash or yield inconsistent results.

As a first illustration, let S_n be a set of $n \geq 4$ points in the Euclidian plane, with no more than two on the same line. The convex hull $\mathcal{C}$ of S_n is the smallest convex polygon enclosing all its elements. A simple method to construct the edges of $\mathcal{C}$ by enumeration is to identify all pairs (a, b) in S_n^2 such that all points in $S_n \setminus \{a, b\}$ lie on the same side of infinite line (ab) (note that imposing no more than two aligned points in S removes special cases here). Although this method is correct from a theoretical point of view, it may fail to yield consistent results in practice: to see this, consider applying it to $n = 4$ nearly aligned points. Because the points may be arbitrarily close to being aligned without being exactly so, the previous test may easily fail on any pair of points, due to inaccuracy in the computations! This failure may seem striking at first glance, but is it really more striking than the impossibility to verify identities such as: $(\sqrt{2})^2 = 2$ or $(1/3) \times 3 = 1$ or $(\cos \theta)^2 + (\sin \theta)^2 = 1$ with floating-point arithmetic?

To get more insight on what goes wrong in this example, consider a, b, c to be 3 distinct points in the Euclidian plane. The "orientation" of the three points, $\mathcal{O}(a, b, c)$ is defined as the sign of the determinant:

$$\begin{vmatrix} x_a & y_a & 1 \\ x_b & y_b & 1 \\ x_c & y_c & 1 \end{vmatrix},$$

which represents the signed volume of the parallelepiped generated by the vectors $(a, 1)$, $(b, 1)$, $(c, 1)$. Intuitively, the three points form a left, right, or null "turn" depending on whether $\mathcal{O}(a, b, c)$ is positive, negative or null. Obviously, $\mathcal{O}(a, b, c)$ and $\mathcal{O}(c, b, a)$ must have opposite signs, but one may easily generate three almost (but not exactly) aligned distinct points a, b, c with floating-point coordinates, which contradict this property. To help solve such inconsistencies, D.E. Knuth [9] suggested a set of axioms fulfilled by the orientation predicates, assuming for the simplicity of proofs that no more than two data points may lie on the same line. He later realized that he had set up the axioms for oriented matroids with rank 3. All his axioms and resulting theorems are contradicted by floating-point configurations, due to inaccuracy.

Inaccuracy introduces inconsistencies in geometric data structures. Among the most basic geometric data structures, some represent point/line or point/plane incidences. Typically a 2D point is described by its cartesian coordinates (x, y), and a line with equation $ax + by + c = 0$ by the triple (a, b, c). The intersection point between 2 lines (a, b, c) and (a', b', c') is easily computed with standard linear algebra. Due to inaccuracy, $\Omega(x_\Omega, y_\Omega)$, the computed intersection point will not lie on these lines (*i.e.*, $ax_\Omega + by_\Omega + c \neq 0$ and $a'x_\Omega + b'y_\Omega + c' \neq 0$): a contradiction between a numerical test and the incidence fact stored in the data structure. Of course, most geometry programmers are aware of this difficulty, and hence, to find out the position of a vertex v relatively to a line D, they first check whether the data structure does not explicitly hold the information "v lies on D"; if not, a numerical orientation test is performed. This two-stage procedure eliminates the more obvious inconsistencies. However, many geometric theorems of projective geometry—or even geometric constructions—imply non-trivial incidences, which such simple "precautions" cannot detect, as we shall now see through five well-known theorems from the field of classical geometry.

Theorem 1 (Harmonic conjugate, Fig. 1, left). *Let A, B, X be 3 distinct aligned points. Let L be any line through X, s any point outside L and ABX. Then the point X' defined by the construction: $a = sA \cap L$, $b = sB \cap L$, $s' = aB \cap Ab$, $X' = ss' \cap AB$, depends neither on L nor on s.*

X' is called the harmonic conjugate of X relatively to A, B. The harmonic conjugate of X' is X.

Theorem 2 (Desargues theorem, Fig. 1, right). *In 2D or 3D, if two triangles abc and ABC are such that aA, bB, cC concur (the two triangles are said to be "perspective"), then homologous sides meet in 3 aligned points, i.e., $ab \cap AB, bc \cap BC, ca \cap CA$ are collinear. The converse is true as well.*

Theorem 3 (Pappus theorem, Fig. 2, left). *In 2D (i.e., in the projective plane), if p_1, p_2, p_3 are three distinct aligned points, and if q_1, q_2, q_3 are three distinct aligned points, then the three intersection points $p_1q_2 \cap p_2q_1$, $p_1q_3 \cap p_3q_1$ and $p_2q_3 \cap p_3q_2$ are aligned as well.*

Theorem 4 (Pascal, Fig. 2, right). *6 coplanar points belong to the same conic if and only if the 3 opposite sides (in any order) meet in 3 aligned points.*

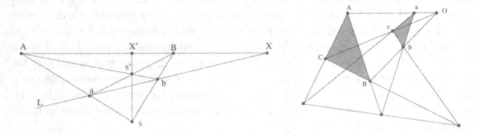

Fig. 1. Left (harmonic conjugates): for 3 given aligned points A, B, X, the point X' does not depend on L nor s. Right: Desargues theorem.

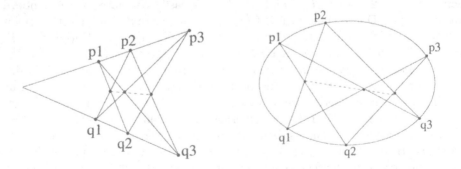

Fig. 2. Pappus' and Pascal's theorems

Theorem 5 (Pouzergues (hexamys)). *In the projective plane, an hexamys is a (possibly concave and self-intersecting) hexagon such that its three opposite sides meet in three aligned points. Then every permutation of an hexamys is also one.*

Pouzergues' theorem may be seen as a particular case of Pascal's theorem with no conic involved. All those geometric theorems are "wrong" when using the floating-point arithmetic, and "true" when using an exact arithmetic. An exact rational arithmetic is sufficient to prove the harmonic conjugate theorem, Pouzergues', Pappus' and Desargues' theorems, assuming the initial coordinates are rational. An algebraic arithmetic is required for Pascal's theorem, if the points on the conic are intersection points between general conics.

3 Classical Methods

3.1 The Epsilon Heuristic

To overcome inaccuracy, the most popular trick used in geometric modelers is the ϵ heuristic. When two floating-point numbers differ by less than a given threshold traditionally called ϵ, they are considered to be equal. The test may be made in an absolute ($|a - b| < \epsilon$) or relative ($|a - b| < \epsilon \times max(|a|, |b|)$) manner. Some modelers use several ϵ values, say one for lengths, another for areas, another for angles, etc.

This heuristic loses the equality transitivity: it is easy to find a, b and c so that $a =_\epsilon b$, $b =_\epsilon c$, but $a \neq_\epsilon c$, with $=_\epsilon$ meaning "equal for the ϵ heuristic": thus inconsistencies remain possible.

Moreover, finding the relevant value(s) for $\epsilon(s)$ is much of a difficult task, depending on the usual range of numbers (itself depending on the applications), and on the format of floating-point numbers: it is common folklore in the CAD-CAM community that the conversion from 32-bits floating-point numbers to 64-bits has required a not so easy ϵ's updating. Of course the ϵ heuristic may fail, and sometimes it does. In practice, it seems to work not so bad and to improve the geometric modelers' robustness.

3.2 The Exact Computation Paradigm

In geometric computations, a lot of effort has been put to design theoretically fast methods, assuming that an exact arithmetic is available for free. These sophisticated methods do not resist to inaccuracy and the induced inconsistencies. They require consistency to work.

Exact rational arithmetics: In the presence of alternatives and tests, geometric methods branch according to the (positive, negative or null) sign of expressions, called predicates: the orientation test of three points in the plane is a typical example. One method to prevent inconsistencies in computations is to take "exact" decisions in the branching tests, by means of an exact arithmetic.

Systematically using such an arithmetic consumes too much time and space resources, and hence various authors have advocated for the use of some sort of 'filtering" (also known as "laziness"):

- First compute guaranteed bounds on the expressions to be tested. Most of the time, those are sufficient to determine the sign of the expression.
- Whenever they are not (*i.e.*, 0 lies within the bounds), use an exact arithmetic to determine the sign of the expression.

An example of such a method is the lazy rational arithmetic [10]: a lazy number is represented by an enclosing interval, and by a definition (either an initial rational number, or the sum, product, opposite, inverse of other lazy numbers). The interval is systematically computed. When it is not sufficient to decide the sign of a number, the definition associated with the number is evaluated using an exact rational arithmetic.

All filter-based solutions use the same basic scheme, they may differ by the way they define aspects such as:

- the exact arithmetic used (remainder number system, *i.e.*, modular arithmetic; strings of digits, ...),
- the evaluation strategy,
- the method for storing the exact values or the definition itself,
- the method for evaluating the bounds (statically at compile time, or dynamically at run time).

Such techniques, routinely used in major geometric applications, for instance CGAL [11], XSC [12], LOOK [13] or LEDA [14] libraries, are, unfortunately, limited to computations in the field of rational numbers. However, it is possible to generalize the lazy (or filter) paradigm if an exact algebraic arithmetic is available, as the "gap arithmetic" used in LEDA::real, CORE and CGAL [15,16,11].

Gap arithmetics: Canny's gap theorem gives a way to *numerically* prove that a number *is* zero: compute a (guaranteed) interval containing it, with width smaller than ϵ_c [17]. As soon as the interval does not contain 0, the number is clearly not 0 and its sign is known. Otherwise, if the interval contains 0 and has width less than ϵ_c, the number can only be 0.

Theorem 6 (Canny's gap theorem). *Let* x_1, $x_2 \ldots x_n$ *be the solutions of an algebraic system of* n *equations and* n *unknowns, having a finite number of solutions, with maximal total degree* d, *with relative integer coefficients smaller or equal to* M *in absolute value. Then, for all* $i \in [1, n]$, *either* $x_i = 0$ *or* $|x_i| > \epsilon_c$ *where* $\epsilon_c = (3Md)^{-(nd^n)}$.

Unfortunately, there are several problems. First, ϵ_c is far much smaller than the ϵ used in geometric modelers; actually ϵ_c is generally much smaller than the smallest positive floating-point number, even in simple examples, hence the need for some "big-float" arithmetic. Second, given such an arithmetic, the computational scheme described here has an exponential cost: an exponential number of digits is needed to prove the nullity of a number because of the nd^n term in Canny's theorem. There is no hope to significantly widen Canny's gap in the worst case, because it is almost reached in the following simple instance: $x_1(Mx_1 - 1) = 0$, $Mx_2 - x_1^2 = 0 \ldots Mx_n - x_{n-1}^2 = 0$. See [18,19,20] for implementations of gap arithmetics or gap theorems, and [21,22] for related root separation bounds.

Pros and cons of the exact computation paradigm: CGAL geometric library relies on filters and lazy rational arithmetic to achieve robustness [11]. CGAL is well-known for its reliability and speed; its Delaunay routine is often used in industry for surface reconstruction from a set of sampling points. This alone stands as a good point of the paradigm. However, the exact approach has important limitations:

- Exact algebraic arithmetics are too slow to be practical. Unfortunately, algebraic numbers are ubiquitous in geometric computations: rotating an object by an angle $k\pi, k \in \mathbb{Q}$, intersecting conics or other algebraic curves, intersecting quadrics or other algebraic (parametric or implicit) surfaces, all those "primitives" introduce algebraic numbers. [23] implemented robust boolean operations between 3D algebraic shapes, but the corresponding program is an order of magnitude too slow.
- In applications such as CAD-CAM, computer graphics and GIS, data are inaccurate; it does not make sense to compute results which are more accurate than data itself; the only justification could be the attempt to make the most fragile algorithms "work"; moreover the exact results are typically rounded to communicate with the rest of the world, which uses only floating-point arithmetics.
- Industrial applications use the floating-point arithmetic to represent geometric objects. Translations from exact representations to floating-point representations and vice-versa are thus essential. Another reason for such "rounding" is that geometric modelers are shape editors, and the algebraic complexity of the edited shape increases with each editing operation. Rounding floating-point geometries to exact (and consistent) geometries is as much difficult as "repairing" inconsistent geometric objects (a situation known as the "polygon soup").

When the cost of exact arithmetics is taken into account, some of these algorithms may become unpracticable, or slower than more rudimentary methods (see section 4), which are more robust and still work with inaccuracy, because they do not propagate inaccurate results.

3.3 Interval Computations

A natural idea to rid geometrical computations of inconsistencies is to resort to some kind of interval computations.

Basic interval arithmetic: The first and most basic interval arithmetic is due to R. Moore [24]. Basically, numbers are represented with intervals, which stand for the uncertainty associated with each one, a notion which was obviously borrowed from the everyday practice of physicists. Interval computations are defined for the sum, difference, product and inverse of intervals. Hence, it is possible to maintain intervals for combinations of, and even simple functions on, elementary data. It is also possible to define interval variants for the exponential ($\exp([a, b]) = [\exp(a), \exp(b)]$), logarithm, sine, cosine functions, etc. For non-monotonous functions, the interval argument is usually decomposed into subintervals on which the function is monotonous.

Following the basic theory, the width of the sum or the difference of two intervals is the sum of the widths of the two added or subtracted intervals. Thus, $X - X$ is not equal to 0: interval arithmetic "loses the dependence between variables". A consequence of this is the wrapping effect: the overestimation of intervals increases with the number of operations.

To restrict such a wrapping effect when evaluating polynomials, a possibility is to use the central evaluation form, as follows: $f(X) \subset f(X_c) + (X - X_c)f'(X)$, where X_c is the center of the interval X; $X - X_c$ is the halfwidth of X; $f'(X)$ is an interval enclosing the derivative of f inside X; it is computed either with the naive arithmetic, or recursively with the central evaluation form. This extends to multivariate polynomials. The analytical definition of f is explicitly required, and may not be considered as an "oracle" or a black box by itself.

Affine interval arithmetic: As mentioned just above, the naive interval arithmetic "loses the dependence between variables"; the affine interval arithmetic was intended to fix this flaw [25,26]. It is designed as a model for "self-validated computations", that keeps track of the first-order correlations between computed and input quantities. Quantities manipulated by the arithmetic are represented by affine expressions of the form:

$$\hat{x} = x_0 + x_1\varepsilon_1 + \ldots + x_n\varepsilon_n,$$

where the ε_i are called the "noise symbols", and the x_i are real numbers. Each noise symbol stands for an independent component of the total uncertainty of the ideal quantity it is associated with. Standard operations are defined over those forms, and affine interval arithmetic may be seen as a generalization of interval arithmetic with richer properties. In particular, it is possible to use the noise symbols to identify the dependency between variables (or even one variable appearing in more complex algebraic expressions, as in $(1 + x)(1 - x)$), and hence to prevent the dynamic interval bounds on expressions from growing as rapidly as they would if a basic interval arithmetic is used. This technique requires good approximations of higher degree expressions with affine forms in the appropriate noise components. This of course is fairly more time-consuming than standard interval arithmetic.

Bernstein-based intervals: In the CAD-CAM and computer graphics communities, since the pioneering works by Bézier and de Casteljau, it is well-known that the properties of the Bernstein basis yield sharp enclosing intervals for polynomials

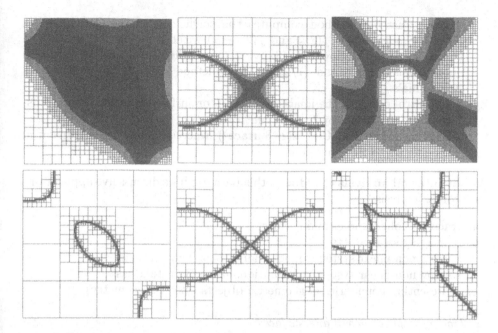

Fig. 3. The naive interval arithmetic is used in the top row, and the Bernstein-based interval arithmetic in the bottom row, for displaying the same three curves. Left column: the curve has equation: $f(x,y) = 15/4 + 8x - 16x^2 + 8y - 112xy + 128x^2y - 16y^2 + 128xy^2 - 128x^2y^2 = 0$ and is displayed in the square $[0,1] \times [0,1]$. Middle column: Cassini's oval. Right column: a random curve with degree 14.

[27,28,29], as Fig. 3 gives visual evidence. This knowledge finally percolated to other communities [30,31,32]. A Bernstein basis solver for polynomial systems written by Mourrain and Pavone is available in GALAAD [33].

The canonical basis for degree d polynomials in t is: $T = (1, t, t^2 \ldots t^d)$. The Bernstein basis is: $B = (B_0^{(d)}(t), B_1^{(d)}(t), \ldots B_d^{(d)}(t))$ where $B_i^{(d)}(t) = \binom{d}{i} t^i (1-t)^{d-i}$.

The conversion between the Bernstein and the canonical bases is a linear mapping, representable by a $(d+1) \times (d+1)$ square matrix M such that: $B = TM$. For instance, for $d = 3$:

$$(B_0, B_1, B_2, B_3) = (1, t, t^2, t^3) \begin{pmatrix} 1 & 0 & 0 & 0 \\ -3 & 3 & 0 & 0 \\ 3 & -6 & 3 & 0 \\ -1 & 3 & -3 & 1 \end{pmatrix}$$

Some remarkable properties of the Bernstein basis are:

- The polynomial lies inside the convex hull of its coefficients z_{ij} in the Bernstein basis. Thus $z = f([0,1],[0,1])$ lies in $[\min_{i,j} z_{ij}, \max_{i,j} z_{ij}]$. This property extends to all dimensions.

- The de Casteljau method computes the coefficients in the Bernstein basis of the polynomials $f(2x)$ and $f(2x - 1)$ without having to refer to the canonical basis; it allows to subdivide the studied interval $[0, 1]$ into two subintervals $[0, 1/2]$ and $[1/2, 1]$. This method extends to multivariate polynomials.
- The control points of the image of a curve (or surface) by some affine transform T are the images by T of the control points of the curve (or surface). *i.e.*, $T(\text{CtrPts}(\text{Surface})) = \text{CtrPts}(T(\text{Surface}))$, where T is any affine transform. This extends to all dimensions.

What interval analysis can do: In this paragraph we discuss five applications of interval analysis:

a. tracing implicit curves and surfaces
b. displaying strange sets
c. solving non-linear systems of equations using Newton method
d. solving non-linear systems of equations using Bernstein basis
e. representing boundaries of geometric objects (interval geometry).

a. Tracing implicit curves and surfaces:
Interval analysis is used in computer graphics to display semi-algebraic sets defined by boolean combinations of polynomial inequalities, in 2D (such as $f(x, y) \geq 0$ and $g(x, y) \geq 0$) or 3D (such as $f(x, y, z) \geq 0$ and $g(x, y, z) \geq 0$). See Fig. 3 for an illustration. Without loss of generality, we present the classical subdivision method used to display 2D curves defined by an implicit equation $f(x, y) = 0$ as follows:

```
Subdivide(f: function, X : interval, Y: interval, depth: int) {
  interval F = f(X, Y);
  draw_rectangle(X, Y);
  if (F contains 0)
  { if (depth==0) fill_rectangle(X, Y)
    else
    {
        interval X1 = [min(X), middle(X)];
        interval X2 = [middle(X), max(X)];
        interval Y1 = [min(Y), middle(Y)];
        interval Y2 = [middle(Y), max(Y)];
        Subdivide(f, X1, Y1, depth-1);
        Subdivide(f, X1, Y2, depth-1);
        Subdivide(f, X2, Y1, depth-1);
        Subdivide(f, X2, Y2, depth-1);
    }
  }
}
```

b. Displaying strange sets:
Interval analysis methods may also be used to compute guaranteed covers of
fractals or strange sets such as Julia sets or the Hénon attractor (see Fig. 4), a
process that we now sketch rapidly:

- The "Hénon mapping" sends the point $(x, y) \in \mathbb{R}^2$ onto $(x', y') = (y + 1 - ax^2, bx)$, where a, b are two parameters; classically, $a = 1.4$ and $b = 0.3$.
- The Hénon set is the set of points whose orbit (set of H-iterates) remains bounded.
- Assume, for the sake of simplicity, that a square bounding box B of the Hénon set is known:
  - Subdivide B in 16×16 square cells;
  - For every cell C, compute an enclosure of $H(C)$, using an interval arithmetic;
  - Each time $H(C)$ overlaps a cell C_i, add an arc $C \to C_i$ in a graph H_G, whose vertices are cells partitioning B.
  - Compute the strongly connected components of H_G: a strongly connected component is *transient* if and only if it contains only one cell C and there is no loop $C \to C$ in H_G; then this cell cannot contain any point in the Hénon set.
  - For non-transient cells C, the graph H_G contains at least one loop $C \to C$ or $C \to C_1 \to \ldots C$, thus C may contain points of the Hénon set.

    Non-transient cells are subdivided, and the method is applied again.
- The recursion stops when an accurate picture is obtained.

This method guarantees a sharp cover of the Hénon set, and shows that the
classical orbit method used to display strange sets may yield erroneous results
(for some values of a, b). Even the need of an initial bounding box may be relaxed,
since the projective plane is bounded (it may be mapped to a sphere). Refer to
[34] for more details.

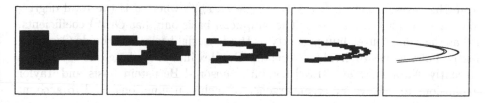

Fig. 4. The Hénon set with increasing $8^2, 16^2, 32^2, 64^2, 1024^2$ resolutions

c. Solving non-linear systems of equations using interval Newton methods:
Let $F(x) = 0$ be a system of n equations in n unknowns. The classical Newton-
Raphson method iterates: $x_{n+1} \leftarrow x_n + F(x_n)F'(x_n)^{-1}$ until convergence. A
variant is the secant method which does not update the inverse of the derivative
F' at each step, and iterates: $x_{n+1} \leftarrow S(x_n) = x_n + F(x_n)J^{-1}$, where J is the
Jacobian of F at x_0.

To find roots of F inside a prescribed initial box X, a natural idea is to compute $S(X)$ for an interval X. Since $S(X) = X+$ something, the width of $S(X)$ is always greater than the width of X, if evaluated with the naive interval arithmetic, which prevents convergence. A solution is to use the central form evaluation to compute $S(X)$, which leads directly to the Krawczyk (or Krawczyk-Moore) operator.

The interval secant method may be described as follows: if $S(X) \subset X$, then S is contractant inside X, thus X contains an isolated root of F, and the secant method will converge to this root starting from any point of X. If $X \cap S(X) = \emptyset$, then X contains no root of F. Otherwise possible roots of F in X can only be in $X \cap S(X)$; if $X \cap S(X)$ is significantly smaller than X, then the method is resumed on $X \cap S(X)$, i.e., $S(X \cap S(X))$ is computed, etc. Otherwise, the former is likely to contain several roots of F, and is bisected: eventually, these bisections separate roots. The combinatorial complexity of the method is due of course to the bisection steps.

The computation of

$$X \cap S(X) = (X_1 \cap S_1(X_1, \ldots X_n), \ldots X_n \cap S_n(X_1, \ldots X_n)),$$

can be optimized using the most recent value of $X_i, i = 0, \ldots k$ when computing $S_k(X)$, and hence by cascading the computations:

$$X_1 \leftarrow X_1 \cap S_1(X_1, \ldots X_n), X_2 \leftarrow X_2 \cap S_2(X_1, \ldots X_n),$$

and so, see [35,36] for more details.

d. *Solving non-linear systems of equations using Bernstein basis:*
The Bernstein-based subdivision method may also be used to solve polynomial systems of equations, as it was done by Patrikalakis [29] and Garloff [30]. These authors use the tensorial Bernstein basis, which has limitations: it is a dense representation (even if the polynomial is sparse in the canonical basis), and has an exponential number of coordinates.

A typical solution is likely to use the simplicial Bernstein basis, as obtained from the development of: $(x_0 + x_1 + x_2 + \ldots x_n)^d$, where d is the total degree, and $x_0 + x_1 + x_2 + \ldots x_n = 1$. The simplicial basis only has $O(n^d)$ coefficients, it has the same convex hull property as the tensorial basis, and the de Casteljau method also allows to divide a Bernstein simplex into two, along one of its edge. Recently, Nataraj *et al.* [31,32] combine tensorial Bernstein basis and Taylor expansions to achieve superconvergence of inclusion functions, and to account for non polynomial functions.

e. *Representing boundaries of geometric objects:*
Quite recently also, several authors (*e.g.*, [37]) suggested to extend the principle of intervals, which enclose boundaries in 1D, to geometric objects in 3D (interval geometry). A boundary surface is thus enclosed inside two polyhedra (typically with rational coordinates). The inside and outside polyhedra may have different topologies, for instance a different number of connected components. Computable Analysis extends to such objects, *i.e.*, it is possible to compute (using increasing computing resource) closer and closer nested approximations.

This approach does not suffer from the incompatibility between, on the one hand, classical boundary representations or geometric algorithms which require to compute the exact sign of numbers: $0, +, -$, and on the other hand interval arithmetics which are intrinsically unable to compute the sign of 0 from an enclosing interval. However interval geometry suffers from the same intrinsic restriction as interval arithmetics: interval arithmetics cannot compute the sign of 0, and interval geometry cannot detect if two shapes are tangent: it can only detect whether they are disjoint or intersect.

In another effort similar to the one of interval geometry, Foufou et $al.$ introduced the fuzzy geometry and proposed to use it to classify surfaces against their intersection status [38]. Geometric entities are replaced by thicken entities. The associated fuzzy intersection algorithm provides a three-state classification of surfaces couples: certainly intersecting, certainly non intersecting and potentially intersecting.

What interval computations cannot do: All interval-based solutions have the intrinsic limitation of being only "half deterministic" in the following sense: if a number is zero, no interval computation can detect it (in finite-time); if a number is either strictly positive or strictly negative, a tight enough interval will find its sign. Thus interval analysis cannot decide nullity. For the same reason, it cannot decide equality ($i.e.$, the nullity of the difference of two equal numbers). This argumentation has been formalized by Computable Analysis [39] as follows: By definition, a number is computable if a finite-time algorithm provides an arbitrarily tight interval enclosing it. For instance, the interval arithmetic provides a stream of (nested) Cauchy intervals. Interval arithmetics which are capable of computing such arbitrarily tight enclosures are called Real Arithmetics. Several implementations have been suggested (and sometimes proven) [40,41].

By definition, a real function f is computable if there is an algorithm which, when given the input numbers $x_1, \ldots, x_n$ (a stream of nested Cauchy intervals), is able to compute a stream of nested Cauchy intervals describing $f(x_1, \ldots, x_n)$.

No interval arithmetic may be more powerful than the Real Arithmetic, an ideal interval arithmetic which does not suffer from practical limitations ($e.g.$, restricted memory). But even Real Arithmetic is intrinsically restricted:

Theorem 7 (Computable Analysis). *Only continuous functions are computable.*

This theorem has considerable implications, which even today have not been fully realized. For instance let us just consider that the sign function is equal to $+1$ for positive numbers, to -1 for negative numbers, and to (say) 0 for zero. This function is discontinuous in 0; since no interval computation may establish that a number is zero, the sign of zero is not computable.

There is a quantitative variant of this theorem: the greater $|f'_{x_i}|$, the sharper the interval for x_i so as to compute $f(x_1, \ldots x_n)$ with a prescribed precision. For a discontinuous function, its slope is infinite at the discontinuity, thus the argument has to be known with infinite precision to compute the function at that point.

In practice, an arbitrarily tight interval cannot be computed, because of storage limitations (and the patience limitation of the user). But the theorem holds even without taking this common-sense argument into account. This theorem has dramatic consequences on geometric computing.

The most basic geometric algorithms typically require to compute non-continuous functions, such as: do two geometric objects intersect or not? does this point lie on the left, on the right, or just on a boundary surface? does this triple of 2D points turn left, right, or are they aligned?

As we have seen, geometric algorithms use "predicates" for branching; the branch is chosen according to the sign of a predicate, such as the orientation predicate. Predicates being discontinuous functions, they are not computable with Real Analysis. Hence, interval computations (which seem to be the only realistic guaranteed way of computing) are not compatible with the most basic geometric algorithms of Computational Geometry, and not compatible with the most basic geometric data structures used to represent incidence between points, curves and surfaces. Nearly all geometric modelers use these data structures.

Stated crudely: no geometric programs computing non-continuous functions, and using floating-point arithmetic, or using interval computations or even Real Arithmetic, may be proven. Actually, they all are wrong. The best they can achieve is to work most of the time (*i.e.*, not fail more often than their competitors).

This limitation of interval arithmetic is the main argument to promote discrete geometry, and stochastic or probabilistic approaches.

4 Probabilistic Approaches

4.1 Solutions at the Arithmetic Level

Probabilistic gap arithmetics: A practicable but only probabilistic method is to compute with some "big-float" library, and to use the ϵ heuristic (see section 3.1) with a small ϵ, *e.g.*, $\epsilon = 10^{-200}$, and hope that life will not be so bad as to produce a counterexample. We are not aware of any report on this kind of experiment.

The stochastic arithmetic: The stochastic arithmetic, implemented in CADNA [42], estimates and limits the propagation of round-off errors of the floating-point arithmetic in scientific software (*e.g.*, simulation of fluid mechanics), and may detect the source of numerical instabilities. A stochastic number is represented by a mean value (a floating-point number) and a variance. Each floating-point operation is performed with n (typically $n = 3$) samples of the stochastic number. A stochastic zero is a number with no significant bits.

While the wrapping effect of interval arithmetics overestimates the interval widths, the stochastic arithmetic gives more realistic confidence intervals, which accounts for the fact that round-off errors often compensate. The stochastic arithmetic is well-adapted to programs with few branchings, where tests are used mainly to detect the convergence of a numerical algorithm.

For geometric computations, the stochastic method will decide the sign of numbers equal to 0 or close to 0 only randomly, thus it takes neither exact nor consistent decisions and it does not seem to make sense to use a stochastic arithmetic for geometric algorithms (*e.g.*, computing a convex hull or a Delaunay triangulation). However, it does make sense to use the stochastic scheme at the geometric level, as we shall soon see.

Zero free arithmetic: There are 2 kinds of exact arithmetics: those (*e.g.*, rational arithmetics) which provide a test against zero (*i.e.*, is a number equal, superior or inferior to 0?), and those that do not (*e.g.*, real computable arithmetics). In the latter class, real numbers are typically represented by streams (potentially infinite lists) of nested Cauchy intervals.

In geometric methods, programmers usually round data to integers or rationals, and then use the first kind of arithmetics. This precludes Computational Geometry from addressing algebraic non-rational problems (*e.g.*, intersection between algebraic curves and surfaces).

The idea of the zero-free exact arithmetic [43] is to round data numbers to algebraically independent, transcendental numbers, by perturbing them with a stream of random digits. Assume that all the tests in geometric programs involve the signs of polynomials: $f(u_1, u_2 \ldots u_n)$ where $u_1, \ldots u_n$ are initial data numbers; if the latter are rounded on algebraically independent numbers, then the only polynomial f which can vanish is the identically zero polynomial. It makes no sense to ask for the sign of this polynomial in an instruction such as `if(0==0) then` .... In consequence, tests never evaluate to zero, and it is possible to use real computable arithmetics. Moreover this method also solves the problem of degeneracies (the zero-free arithmetic may be seen as a generalization of the so-called "simulation of simplicity" (*SoS*) technique suggested by Edelsbrunner and Mücker [44], with the major difference that *SoS* requires an arithmetic capable of exactly testing the sign of numbers.): three points will never be aligned, four points will never be coplanar nor cocyclic, etc. Finally, this solution allows geometric methods to address problems requiring non-rational arithmetics: algebraic arithmetics, or even transcendental numbers. It becomes possible to use numbers defined by a convergent algorithm, which computes roots of polynomial systems for instance, as long as we are sure that all tests involve algebraically independent numbers.

This assumption is the cornerstone of the zero-free arithmetic, and its main weakness: some geometric programs allow to derive geometric objects or numbers (coordinates for intersection points) from the initial (perturbed) ones, with some geometric constructions. Clearly, the former algebraically depend on the latter. For instance, cutting one (perturbed) initial line with three other (perturbed) initial lines produces three aligned points; the orientation test for these three points will not terminate: the related determinant is zero. A lot of geometric theorems (Pappus, Desargues) allow to construct less trivial alignments, and occurrences of zero. At the other end of the spectrum, computing $x - x$, where the two instances arise from different computation contexts (*e.g.*, the dependance of the variables has been lost as in interval arithmetic), is null but not "identically null".

Combining intervals and randomized nullity tests: The previous method does not terminate in case of a null number. In order to circumvent this problem, it is possible to combine it with a randomized nullity test [45,46,47,48]. For instance, assuming all numbers involved are rational, a number is probably null if its interval is sharp enough and contains zero, and the number's hashed value (result of a modular computation modulo a large prime) is zero. This method is efficient and simple in the rational case, but is less appealing in the algebraic case [47,49,47,50].

Probabilistic tests can also be used for polynomials and not only for numbers. J.T. Schwartz [45] introduced this method to test algebraic identities: if a polynomial (*e.g.,* the determinant of a square matrix with polynomial entries) vanishes when evaluated at random values of its variables, it is likely identically null. This test is only probabilistic, but extremely fast. This probabilistic principle, called *proof by example*, is also used for probabilistic proofs of geometric theorems [51,52,53]. It has been extended beyond polynomials [49,47,50], *e.g.,* Tulone *et al.* [49] extend it to radical expressions which occur in ruler and compass geometric constructions and related geometric theorems, such as Pascal's for a circle.

This probabilistic test may be made deterministic in several ways. For the sake of simplicity, consider a polynomial in one variable: if an upper bound d of the degree is known, and if the polynomial vanishes in $d + 1$ distinct sample values, then it may only be the zero polynomial; or, if we have an upper bound for the magnitude of the coefficients, we can also use some formula [20] to compute an upper bound for the magnitude of the root module, or a lower bound for the magnitude of the root inverse, so if the polynomial vanishes for a number outside these bounds, then it can only be the zero polynomial [51]. This principle for univariate polynomials may be extended to the multivariate case, and always with an exponential cost.

Clarkson and Shor used random sampling for several geometric algorithms and showed that random subsets can be used optimally for divide-and-conquer algorithms and for bounds computations for incremental building of geometric structures [54,55].

4.2   Solutions at the Geometric Level

The motion planning problem: In robotics, the motion planning problem (also called the piano mover's problem) consists in finding a trajectory for a robot, which moves from a starting position to a final one in an environment cluttered with obstacles. The robot is represented by a point in a configuration space; the configuration space is the usual Euclidean space (which allows to describe the location of some origin point of the robot), augmented with all parameters which describe the spatial orientation (3 angles for a rigid body in 3D) and the configuration (one angle for each articulation of the robot, one length for each jack) of the robot. Obstacles – and the constraints of avoiding self-intersection of the robot – forbid some areas of the configuration space. There is a trajectory for the robot if its initial and final configurations lie in the same connected component of the configuration space. Typically, the routine that decides whether a given point in the configuration space is collision-free or not, returns in fact a signed "interpenetration depth" or some kind of signed minimal distance to the closest obstacle. In the first stages of the algorithm, interpenetration is tolerated in order to find a coarse path rapidly. In the final stages, the acceptable tolerance is gradually reduced.

This problem is decidable, and exact and deterministic algorithms from computer algebra (say the Collin's Cylindrical Algebraic Decomposition, and its variants and optimizations) solve it by computing an explicit representation of the feasible part (the part not forbidden by obstacles and self-avoidance constraints of the robot) of the configuration space. However, such approaches turn out to be definitively not practical due to their high combinatorial costs, to their lack of robustness (these methods do not resist inaccuracy and degeneracy), and also to the high dimension of the configuration space ($\approx$ 200 for a human body, $\approx$ 50 after simplification; from 10 to 20 for a simple robot) occurring in industrial problems.

By the end of the 1980's some roboticians broke away from this exact and deterministic approach. They no more compute an explicit representation of the feasible configuration space. Instead, they randomly sample the configuration space, keeping only collision-free samples; they build a graph, called the *road map*, where close collision-free samples of the configuration space are linked if the line segment (in configuration space) connecting the two samples is collision-free, or, more precisely, if all samples are collision-free when sampling regularly the line segment (the detection of collision for a point in the configuration space is simpler than for a segment). In this approach, finding a trajectory reduces to finding a path in the road map graph, with a standard algorithm like Dijkstra's. Once a path has been found, it may be refined in several ways, for instance by trying to connect two non-contiguous vertices of the path by a segment and resampling more densely a neighborhood of the path.

This method is neither deterministic nor exact; it may fail to find complicated paths in very cluttered environments. Its completeness is probabilistic: it will find a solution (when there is one) with probability one if it runs indefinitely; the probability of failure decreases exponentially with the running time.

This approach is a technological breakthrough as it solves in interactive running times problems of industrial size which are completely out of reach of the deterministic exact methods; moreover it is extremely robust against inaccuracy or degeneracy: for deciding if a sample point of the configuration space is collision-free or not, it uses the standard floating-point arithmetic. We refer the reader to Laumond's book for more details [56].

Roadmap and topological computations: A similar approach is used in [57] to compute (approximated) shortest circuits with a prescribed topology ($\approx$ shape) on a given surface. Actually, the authors started by trying to use mesh representations provided by some industrial software; but these meshes turned out to be inconsistent, containing self intersecting triangles, and gaps (holes between triangles), which means that some sections of the meshes were not even connected. They then decided to sample the mesh, and build a graph (a roadmap) where each edge connects two close enough samples. They also assume that each triangle (three pairwise-connected vertices) in this roadmap corresponds to a triangular patch on the sampled surface: this assumption is sufficient for roadmaps to allow the topological computations required to decide the equivalence of two circuits on a surface with holes.

Topological computations are usually done from some simplicial complexes such as triangulated meshes; these sophisticated data structures are terribly fragile – it is the reason why CGAL requires exact computations. Roadmaps are not triangulations, and they are sufficient to perform topological computations. Moreover, roadmaps are rudimentary and robust: they use floating-point computations without problems.

No motion planning algorithm uses topological computations yet, as far as we know. Such computations can detect several kinds of paths or circuits (*e.g.,* passing at the left or at the right of an obstacle) in the configuration space.

Note: the length of a line through rational vertices is a sum of square roots of rational numbers [58]; this length is an algebraic number with degree 2^{n-1} if there are n vertices. A truly exact algorithm, which exactly computes these lengths (in order to compare them when they are very close) has thus an exponential running time. For this reason, even proclaimed "exact" methods do not use exact computations of lengths.

The radiosity problem: The radiosity problem [59,60] consists in computing photo-realistic images of virtual scenes. Monte Carlo methods simulate the propagation of light by following samples of photons, from the light sources and their reflections or refractions in the scene, until they are absorbed by a surface. Counting the number of absorbed or reflected photons on each surface patch of the scene gives an estimation of the radiosity. Monte Carlo methods do not require an explicit representation of surfaces bounding the objects in the scene, but only a procedure capable of computing the intersection between a line and geometric objects (and the normal to the surface at the intersection point); computing this intersection is an easily to solve one-dimensional problem. Monte Carlo methods are extremely robust: they never fail because of inaccuracy.

Ray tracing [59] is another method used for scene rendering; it follows light rays, *i.e.,* photons, but starts from the (virtual) eye. Like Monte Carlo methods, ray tracing does not need an explicit representation of the scene (similarly the roadmap method for the motion planning problem does not need an explicit geometric representation of the configuration space). It uses the same intersection procedure as Monte Carlo methods. Actually, ray tracing and Monte Carlo methods are combined: the former accounts for specular (mirror-like) light reflections, and the latter accounts for diffusion and scattering of light.

Numerous deterministic methods were proposed to compute more or less realistic images of virtual scenes. They rely on an explicit representation of the boundary of objects in the scene, typically approximating meshes (so they were not even exact). As these deterministic methods use sophisticated data structures (*e.g.,* boundary representations, visibility graphs), they are terribly complicated to implement, and fragile. Note that ray tracing and Monte Carlo radiosity are roughly as old as the deterministic methods; the increasing power of computers has made them practical over the last ten years.

5   Conclusion: Is Randomness the Problem or the Solution?

The inaccuracy of the floating-point arithmetic is often considered as a source of random noise. The latter perturbs computations, introduces inconsistencies in deterministic geometric algorithms and causes strange behaviors such as crashes or infinite loops. Thus randomness is a problem for deterministic algorithms.

Worse even, exact computations turn out to be intractable, and not relevant for real-world applications (except in some very restricted cases). Randomness kills deterministic geometric algorithms.

Probabilistic geometric methods are simpler, very robust, and become relevant and tractable with the increasing power of computers: randomness may very well be the solution to randomness.

References

1. Goldberg, D.: What every computer scientist should know about floating-point arithmetic. ACM Computing Surveys 23, 5–48 (1991)
2. Schirra, S.: Precision and robustness in geometric computations. In: van Kreveld, M., Nievergelt, J., Roos, T., Widmayer, P. (eds.) CISM School 1996. LNCS, vol. 1340, Springer, Heidelberg (1997)
3. Hoffmann, C.M.: Robustness in geometric computations. Journal of Computing and Information Science in Engineering 1, 143–156 (2001)
4. Yap, C.: Robust geometric computation. In: Goodman, J.E., O'Rourke, J. (eds.) Handbook of Discrete and Computational Geometry, pp. 927–952. CRC Press, Boca Raton (2004)
5. Keyser, J.: Robustness issues in computational geometry. Technical report, Comp. 234 Final Paper, Duke University (1997)
6. Sugihara, K., Iri, M.: A solid modelling system free from topological inconsistency. Journal of Information Processing 12, 380–393 (1989)
7. Sugihara, K., Iri, M.: A robust topology-oriented incremental algorithm for voronoi diagrams. IJCGA 4, 179–228 (1994)
8. Sugihara, K.: A robust and consistent algorithm for intersecting convex polyhedra. Computer Graphics Forum 13, 45–54 (1994)
9. Knuth, D.E.: Axioms and Hulls. LNCS, vol. 606. Springer, Heidelberg (1992)
10. Michelucci, D., Moreau, J.M.: Lazy arithmetic. IEEE Transactions on Computers 46, 961–975 (1997)
11. Fabri, A., Giezeman, G.J., Kettner, L., Schirra, S., Schönherr, S.: The CGAL kernel: A basis for geometric computation. In: Lin, M.C., Manocha, D. (eds.) FCRC-WS 1996 and WACG 1996. LNCS, vol. 1148, pp. 191–202. Springer, Heidelberg (1996)
12. Klatte, K., Kulisch, U., Lawo, C., Rausch, M., Wiethoff, A.: C-XSC, A $C++$ class library for extended scientific computing. Springer, Heidelberg (1993)

146   D. Michelucci, J.M. Moreau, and S. Foufou

13. Funke, S., Mehlhorn, K.: LOOK – a lazy object-oriented kernel for geometric computations. In: Proceedings 16th Annual ACM Symposium on Computational Geometry, Hong-Kong, pp. 156–165. ACM Press, New York (2000)
14. Mehlorn, K., Naher, S.: LEDA: A platform for combinatorial and geometric computing. Communications of the ACM 38, 96–102 (1995)
15. Mehlhorn, K., Naher, S.: The LEDA Platform for Combinatorial and Geometric Computing, 1018 pages. Cambridge University Press, Cambridge (1999)
16. Karamcheti, V., Li, C., Pechtchanski, I., Yap, C.: A core library for robust numeric and geometric computation. In: Proceedings 15th Annual ACM Symposium on Computational Geometry, pp. 351–359. ACM Press, New York (1999)
17. Canny, J.: The complexity of robot motion planning. M.I.T. Press, Cambridge (1988)
18. Pion, S., Yap, C.: Constructive root bound method for k-ary rational input numbers. In: Proc. 18th ACM Symp. on Computational Geometry, ACM Press, San Diego, California (2003)
19. Li, C., Yap, C.: A new constructive root bound for algebraic expressions. In: 12th ACM-SIAM Symposium on Discrete Algorithms (SODA) (2001)
20. Mignotte, M., Stefanescu, D.: Polynomials: An algorithmic approach. Discrete Mathematics and Theoretical Computer Science Series, vol. XI. Springer, Heidelberg (1999)
21. Burnikel, C., Fleischer, R., Mehlhorn, K., Schirra, S.: A strong and easily computable separation bound for arithmetic expressions involving square roots. In: Proceedings of the eighth annual ACM-SIAM symposium on Discrete algorithms table of contents, New Orleans, Louisiana, United States, pp. 702–709 (1997)
22. Scheinerman, E.R.: When close enough is close enough. American Mathematical Monthly 107, 489–499 (2000)
23. Keyser, J., Culver, T., Foskey, M., Krishnan, S., Manocha, D.: ESOLID - a system for exact boundary evaluation. Computer-Aided Design (Special Issue on Solid Modeling) 36, 175–193 (2004)
24. Moore, R.: Interval Analysis. Prentice Hall, Englewood Cliffs (1966)
25. Andrade, M.V.A., Comba, J.L.D., Stolfi, J.: Affine arithmetic. In: Abstracts of the International Conference on Interval and Computer-Algebraic Methods in Science and Engineering (INTERVAL 1994), St. Petersburg (Russia), pp. 36–40 (1994)
26. de Figueiredo, L.H., Stolfi, J.: Affine arithmetic: Concepts and applications. Numerical Algorithms 37, 147–158 (2004)
27. Farin, G.: Curves and Surfaces for Computer Aided Geometric Design. Academic Press, London (1990)
28. Hu, C.Y., Patrikalakis, N., Ye, X.: Robust interval solid modelling. part 1: Representations. Part 2: Boundary evaluation. CAD 28, 807–817, 819–830 (1996)
29. Sherbrooke, E.C., Patrikalakis, N.: Computation of the solutions of nonlinear polynomial systems. Computer Aided Geometric Design 10, 379–405 (1993)
30. Garloff, J., Smith, A.P.: Investigation of a subdivision based algorithm for solving systems of polynomial equations. Journal of nonlinear analysis: Series A Theory and Methods 47, 167–178 (2001)
31. Nataraj, P.S.V., Kotecha, K.: Global optimization with higher order inclusion function forms part 1: A combined Taylor-Bernstein form. Reliable Computing 10, 27–44 (2004)
32. Nataraj, P.S.V., Kotecha, K.: Higher order convergence for multidimensional functions with a new Taylor-Bernstein form as inclusion function. Reliable Computing 9, 185–203 (2003)

33. Mourrain, B., Rouillier, F., Roy, M.F.: Bernstein's basis and real root isolation. Technical Report 5149, INRIA Rocquencourt (2004)
34. Michelucci, D., Foufou, S.: Interval based tracing of strange attractors. International Journal of Computational Geometry and Applications 16, 27–39 (2006)
35. Kearfott, R.: Rigorous Global Search: Continuous Problems. Kluwer Academic Publishers, Dordrecht (1996)
36. Hansen, E.R., Walster, G.W.: Global Optimization Using Interval Analysis. Marcel Dekker, New York (2003)
37. Edalat, A., Lieutier, A.: Foundation of a computable solid modelling. Theoretical Computer Science 2, 319–345 (2002)
38. Foufou, S., Brun, J., Bouras, A.: Surfaces intersection for solid algebra: A classification algorithm. In: Strasser, W., Klein, R., Rau, R. (eds.) Proc. Theory and Practice of Geometric Modeling 1996, Tubingen, Germany, Springer, Heidelberg (1996)
39. Weihrauch, K.: Computable Analysis An Introduction. Springer, Heidelberg (2000)
40. Boehm, H.J., Cartwright, R., Riggle, M., O'Donnell, M.: Exact real arithmetic: a case study in higher order programming. In: Proc. ACM Conference on Lisp and Functional Programming, pp. 162–173 (1986)
41. Lester, D., Gowland, P.: Using pvs to validate the algorithms of an exact arithmetic. Theoretical Computer Science 291, 203–218 (2003)
42. Vignes, J., Alt, R.: An efficient stochastic method for round-off error analysis. In: Accurate Scientific Computations, pp. 183–205 (1985)
43. Michelucci, D., Moreau, J.M.: ZEA – a zero-free exact arithmetic. In: Proceedings 12th Canadian Conference on Computational Geometry, Fredericton, New Brunswick, pp. 153–157 (2000)
44. Edelsbrunner, H., Mücke, E.P.: Simulation of simplicity: a technique to cope with degenerate cases in geometric algorithms. ACM Trans. Graph 9, 66–104 (1990)
45. Schwartz, J.: Fast probabilistic algorithms for verification of polynomial identities. J. ACM 4, 701–717 (1980)
46. Agrawal, A., Requicha, A.G.: A paradigm for the robust design of algorithms for geometric modeling. In: Computer Graphics Forum (EUROGRAPHICS 1994), vol. 13, pp. C–33–C–44 (1994)
47. Monagan, M., Gonnet, G.: Signature functions for algebraic numbers. In: Proc. ISSAC, pp. 291–296. ACM Press, New York (1994)
48. Benouamer, M., Jaillon, P., Michelucci, D., Moreau, J.: Hashing lazy numbers. Computing 53, 205–217 (1994)
49. Tulone, D., Yap, C., Li, C.: Randomized zero testing of radical expressions and elementary geometry theorem proving. In: International Workshop on Automated Deduction in Geometry (ADG 2000) (2000)
50. Gonnet, G.H.: New results for random determination of equivalence of expressions. In: SYMSAC 1986: Proceedings of the fifth ACM symposium on Symbolic and algebraic computation, pp. 127–131. ACM Press, New York (1986)
51. Hong, J.: Proving by example and gap theorem. In: I.C.S. (ed.): 27th symposium on Foundations of computer science, Toronto, Ontario, 107–116 (in press,1986)
52. Kortenkamp, U.: Foundations of Dynamic Geometry. PhD thesis, ETH Zurich, Institut fur Theoretische Informatik (1999)
53. Foufou, S., Jurzak, J.P., Michelucci, D.: Numerical decomposition of geometric constraints. In: Proc. ACM Conference on Solid and Physical Modeling, pp. 143–151 (2005)

54. Clarkson, K.L., Shor, P.W.: Applications of random sampling in computational geometry, II. Discrete and Computational Geometry 4, 387–421 (1989)
55. Clarkson, K.L.: New applications of random sampling in computational geometry. Discrete and Computational Geometry 2, 195–222 (1987)
56. Laumond, J.P. (ed.): Robot Motion Planning and Control. Lecture Notes in Control and Information Science. Springer, Heidelberg (1998)
57. Michelucci, D., Neveu, M.: Shortest circuits with given homotopy in a constellation. In: 9th ACM Symp. Solid Modeling and Applications, pp. 297–302 (2004)
58. Choi, J., Sellen, J., Yap, C.: Approximate Euclidean shortest path in 3-space. Int'l. J. Computational Geometry and Applications 271–295 (1997); Journal special issue. Also in 10th ACM Symposium on Computational Geometry (1994)
59. Glassner, A.: An Introduction to Ray Tracing. In: Glassner, A. (ed.), Academic Press, London (1989) ISBN 0-12-286160-4
60. Glassner, A.S.: Principles of Digital Image Synthesis. Morgan Kaufmann Publishers Inc., San Francisco (1994)

Topological Neighborhoods for Spline Curves: Practice & Theory

Lance Edward Miller, Edward L.F. Moore, Thomas J. Peters,
and Alexander Russell

Department of Computer Science & Engineering,
University of Connecticut,
Storrs, CT 06269-2155
tpeters@cse.uconn.edu

Abstract. The unresolved subtleties of floating point computations in geometric modeling become considerably more difficult in animations and scientific visualizations. Some emerging solutions based upon topological considerations for curves will be presented. A novel geometric seeding algorithm for Newton's method was used in experiments to determine feasible support for these visualization applications.

1 Computing the Pipe Surface Radius

Parametric curves have been shown to have a particular neighborhood whose boundary is non-self-intersecting [9]. It has also been shown that specified movements of the curve within this neighborhood preserve the topology of the curve [12, 13], as is desired in visualization. This neighborhood is defined by a single value, which is the radius of a pipe surface, where that radius depends on curvature and the minimum length over those line segments which are normal to the curve at both endpoints of the line segment [9]. Since computation of curvature is a well-treated problem, the focus of this paper is efficient and accurate floating point techniques to compute the other dependency for that radius.

Definition 1. *For a non-self-intersecting, parametric curve c, where*

$$c : [0, 1] \rightarrow \mathbb{R}^3,$$

and for distinct[1] values $s, t \in [0, 1]$, then the line segment $[c(s), c(t)]$ is doubly normal if it is normal to c at both of the points $c(s)$ and $c(t)$.

To avoid unnecessary complications with computing derivatives, only curves with regular parameterization [7] are considered.

Definition 2. *The* global separation *is the minimum over the lengths of all doubly normal segments. (For compact curves, this minimum has been shown in be positive [10].)*

---

[1] If the curve is closed, the s and t should be distinct values in $[0, 1)$.

P. Hertling et al. (Eds.): Real Number Algorithms, LNCS 5045, pp. 149–161, 2008.

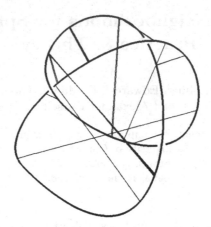

Fig. 1. Many doubly normal segments exist on this curve

An example cubic B-splines curve is given in Figure 1, with

1. control points: (0.0 0.0 0.0) (-1.0 1.0 0.0) (4.5 5.5 2.0) (5.0 -1.0 8.5)
   (-1.5 2.5 -4.5) (4.5 6.0 8.5) (3.5 -3.5 0.0) (0.0 0.0 0.0), and
2. knot vector: {0 0 0 0 0.2 0.4 0.6 0.8 1 1 1 1}

For this curve, there exist many doubly normal segments, as shown in Figure 1.
The problem is how to efficiently find all these doubly normal segments, and
then find the pair which represents the global separation distance, denoted as σ.
A pair of distinct points at parametric values s and t on a curve will be endpoints
of a doubly normal segment if they satisfy the two equations [9] for $s, t \in [0, 1]$:

$$[c(s) - c(t)] \cdot c'(s) = 0 \tag{1}$$

$$[c(s) - c(t)] \cdot c'(t) = 0. \tag{2}$$

In principle, the system given by Equations 1 and 2 could be solved algebraically by writing them in their power basis form, but this approach results
in well-known algorithmic difficulties [14]. Hence, alternative techniques will be
presented. For the software infrastructure available to these authors, it was convenient to convert the B-spline curve into a composite Bézier curve by the usual
technique of increasing the multiplicity of each interior knot to 3. This produces
5 subcurves (depicted in Figure 2 by differing line fonts), with control points:

- (0 0 0) (-1 1 0) (1.75 3.25 1) (3.21 3.29 2.58),
- (3.20 3.29 2.58) (4.67 3.33 4.17) (4.83 1.17 6.33) (3.83 0.67 5.25),
- (3.83 0.67 5.25) (2.83 0.17 4.17) (0.67 1.33 -0.17) (0.58 2.5 -0.17),
- (0.58 2.5 -0.17) (0.5 3.67 -0.17) (2.5 4.83 4.17) (3.25 3.04 4.21),
- (3.25 3.04 4.21) (4 1.25 4.25) (3.5 -3.5 0) (0 0 0).

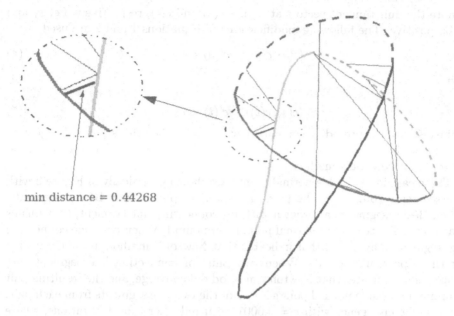

min distance = 0.44268

Fig. 2. Newton's method

Newton's method for two variables [11] was applied to Equations 1 and 2. The numerical experiments reported on prototype code suggest that this approach could be sufficiently rapid to support scientific visualization. These experiments were performed on a 64-bit AMD processor with Red Hat Linux Fedora Core 2 and OpenGL with double buffering. As always, the integration with a specific graphics subsystem is highly dependent upon the underlying architecture, and incorporation of this code on any platform would require further development and experimentation.

As is typical, the 'art' required for the successful use of Newton's method is highly dependent upon the determination of reasonable initial estimates, within the following standard formulation

$$\begin{bmatrix} s_{n+1} \\ t_{n+1} \end{bmatrix} = \begin{bmatrix} s_n \\ t_n \end{bmatrix} - J^{-1}(s_n, t_n) \begin{bmatrix} f(s_n, t_n) \\ g(s_n, t_n) \end{bmatrix}, n = 0, 1, \dots \tag{3}$$

until $|J^{-1}(s_n, t_n)[f(s_n, t_n)\ g(s_n, t_n)]^T|$ is less than some $\epsilon > 0$, where $J^{-1}(s_n, t_n)$ is the inverse Jacobian matrix.

A viable approach to this art is presented and verified on an illustrative example. The general idea is to take finitely many points on each subcurve and consider all line segments between each pair of points as a candidate for being doubly normal. Many of these segments can be excluded from further consideration by an easy culling technique based upon a lack of normality at one end point or the other.

Let $\langle c(s), c(t) \rangle$ denote the vector of unit length, formed by taking the vector between $c(s)$ and $c(t)$ and dividing that vector by its norm. Let $c'(s)$ and $c'(t)$

denote the unit tangent vectors at points $c(s)$ and $c(t)$, respectively. Let ϵ_1 and ϵ_2 be positive. The following modifications of Equations 1 and 2 are used

$$\langle c(s), c(t) \rangle \cdot \ c'(s) < \epsilon_1, \tag{4}$$

and

$$\langle c(s), c(t) \rangle \cdot \ c'(t) < \epsilon_2. \tag{5}$$

If the result of the preceding comparisons[2] are false, then this segment is rejected. Otherwise, it is sufficiently close to being doubly normal to serve as an initial estimate for Newton's method.

These candidate double normal points are shown graphically in Figure 2 with line segments connecting the pairs of candidate points from Bézier segments. When Bézier segments are shown with no connecting line segment, that means that no candidate doubly normal points were found. When only one connecting line segment is depicted, that indicates that Newton's method did not converge for those particular points. When two pairs of connecting line segments are shown, that indicates that Newton's method did converge, and the resulting pair of minimum double normal points is one of the two line segments from each pair. Typically, convergence with $\epsilon = 0.0001$ occurred after 3 or 4 iterations, where similar behavior was corroborated in independently implemented code [2]. Note that Figure 2 depicts the same curve as in Figure 1, but now the curve is rotated about the y-axis to get a better view of doubly normal points, with σ illustrated in the zoomed-in section of Figure 2.

Table 1 summarizes experimental work completed. Tests 1 - 3 report on a naive, direct approach. This relies purely upon the limiting notion that sufficiently many approximation pairs will produce a list that contains a reasonable estimate for σ. This produces the reliable estimates shown in both Tests 1 and 2, but at prohibitively slow performance for visualization applications. Furthermore, Test 3 shows that further coarsening on the partition results in both poor estimates for σ *and* unacceptable performance. Alternatively, Test 4 shows that Newton's method produces a reliable estimate of σ with acceptable performance over a very coarse partition [3]. It should also be noted that the timing for the Newton's implementation is a very rough estimate and that the prototype code is not fully optimized, so further efficiencies could be expected. Even with these disclaimers, the time reported is encouraging for scientific visualization purposes.

2   Guaranteeing a Lower Bound

The estimate of σ produced by Newton's Method can be done quickly, but it could easily be an overestimate of σ. In this section we show how to efficiently

---

[2] The possibility of choosing different values for ϵ_1 and ϵ_2 is left as a user-option and is fully permissible within the theory presented. In practice, these values may often be chosen to be the same.

[3] As a verification of the Newton's code implemented, the value of σ for this experimental curve was corroborated by an independently created code [2].

Table 1. Estimating σ

Test #	Method	Partition Size, n	Time(s)	σ
1	Direct	10,000	85	0.44268
2	Direct	2,000	6	0.44268
3	Direct	1,000	2	0.91921
4	Newton	10	.0004	0.44268

determine a *guaranteed approximation* to the length of the shortest ϵ-nearly doubly normal line segment, a quantity we call $\sigma(\epsilon)$. (This is defined precisely below.) Note that $\sigma(\epsilon) \leq \sigma(0) = \sigma$, and that in order for this to be a guaranteed approximation of σ, one would have to establish a relationship between σ and $\sigma(\epsilon)$. However, if s is a good multiplicative approximation to $\sigma(\epsilon)$ in the sense that $\alpha^{-1} \leq s/\sigma(\epsilon) \leq \alpha$ (for some small $\alpha > 1$), then certainly $\alpha^{-1}s$ is a guaranteed lower bound on σ.

2.1 Partitioning by Taylor's Theorem

Given $\epsilon > 0$, the algorithm presented in this section depends on a subroutine PIPE(δ) that returns a PL approximation to c. Specifically, PIPE is called with a parameter δ and computes a PL approximation of c for which

- the Hausdorff distance between c and the PL approximation is bounded above by $\delta/2$, and
- the PL edges "ϵ-approximate" the tangents of c associated with this edge.

To do this for the curve c, the subroutine PIPE will determine a uniform partition of the parametric interval $[0, 1]$ by the increasing sequence of points

$$0 = s_0, s_1, \ldots, s_\ell = 1.$$

Then a PL approximation to the c is created by connecting the interpolant points

$$c(s_0), c(s_1), \ldots c(s_\ell).$$

Both conditions can be met by invoking Taylor's Theorem [6]. Taylor's Theorem is stated as follows. For $f : \mathbb{R} \to \mathbb{R}$ and $n > 0$, suppose that $f^{(n+1)}$ exists for each x in a non-empty open interval $I \subset \mathbb{R}$ containing a. For each $x \neq a$ in I, there exists t_x strictly between a and x such that

$$f(x) = f(a) + f'(a)(x - a) + \ldots + \frac{f^n(a)}{n!}(x - a)^n + r_n(x),$$

where

$$r_n(x) = \frac{f^{(n+1)}(t_x)}{(n + 1)!}(x - a)^{n+1}.$$

Note that this statement of Taylor's Theorem is for the univariate case into $\mathbb{R}$, whereas the present application is to the map $c : [0, 1] \to \mathbb{R}^3$, a univariate function into $\mathbb{R}^3$. However, the x, y and z components can be treated independently as functions into $\mathbb{R}$.

Definition 3. *For any compact set $K \subset [0,1]$ and any continuous function $f : K \rightarrow \mathbb{R}^3$, and any $t \in K$, denote the components of f as $f_x(t), f_y(t)$ and $f_z(t)$. Then the max norm of $f(t)$ is denoted as $\|f(t)\|_{max}$, with*

$$\|f(t)\|_{max} = max\{f_x(t), f_y(t), f_z(t)\}.$$

Condition 1. PL Approximation within $\delta/2$: This part discusses the creation of a PL approximant of c that is within $\delta/2$ of c.

Since only C^2 functions defined on the compact set $[0,1]$ are considered, there is a maximum positive value for $\|c'(t)\|_{max}$, denoted as M_0. Recall that $c'(t)$ is non-zero. Then for any $t \in [t_0, t_1]$, (when $|t_1 - t_0|$ is sufficiently small), a straightforward application of Taylor's Theorem to the x component of $c(t)$, denoted as $c_x(t)$ would give,

$$c_x(t) = c_x(t_0) + E_x(t^*)$$

for some $t^* \in [t_0, t]$, where

$$E_x(t^*) = (t - t_0)c'_x(t^*),$$

with $E_x(t^*)$ playing the role of $r_1(x)$ above. Clearly, this can be done in each component. Then, since the final intent is to use the Euclidean norm on the vector-valued c, denoted as $\|c(t) - c(t_0)\|$, an elementary algebraic argument shows that the component-wise inequalities can be combined to yield

$$\|c(t) - c(t_0)\| \leq (t_1 - t_0)\sqrt{3}M_0.$$

Observe then that if $|s_{i+1} - s_i| \leq \delta/(2\sqrt{3}M_0)$ for each i, the curve c and this PL approximation are nowhere more than $\delta/2$ apart, as desired.

Note that this analysis only applies to a single curve, and recall that a curve c can be composed of many Bézier sub-curves. Suppose there are j many sub-curves. Then, the Taylor's theorem analysis must be applied to each of the j-many sub-curves.

Condition 2. Guaranteeing Good Local Tangent Approximations: This is analogous to the preceding argument. Suppose the curvature is positive somewhere. If not, the curve is the trivial case of a straight line. Let M_1 denote the maximum value of $\|c''(t)\|_{max}$, and let μ_0 denote the minimum value of $\|c'(t)\|_{max}$. A similar application of Taylor's Theorem yields,

$$\|c'(t) - c'(t_0)\| \leq |t_1 - t_0|\|c''(t^*)\| \leq (t_1 - t_0)\sqrt{3}M_1.$$

Let θ_t denote the angle between $c'(t_0)$ and $c'(t)$. Then,

$$|sin(\theta_t)| \leq \frac{\|c'(t) - c'(t_0)\|}{\|c'(t)\|}.$$

For a sufficiently small value of ϵ chosen to be greater than 0, the arcsine function is monotonically increasing on $[-\epsilon/4, \epsilon/4]$. Therefore, to show that $sin(\theta_t) < sin(\epsilon/4)$ over that interval, it is sufficient to have $|\theta_t| < \epsilon/4$, yielding

$$|sin(\theta_t)| \leq \frac{\|c'(t) - c'(t_0)\|}{\|c'(t)\|} \leq (t_1 - t_0)\frac{M_1}{\mu_0}.$$

Observe then that if $|s_{i+1} - s_i| \leq sin(\epsilon/4)\mu_0/M_1$ for each i, the angular deviation along the curve will be bounded as desired.

The subroutine $\text{PIPE}(\delta)$, then, returns the PL approximation obtained by uniformly dividing the interval so that each

$$|s_{i+1} - s_i| \leq min\left(\frac{sin(\epsilon/4)\mu_0}{M_1}, \frac{\delta}{2\sqrt{3}M_0}\right).$$

2.2   Lower Bound for $\sigma(\epsilon)$

The introduction, here, of the terminology "ϵ-nearly doubly normal" is similar to the conditions previously set forth for the seeds for Newton's Method, as expressed in Equations 1 and 2 in Section 1.

Let $c(s_\sigma)$ and $c(t_\sigma)$ be two distinct points of c such that $d(c(s_\sigma), c(t_\sigma)) = \sigma$. Consider those circumstances, where for sufficiently small positive ϵ there exist $\tilde{s}_\sigma, \tilde{t}_\sigma \in [0,1]$ such that the the normal planes P_1 and P_2 at $c(\tilde{s}_\sigma)$ and $c(\tilde{t}_\sigma)$, respectively, are distinct and intersect in a line near to the segment connecting $c(s_\sigma)$ and $c(t_\sigma)$ such that ν is a point on $P_1 \cap P_2$ which minimizes the sum $d(c(\tilde{s}_\sigma), \nu) + d(c(\tilde{t}_\sigma), \nu)$ and such that the angle ϕ formed between the segments connecting $c(\tilde{s}_\sigma)$ to ν and ν to $c(\tilde{t}_\sigma)$ is between $\pi - \epsilon$ and π. An illustration is shown in Figure 3, where $a = d(c(\tilde{s}_\sigma), \nu)$ and $b = d(c(\tilde{t}_\sigma), \nu)$ denote the lengths along the indicated line segments.

Any two points $c(s)$ and $c(t)$ are said to be ϵ-*nearly doubly normal* if

$$(c(s) - c(t)) \cdot c'(s) = 0 \quad \& \quad (c(s) - c(t)) \cdot c'(t) = 0,$$

or

$$\pi - \epsilon < \phi < \pi.$$

The triangle inequality gives $d(c(\tilde{s}_\sigma), c(\tilde{t}_\sigma)) \leq a + b$, and that $a + b \leq \sigma$. The algorithm described will estimate the global separation distance using approximations of $d(c(\tilde{s}_\sigma), c(\tilde{t}_\sigma))$. Since $d(c(\tilde{s}_\sigma), c(\tilde{t}_\sigma)) \leq \sigma$, the estimate produced, denoted as $\sigma(\epsilon)$ (defined immediately, below) will also be shown be no more than σ. The value $\sigma(\epsilon)$ (See Figure 4) is defined over any two ϵ-nearly normal points $c(t), c(s)$ with $t \neq s$,

$$\sigma(\epsilon) = min_{\{c(t),c(s)\}}\{d(c(t), c(s)).\}$$

The transition to providing an estimate of the more conservative value $\sigma(\epsilon)$ rather than trying to directly approximate σ is motivated by the following example. Let α be a planar C^∞ curve containing an arc of the unit circle with

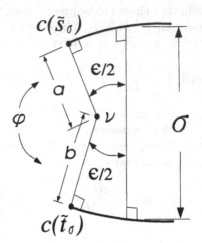

Fig. 3. The points $c(\tilde{s}_\sigma)$ and $c(\tilde{t}_\sigma)$ are ϵ-nearly doubly normal

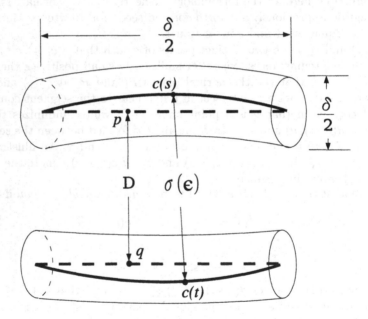

Fig. 4. The points $c(s)$ and $c(t)$ on the curve segments inside each cylinder are ϵ-nearly doubly normal, and D is the distance between the PL segments that approximates the curve segments

arc-length strictly less than π, but where α has its minimum separation distance being much greater than 2 and found elsewhere on the curve. For any algorithm that attempts to approximate σ by focusing upon pairs of points that were nearly normal within some fixed tolerance, there would always be

---

Global Separation Distance Estimate Algorithm

Input: A spline curve c & ϵ.
   0. Initialize $\delta = 1$ and ω for upper precision bound.
   1. Initialize $A(\epsilon) = 0$.
   2. Apply PIPE(δ) to create a PL approximation.
   3. Find pw-distances, $d(e_i, e_j)$
      with points p, q that realize $d(e_i, e_j)$.
   4. If p and q are ϵ-nearly doubly normal
      retain $d(e_i, e_j)$ for further consideration,
      Else discard.
      Let $D = min(d(e_i, e_j))$ over the remaining points.
   5. If $D \geq 4\delta$
      $A(\epsilon) = D - 2\delta$
      Else $\delta = \delta/2$, and Go to Step 1.
Output: $A(\epsilon)$ = estimate for global separation distance.

Fig. 5. General algorithm for estimating the global separation distance

some input curve like α which would return some value near 2, since this arc-length can be made arbitrarily close to π.

The value $\sigma(\epsilon)$ is now accepted as a good estimate of σ, and the focus shifts to approximating $\sigma(\epsilon)$, recalling that $\sigma(\epsilon) \leq \sigma$. Then, the algorithm below in Figure 5 will return an approximation $A(\epsilon)$ of $\sigma(\epsilon)$, with the following two guarantees:

- $A(\epsilon) \leq \sigma(\epsilon) \leq \sigma$, and
- $A(\epsilon) > (\sigma(\epsilon))/2$.

Recall that the previous Taylor analyses guarantees that the result of PIPE(δ) satisfies the following three conditions:

- the length of each cylinder is strictly less than $\delta/2$,
- the radius of each cylinder is strictly less than $\delta/2$, and
- the angular deviation between tangents on the curve segments in each cylinder is strictly less than $\epsilon/4$.

The value for δ is initialized to 1. (It can be assumed that the curve has been normalized so that it lies in a sphere of radius 1 (Note that this also makes $\sigma(\epsilon) < 1$ for all $\epsilon > 0$, which is invoked later).) The resultant estimate $A(\epsilon)$ is then tested for validity (See algorithm in Figure 5), and failure results in halving the value of δ, repeating the iterations until a valid value is obtained. In this way, the overall algorithm is logarithmic in $1/\sigma(\epsilon)$.

2.3   Termination and Satisfactory Value

A termination, let $\check{D}$ be defined as the distance between two PL segments that approximate the curve segments in which $\sigma(\epsilon)$ is actually realized. Note that $D \leq \check{D} \leq \sigma(\epsilon) + \delta$, since the radius of the cylinders shown in Figure 4 is at most $\delta/2$, as given previously by Taylor's analysis.

The algorithm will terminate when $2\delta < \sigma(\epsilon)$. Several applications of the triangle inequality in Figure 4 show that $D \geq \sigma(\epsilon) - 2\delta$, or equivalently $D + 2\delta \geq \sigma(\epsilon)$, yielding

$$\frac{D}{\sigma(\epsilon)} \geq \frac{D}{D + 2\delta} = \frac{1}{1 + \dfrac{2\delta}{D}} \geq \frac{1}{1 + \dfrac{2\delta}{4\delta}} = 2/3.$$

Hence, $D \geq (2/3)\sigma(\epsilon)$ and $D \leq \sigma(\epsilon) \leq \sigma$, as desired.

The global separation distance algorithm in Figure 5 assumes the existence of a geometric distance predicate $d(e_i, e_j)$ between two line segments, e_i and e_j, which returns:

– the distance $d(e_i, e_j)$ between the two line segments, and
– the points p and q on e_i and e_j, respectively, where that distance is realized.

2.4   Asymptotic Time Bound

The time taken to approximate the global separation distance by this algorithm is quadratic in the bounds derived earlier for the Taylor's analysis. The final bound $\sigma(\epsilon)$ is computed within an *a priori* upper bound on the total number of subdivisions required as is standard practice [8]. As the algorithm is guaranteed to terminate when $\delta < \sigma(\epsilon)/2$ and δ is halved during each iteration, no more than $O(\log \sigma(\epsilon)^{-1})$ calls to PIPE are invoked. In the worst case, checking validity for a given PL approximation produced by PIPE takes quadratic time in the number of edges. The total time is thus no more than

$$O\left(\log(1/\sigma(\epsilon)) \max\{ \ (\frac{M_0}{\delta})^2, \ (\frac{M_1}{\mu_0})^2 \ \}\right).$$

2.5   Example Analysis

For the curve already used, the values for the indicated parameters, above, were computed using the Maple computer algebra system as

– $M_0 = 14.9$,
– $\mu_0 = 3.4$,
– $M_1 = 21.9$.

Then an easy analysis shows that the number of subintervals generated for each sub-curve is 2048, which is consistent with the empirical findings in Table 1, where approximately 2000 sampled points per sub-curve produced an acceptable approximation. However, this algorithm for $\sigma(\epsilon)$ provides the additional information that $\sigma(\epsilon)$ is a lower estimate and is truly close in the precisely defined sense given in Section 2.3. Of course, the input numerical parameters between the two algorithms would cause ·slight variances, but the agreement within this order of magnitude comparison is of interest.

2.6   Open Issues for Future Work

Within the Taylor's analysis performed, a well-defined lower estimate is established. Neither of the two algorithms presented (based upon Newton's Method or Taylor's Theorem) can effectively preclude output of a value that might be generated due purely to local properties of the curve. In practice, though, this is not quite as problematic as it may first appear. Recall that the purpose in estimating σ was to find the *global* factor that contributed to determining the radius of the neighborhood around the curve, while the *local* factor was in terms of curvature. A minimization is taken over those two factors to determine the radius. So, if either of the algorithms presented here returns a minimum value that is reflective of local properties, then this may suggest that the curvature is the determining factor for the radius. In those cases where curvature is the determining factor, then one need not even estimate σ, but these authors know of no *a priori* way to discriminate these cases, in order to avoid unnecessary computations. Resolving this issue remains beyond the scope of the present article but it is of interest for future investigation, both

- *experimentally*, with the algorithms discussed on more examples, and
- *theoretically*, by examination of adaptive skeletal structures, such as the medial axis [1], but also inclusive of more recent alternatives [3, 4, 5].

Extensions to higher dimensional geometric elements appear to be possible, but remain the subject of future work.

3   Experimental Observations

The curves here were assumed to be C^2. While this is sufficient for Newton's method, it is clearly *not* necessary to have the curve be C^2 *globally* . Clearly, Newton's method is local, so it it will be sufficient to have the C^2 condition *locally* within neighborhoods of the seeds. This is shown in Figure 6. This composite cubic Bézier curve has a point of non-differentiability at the top, where the three segments are shown in differing line fonts. Yet Newton's method easily and quickly estimates σ as 1.52, using only 10 partitioning points per sub-curve. This value of 1.52 was verified by the direct method discussed in Section 1. In Figure 6, two line segments are shown, with the thinner font indicating the seed and the thicker font denoting the converged value from Newton's method.

This composite Bézier curve has three segments and its control points are

- (0, 0.5, 0), (0.75, -1, 0), (0.83, -1.67, 0), (0.72, -2.11, 0),
- (0.72, -2.11, 0), (0.5, -3, 0), (-0.5, -3, 0),
- (-0.72, -2.11, 0), (0.83, -1.67, 0), (0.75, -1, 0), (0, 0.5, 0).

Fig. 6. A composite Bézier curve with a non-differentiable point

Fig. 7. A Bézier curve with a cusp

When the bound on the angular deviation of **Condition 2** of Subsection 2.1 is $\epsilon = 0.1$ (as done Subsection 2.5), the Taylor's analysis yields a value of $\ell = 10$, indicating $2^{\ell} = 1024$ partition points, far in excess of the 10 used here for Newton's method.

Of course, care must still be exercised in using Newton's method, as shown in Figure 7. Here there is a cusp at the top and the control polygon is shown. Using a very fine sampling relative to Inequalities 4 and 5, results in accepting a seed that is far into the cusp. Under Newton's method such a seed converges to an estimate of zero for σ. For the particular curve σ does equal zero, but the curve shown could be merely a subset of a much larger closed curve having a non-zero value for σ, meaning that this zero estimate would be inappropriate. Note that the algorithm for $\sigma(\epsilon)$ would specifically detect this difficulty by its check on the magnitude of the derivatives, thereby identifying this unbounded derivative and terminating the algorithm. Similar checks should also be incorporated into any practical code for Newton's method in this application. This example provides further motivation for studying the trade-offs regarding local and global properties, as mentioned in Subsection 2.6.

4 Conclusion

Newton's method in two variables, when implemented with some novel geometric seeding techniques, provides an approach that is promising for preservation of topological characteristics during scientific visualization. Experiments and an alternative theoretical analysis, based upon Taylor's Theorem, are presented.

Acknowledgements. The authors were partially supported by NSF grants DMS-9985802, DMS-0138098, CCR-022654, CCR-0429477 and/or by an IBM Faculty Award. All statements here are the responsibility of the authors, *not* of the National Science Foundation *nor* of IBM. The authors thank the Dagstuhl Seminar organizers and the Dagstuhl staff for providing the intellectually stimulating environment for refinement of these ideas, which were initially based upon the dissertation of E. L. F. Moore.

References

1. Amenta, N., Peters, T.J., Russell, A.C.: Computational topology: ambient isotopic approximation of 2-manifolds. Theoretical Computer Science 305, 3–15 (2003)
2. Bisceglio, J.: Personal communication. justin.bisceglio@gmail.com (October 2005)
3. Damon, J.: On the smoothness and geometry of boundaries associated to skeletal structures, i: sufficient conditions for smoothness. Annales Inst. Fourier 53, 1941–1985 (2003)
4. J. Damon.: Determining the geometry of boundaries of objects from medial data (pre-print, 2004)
5. J. Damon.: Smoothness and geometry of boundaries associated to skeletal structures, II: geometry in the Blum case (pre-print, 2004)
6. Ellis, R., Gullick, D.: Calculus with Analytic Geometry, 3rd edn., Harcourt Brace Jovanovich (1986)
7. Farin, G.: Curves and Surfaces for Computer Aided Geometric Design: A Practicle Guide, 2nd edn. Academic Press, San Diego (1990)
8. Lutterkort, D., Peters, J.: Linear envelopes for uniform B–spline curves. In: Curves and Surfaces, St Malo, pp. 239–246 (2000)
9. Maekawa, T., Patrikalakis, N.M.: Shape Interrogation for Computer Aided Design and Manufacturing. Springer, New York (2002)
10. Maekawa, T., Patrikalakis, N.M., Sakkalis, T., Yu, G.: Analysis and applications of pipe surfaces. Computer Aided Geometric Design 15, 437–458 (1998)
11. Mathews, J.H.: Numerical Methods for Computer Science, Engineering and Mathematics. Prentice-Hall, Inc., Englewood Cliffs (1987)
12. Moore, E.L.F.: Computational Topology of Spline Curves for Geometric and Molecular Approximations. PhD thesis, The University of Connecticut (2006)
13. Moore, E.L.F., Peters, T.J., Roulier, J.A.: Preserving computatational topology by subdivision of quadratic and cubic Bézier curves (to appear)
14. Piegl, L., Tiller, W.: The NURBS Book, 2nd edn. Springer, Heidelberg (1997)

Homotopy Conditions
for Tolerant Geometric Queries*

Vadim Shapiro

Mechanical Engineering and Computer Sciences
University of Wisconsin – Madison, Wisconsin 53706 USA
vshapiro@engr.wisc.edu

Abstract. Algorithms for many geometric queries rely on representations that are comprised of combinatorial (logical, incidence) information, usually in a form of a graph or a cell complex, and geometric data that represents embeddings of the cells in the Euclidean space E^d. Whenever geometric embeddings are imprecise, their incidence relationships may become inconsistent with the associated combinatorial model. Tolerant algorithms strive to compute on such representations despite the inconsistencies, but the meaning and correctness of such computations have been a subject of some controversy.

This paper argues that a tolerant algorithm usually assumes that the approximate geometric representation corresponds to a subset of E^d that is homotopy equivalent to the intended exact set. We show that the Nerve Theorem provides systematic means for identifying sufficient conditions for the required homotopy equivalence, and explain how these conditions are used in the context of geometric and solid modeling.

1 Queries on Combinatorial Representations

1.1 Queries on Combinatorial Data Structures

Shapes, configurations, and other geometric objects in computational geometry and geometric modeling may be represented implicitly by a system of predicates, or *combinatorially*. A distinguishing feature of a combinatorial representation is that it includes an explicit data structure to represent logical incidence between a finite collection of simpler 'primitive' objects. In geometric applications, a combinatorial representation also includes some representation of geometry that embeds these primitive objects into (typically) Euclidean space E^d. Practitioners often refer to the two parts of such a representation as 'topology' and 'geometry' respectively. Examples of combinatorial representations include arrangements of hyperplanes, triangulations, polyhedra, boundary representations in solid modeling, Voronoi diagrams, and many others.

This paper deals exclusively with combinatorial representations. Without loss of generality, we will assume that logical incidence relationships in a combinatorial representation are stored as an *abstract* cell complex K, which is essentially a

---

* Based on the talk at the Dagstuhl Seminar on Reliable Implementation of Real Number Algorithms: Theory and Practice, January 8-13, 2006.

P. Hertling et al. (Eds.): Real Number Algorithms, LNCS 5045, pp. 162–180, 2008.

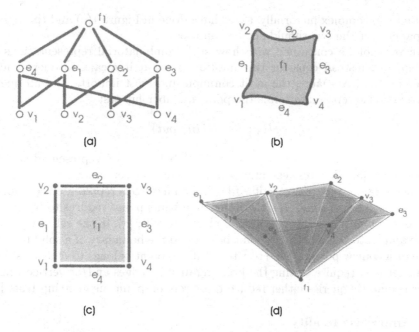

Fig. 1. Combinatorial representation for a quadrilateral: (a) graph representation of an abstract complex K; (b) geometric realization $|K|$ is a union of embedded cells; (c) geometric representation of the abstract complex K in (a); (d) nerve $\mathcal{N}K$ of the complex K consists of four solid tetrahedra

partial ordering on a collection of k-cells (vertices, edges, faces, etc.). For convenience, the ordering may be defined in terms of subsets of 0-cells and represented by a graph, with arcs indicating the containment relation. Thus, a 1-cell is a collection (usually a pair) of 0-cells; 2-cell is a collection of (bounding) 1-cells, and so on. Figure 1(a) shows a typical abstract cell complex for a combinatorial representation of a quadrilateral, consisting of four vertices (0-cells) v_i, four edges (1-cells) e_i, and one face (2-cell) f_1. Since higher dimensional cells contain all incident lowerdimensional cells, by definition, all cells are closed. Additional conditions on K may be imposed depending on specific use, such as in triangulations, Voronoi diagrams, and boundary representations in solid modeling. For example, a boundary representation in solid modeling requires that the abstract complex is an orientable k-cycle[1,2]. Thus, topology of the boundary representation for the quadrilateral would be the abstract cell complex in Figure 1, but without the 2-cell f_1.

Each p-cell $\sigma^p \in K$ is embedded (or realized) geometrically as a set $|\sigma^p|$ in E^d either explicitly (for example, vertex coordinates, curve or surface parameterization), implicitly (for example, as the intersection of embedded higher dimensional cells $|\sigma^{p+1}|$), or procedurally (for example, as the interpolation of lower-dimensional cells $|\sigma^{p-1}|$). The union of all embedded cells of K is usually called a geometric realization or underlying space of K and is denoted by $|K|$. A (non-unique) geometric realizations of the abstract complex for quadrilateral is shown in Figure 1(b). Geometric realizations can be used to represent the abstract cell complex pictorially, as we have done in Figure 1(c) and throughout the paper, but the two should not be confused.

We will not be concerned with how such combinatorial representations are created, but instead consider the question of what it means to query such a representation. Arguably, the most common and most important of all queries on a geometric representation is the point membership test

$$\Pi_S : R^3 \rightarrow \{in, \ out\}$$

to determine whether a given point $p \in R^3$ belongs to a represented set S of points or not[3]. This test essentially implements the set charateristic function, thus defining the *induced* set S itself [4]. When a set S is represented combinatorially as a subset of some $|K|$, the point membership test reduces to testing the candidate point p against individual embedded cells, $|\sigma|$, in the cell complex K. For example, to test whether a point belongs to the boundary of a solid (or a triangle or a convex polygon), we test to see if the point belongs to one of its faces; the latter may require testing the point against its edges and/or vertices. Many other geometric queries either reduce to or rely on point membership tests [2].[1]

1.2 Imprecise Reality

The above description of a combinatorial structure is typical in literature but is often misleading for practical applications, because embeddings of cells in the cell complex K are usually known only with some degree of uncertainty. The reasons for this vary, but there are several well known generic causes that are summarized below and illustrated in Figure 2 for the combinatorial representation of the quadrilateral boundary:

- Geometric data for scientific and engineering models comes from a variety of sources and is intrinsically imprecise; this is obviously true for measured and reconstructed models, but is also the case for models that are transferred between various systems using some neutral exchange format [6]. Figure 2(a) shows a typical geometric data obtained from an inexact source.
- Embedding may be exact, but point membership test requires computation of quantities that are computed only approximately, for example, because of round-off errors in floating point computations. Even the simplest and the most common point membership test against a linear halfplane is guaranteed to be correct only for points that are at least some distance δ away from the line segment bounding the halfspace [7]. Figure 2(b) illustrates typical 'safe' zones where the point membership is guaranteed to be correct.
- Geometric representations are often based on incomplete spaces – by design. For example, rational parametric curves and surfaces are standard in most

---

[1] The point membership test as used in this paper should not be confused with the point membership classification (PMC) function. PMC also requires distinguishing the points on the boundary from those in the interior of the set[5]; however, most PMC implementations rely on point membership testing in one form or another[2].

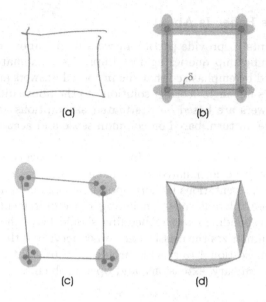

Fig. 2. Well known causes of imprecision: (a) data comes from imprecise source; (b) point membership is limited by finite resolution; (c) incomplete representation spaces lead to redundant representations; (d) containment sets in multi-resolution and subdivision methods

computer-aided design systems, but their transformations and intersections may not admit rational parameterizations and must be approximated by independent embeddings that are logically redundant[8]. For example, in Figure 2(c), both vertices and edges (with their end points) are embedded independently.

– Multi-resolution and subdivision methods for representing curves, surfaces, and shapes are increasingly popular in geometric modeling due to their attractive computational properties. By their very nature, such methods represent the geometry only approximately, even when the exact point set is well defined in the limit [9]. For example, the edges in Figure 2(d) are contained inside a hierarchy of convex regions. See [10] for many additional examples.

Regardless of the source of the imprecision, we are faced with an undeniable reality: point membership testing against embedded cells $|\sigma|$ or embedding of their complex $|K|$ is usually not computable exactly. What is the meaning of such a combinatorial representation? Without proper semantics, correctness of most geometric queries, and point membership queries in particular, on such combinatorial representations are compromised. Two difficulties are immediately apparent. First, it is not clear what it means to test a point against an imprecise embedding of a cell σ. Secondly, depending on the answer to the first question, the union of $|\sigma|$ may or may not (and usually does not) form a valid realization of the cell complex K. Then, under what conditions point membership testing against $|K|$ make sense and reduces to tests against individual cells $|\sigma|$?

1.3   What This Paper Is About

This paper attempts to provide partial answers to the above questions, in the case of point membership queries against imprecise combinatorial representations. But it should be emphasized that the proposed answers are not a panacea, and indeed there is no single "correct solution" to the above difficulties. Rather, the proposed answers are based on postulated assumptions about such representations that are, in turn, based on common sense and accumulated practical experience.

The rest of the paper is as follows. In section 2, we argue that most imprecision problems may be formulated by replacing exact geometric embeddings in combinatorial representations by corresponding tolerant zones. We postulate properties for these tolerant zones, including the requirement that a tolerant zone and the corresponding exact embedding should be of the same homotopy type. Section 3 summarizes the classical notion of nerve and the Nerve Theorem, that are applied in Section 4 to specific problems of tolerant modeling. Section 5 concludes with summary, extensions, and open problems.

2   Tolerant Zones in a Complex

2.1   To Tolerate Imprecision

The difficulty in proposing a standard notion of tolerant point membership query on imprecise representations lies in the ambiguity of the term "imprecision" and the diversity of its sources. But let us consider the hypothetical task of classifying a point against any one of the imprecise representations illustrated in Figure 2.

When the data comes from an imprecise source as in Figure 2(a), or is known to be redundantly approximated as in Figure 2(c), it makes little sense to test points against the given embeddings because the union of the embedded cells $|\sigma|$ is clearly not a valid complex K. Two approaches have been proposed in such situations[4]: either perturbing the original embeddings slightly to obtain a valid cell complex, or classifying points within some small but finite *distance* δ of $|\sigma|$ as being *in* $|\sigma|$. The latter approach would have an effect of "thickening" the individual cell embeddings by δ in a hope that the adjacent cells connect to each other as intended. In situations illustrated in Figure 2(b) and (d), perturbations are not helpful at all, since the cause of imprecision is the resolution of the point membership algorithm itself. But in these cases too, points in the vicinity of the embedding $|\sigma|$, which may or may not be known, are considered to belong to the embedded set.

We conclude that in most cases, to tolerate imprecision means that a candidate point p is tested not against the set $|\sigma|$, but against some enlarged set Z_σ^δ associated with the cell in question. We will refer to this set Z_σ^δ as a *tolerant zone*[2] of cell σ of "size" δ. In general, δ can be a function of the embedding and vary spatially, but in practice δ is often used as an upper bound on the distance from the

---

[2] Since the zone is associated with a set $|\sigma|$ and not an abstract cell σ, it would be more accurate (and more awkward) to use $Z_{|\sigma|}^\delta$ to denote the zone.

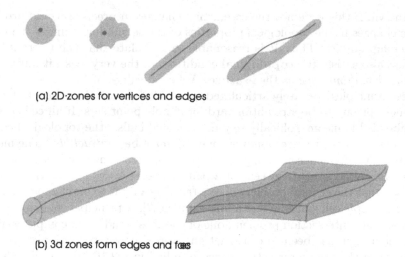

(a) 2D-zones for vertices and edges

(b) 3d zones form edges and faces

Fig. 3. Tolerant zones Z_σ^δ for individual cells $\sigma \in K$. Note that a tolerant zone may not contain the exact embedding $|\sigma|$, which may or may not be known.

embedding $|\sigma|$, which may or may not be known. Throughout the paper, we may drop either δ or σ from Z_σ^δ when their meaning is clear from the context. Typical tolerant zones for point, edges, and faces are illustrated in Figure 3.

The notion of a tolerant zone appears explicitly in many proposed approaches to dealing with imprecision, for example, in [11,12,13,14]. The zones are also implied by epsilon predicates in [15], box and ball covers of boundaries [16,4], and surface thickening[17]. Zones that are defined by maximum distances to the corresponding cells can be constructed in terms of offsets or sweeps (Minkowski sums); for B-splines, the zones are naturally associated with their control nets and polyhedral enclosures [9,10]. The notion of a tolerant zone is also present implicitly even when the embeddings are known exactly and may be perturbed to create a valid cell complex K. Different perturbation techniques are described in [18,19] and in [20], but all such techniques share the assumptions that the perturbed embeddings (usually curves and/or surfaces) lie in the vicinity of the original embeddings and preserve their topological form in some well defined sense.

We could now say that a point membership query against set X is δ-*tolerant* iff it returns *in* for every point in the tolerant zone Z_X^δ, and *out* for all other points. For this definition to make sense, we have to be clear about the *assumed* properties of the tolerant zones. Based on the above discussion, it is clear that a tolerant zone must satisfy some **metric** condition: the size of zone Z_X^δ is bounded by δ. It may be interpreted to mean that every point in the zone is at most δ away from X, or that the zone Z_X^δ may be covered by an infinite family of balls of radius δ, or perhaps in another way depending on a specific situation. A second condition becomes apparent when we consider what happens to the tolerant zone with increased precision. As $\delta \to 0$, the tolerant zone Z_X^δ must gradually shrink and deform onto the exact set $X = Z_X^0$, if that exact set were known. If this were not the case, small changes in precision could produce drastic changes in the topological properties of the sets induced under the point membership queries. Thus, it is reasonable to postulate that the tolerant zone Z_X^δ must also satisfy the **topological** condition: at the very least, it must be of the *same homotopy type* as the exact set X for all values of δ.

These principles are rarely articulated, probably because the concept of tolerant zones appears to be straightforward for simple point sets. If all cells $\sigma \in K$ are embedded homeomorphically to p-dimensional balls,[3] the topological condition implies that their corresponding zones Z_σ^δ must be *contractible*.[4] The metric conditions are often chosen to imply that the zones are either (unions of) convex sets or Minkowski sums (a sweep) of a ball and some known geometry. That is not to say that the general problem of constructing a tight tolerant zone is trivial, even for a simple cell embedded as a line segment [7]. A principal challenge in the surface-surface intersection problem, one of the most studied in computer-aided geometric design, has been to construct an approximate intersection curve that is isotopic to the true intersection curve (studied in [22,16,23] and elsewhere). Requiring that the tolerant zone of a curve segment is of the same homotopy type is a weaker condition that is easier to obtain and is often used to establish the isotopy condition [17,23].

2.2 Collection of Tolerant Zones

Let us now suppose that an implemented point membership test is δ-tolerant against each embedded cell σ corresponding to a collection of cells $\{\sigma\}$ that form a cell complex K. A usual implementation of the point membership query against the embedded complex $|K|$ returns *in* if and only if the point in question classifies *in* with respect to one of the embedded cells $|\sigma|$, $\sigma \in K$. This seems reasonable since the union of all tolerant zones $\bigcup Z_\sigma$ is an approximation of some exact realization $|K| = \bigcup |\sigma|$ (which may or may not be known). But is such a query δ-tolerant for some δ determined from the tolerant zones of the individual cells? In particular, is it true that the union of the tolerant zones $\bigcup Z_\sigma$ for the cells in the complex is a combined tolerant zone Z_K^δ for the complex? The examples in Figure 4 easily show that this is not the case. The metric condition is satisfied trivially, but the topological condition is problematic. Both examples in Figure 4 fail the topological condition, each for different reasons, because the union of the cell zones is not homotopy equivalent to the embedding $|K|$ of the original cell complex K. In this case, K is the abstract cell complex associated with a quadrilateral's boundary $|K|$. Indeed, we usually do not know what $|K|$ is, but its intended topological properties are completely defined by the abstract cell complex K.

Intuitively, the solution to the difficulty appears to be straightforward: if $\bigcup Z_\sigma$ were to form a tolerant zone Z_K for the whole complex, then the tolerant zones

---

[3] This is not always the case. For example, computer-aided design systems often rely on geometric cell complexes built from submanifold cells of arbitrary genus [21].

[4] By definition, contractible spaces are homotopy equivalent to a point; in particular, they must be connected and contain no holes.

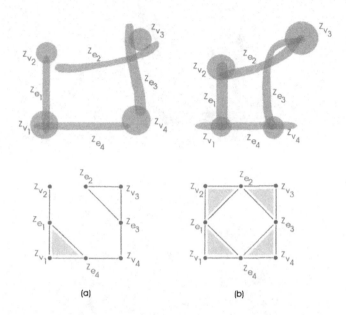

Fig. 4. The unions of tolerant zones $\bigcup Z_\sigma$ (shown above) and the corresponding nerves $\mathcal{N}\{Z_\sigma\}$ (shown below) associated with vertices and edges in a quadrilateral's boundary. (a) The intersections of zones $\{Z_\sigma\}$ in three out of four corners do not correspond to incidence of cells in the quadrilateral's boundary; (b) Zones intersections are in one-to-one correspondence with incidences in the boundary, but the union of the zones is not homotopy equivalent to the quadrilateral's boundary.

Z_σ must maintain the same incidence relationships as the corresponding cells σ have in K. In other words, this requires that for any two cells $\sigma_i, \sigma_j \in K$, the tolerant zones $Z_{\sigma_i} \cap Z_{\sigma_j} \neq \emptyset$ if and only if $\sigma_i \cap \sigma_j \neq \emptyset$. Indeed, this is one of the conditions proposed in [12] as a requirement for "consistency" of a polyhedral boundary representation model. Note that this condition is satisfied by the collection of zones in Figure 4(b), but is not sufficient because tolerant zones are not convex and their intersections are not contractible. A different set of conditions on the tolerant zones proposed in [13] suggests that only *some* of these intersections are required, but connected components of all such intersections must be contractible. Both authors proposed additional metric and containment conditions on the zones, without formal justification. Consider the example in Figure 5. Clearly the union of shown tolerant zones $\bigcup Z_\sigma$ is homotopy equivalent to the boundary of the rectangle, even though the relationships between the zones fail the previously proposed conditions. Specifically, the conditions proposed in [12] are violated in the upper left corner, where the zones Z_{e_1} and Z_{e_2} of the incident edges do not intersect; the conditions proposed in [13] are violated in the lower left corner where the vertex zone Z_{v_1} is not large enough to contain the intersection of the two incident edge zones. These and other proposals for the notion of topological consistency rely on heuristic arguments and seemingly contradictory conditions, largely because the notion itself is not well defined [14,24,25].

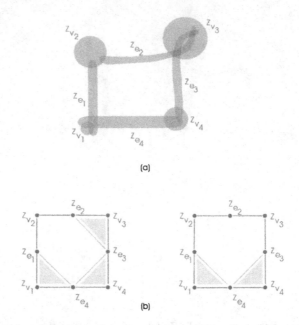

(a)

(b)

Fig. 5. (a) The union of tolerant zones $\bigcup Z_\sigma$ is homotopy equivalent to the boundary of a rectangle, but fails the conditions proposed in [12] (upper left corner) and [13] (lower left corner); (b) Full nerve (left) and inclusion-reduced nerve (right) for the collection of zones in (a)

However, if the notion of homotopy equivalence is accepted as a guiding principle for topological consistency, then it becomes possible to derive sufficient conditions for consistency and to establish a formal relationship between various heuristic arguments. The rest of this paper shows how this can be accomplished using the Nerve Theorem[26].

3 Homotopy Via Nerves

3.1 Nerve of Collection of Sets

The differences between various examples in Figures 4 and 5 can be understood in terms of intersections between the tolerant zones Z_σ for various cells $\sigma \in K$. The example in Figure 4(a) does not satisfy the homotopy condition because some zones of incident edges do not intersect. All required zone intersections are present in Figure 4(b), but the intersections themselves are problematic, and it may not be clear how to differentiate this case from the example in Figure 5. Intuitively, we want to know *which* zones intersect and *how* they intersect, so that the union of zones $\bigcup Z_\sigma$ is homotopy equivalent to $|K|$. We will use the concept of nerve to order all possible intersections between the zones, and use the Nerve Theorem to enforce the required homotopy.

Given a finite collection of (closed) sets $\{X\}$, the **nerve** $\mathcal{N}\{X\}$ is an abstract simplicial complex constructed as follows. Let 0-simplex $\langle X_i \rangle$ be a vertex

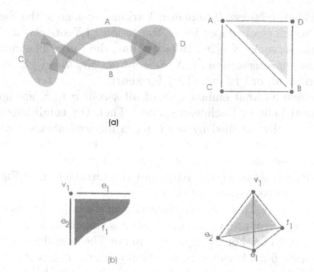

(a)

(b)

Fig. 6. Collections of closed sets and their corresponding nerves: (a) The nerve $\mathcal{N}\{A, B, C, D\}$ is a two-dimensional complex determined by which sets intersect each other, but says nothing about the topological properties of intersections; (b) a nerve of a typical two-dimensional "corner" is a solid tetrahedron

corresponding to the set X_i in the collection. A 1-simplex $\langle X_i, X_j \rangle$ is associated with every pair of sets X_i, X_j whose intersection is non-empty. More generally, k-simplices in the nerve $\mathcal{N}\{X\}$ are defined by collections of $k+1$ vertices corresponding to $k+1$ sets $\{X_j\}$ such that $\bigcap_j X_j \neq \emptyset$. It is easy to see that the collection of such simplices is a complex, since every sub-collection of intersecting sets is non-empty. Two examples of nerves are shown in Figure 6. Figure 6(a) shows the nerve for the collection of four sets. Observe that the dimension of the nerve corresponds to the maximum number of sets in the collection with non-empty intersection, which is three in both of these examples. In a cell complex K, the maximum number of intersections occurs at the vertices, where all higher dimensional cell intersect. For example, consider a typical vertex neighborhood in Figure 6(b) where a vertex, two edges, and one face intersect. The corresponding nerve will contain a 3-simplex which can be visualized as a tetrahedron. The nerve of the abstract complex K for quadrilateral (with interior) is shown in Figure 1(d). Nerves for collections of tolerant zones in Figures 4 and 5 are shown next to them.

3.2 Nerve Theorem

The concept of a nerve is apparently due to Alexandrov[27], and has become a standard tool in algebraic topology[28], combinatorics [26], and computational topology [29]. We will use nerves to enumerate, represent, and analyze the intersections between the tolerant zones Z_σ, and to compare them to the intersections implied by the corresponding abstract cell complex K. The main tool in our analysis is the **Nerve Theorem**. Various versions of the Nerve Theorem correlate the topological properties of the union $\bigcup X$ of sets in the collection and those of its nerve $\mathcal{N}\{X\}$ [26]. In particular, the Nerve Theorem states that, *if every non-empty intersection $\bigcap_j X_j$ is contractible, then $\bigcup X$ and $\mathcal{N}\{X\}$ are homotopy equivalent*, or $\bigcup X \simeq \mathcal{N}\{X\}$ for short.

Let us now assume that embeddings of all k-cells $\sigma \in K$ are homeomorphic to k-dimensional balls in Euclidean space.[5] Then the conditions of the Nerve Theorem are trivially satisfied by the cells in any cell abstract complex, since the intersection between any two cells in the complex is either empty or another cell. It is easy to see that the nerve of Figure 1(d) can be continuously deformed onto the quadrilaterals in Figure 1(b), and the tetrahedron in Figure 6(b) can be collapsed on the corresponding vertex.

Applying the Nerve Theorem to a collection of zones $\{Z_\sigma\}$ is more interesting. By assumed topological properties, every zone is contractible, but their intersections do not have to be. For example, the nerve Theorem does not apply to the example in Figure 6(a), because intersection of the two sets $A \cap B$ is a disconnected set with two components. In this case, the nerve $\mathcal{N}\{A, B, C, D\}$ is not homotopy equivalent to $A \cup B \cup C \cup D$. Similarly, the Nerve Theorem cannot be applied to the collection of zones in Figure 5, because the intersection of zones $Z_{e_2} \cap Z_{e_3}$ is not contractible. Nevertheless in this particular case, the nerve is homotopy equivalent to the union of zones, by inspection. On the other hand, the conditions of the Nerve Theorem are satisfied by every collection of zones in Figure 4(a), and indeed it is easy to see that in each case, the nerve is homotopy equivalent to the corresponding union of zones.

4   Tolerant Complex from Tolerant Zones

We are now ready to tackle the original question posed in Section 2.2, namely under what conditions the union of tolerant zones $\bigcup Z_\sigma$ is in fact a tolerant zone Z_K for the embedding of the abstract complex $|K|$? There are two apparent difficulties. First, we do not really know the embedding $|K|$ but only the abstract complex K. Secondly, as we saw above on the examples in Figure 4 and 5, it is not clear that the Nerve theorem actually helps to distinguish the tolerant embedding of K. We address both of these difficulties below.

4.1   Sufficient (and Almost Necessary) Conditions

The first of the above difficulties is resolved by applying the Nerve theorem twice: once to the abstract cell complex K and second time to the collection of the tolerant zones Z_σ, $\sigma \in K$. The key observation is that we really do not need to know the embedding $|K|$, because by definition, any exact embedding must be homotopy equivalent to the abstract complex K. Therefore, by the Nerve theorem, $|K| \simeq \mathcal{N}K$, where $\mathcal{N}K$ stands for the nerve of the abstract complex K.

---

[5] For more general embeddings, other versions of the Nerve Theorem relate k-connectedness of the nerve to that of the union of the sets in the collection [26].

Applying the Nerve theorem to the collection of tolerant zones $\{Z_\sigma\}$, we observe that contractibility of all non-empty intersections between the tolerant zones is a sufficient condition for homotopy equivalency between the union of all zones and their nerve, i.e. $\bigcup Z_\sigma \simeq \mathcal{N}\{Z_\sigma\}$. Combining the two observations together, we arrive at sufficient conditions for a complex of zones to be tolerant.

Theorem 1 (Twin-Nerve). *Let $\{Z_\sigma\}$ be a collection of tolerant zones associated with abstract cells $\sigma \in K$, such that every non-empty intersection $\bigcap Z_{\sigma_i}$ is contractible. Then $\bigcup Z_\sigma$ is a tolerant zone Z_K for the complex K if and only if $\mathcal{N}K \simeq \mathcal{N}\{Z_\sigma\}$.*

To show that the theorem is true, we only need to show that the homotopy equivalence of the two nerves implies that $|K|$ and $\bigcup Z_\sigma$ are of the same homotopy type as well, and vice versa. This follows immediately by applying the Nerve theorem twice as discussed above:

$$|K| \simeq \mathcal{N}K \simeq \mathcal{N}\{Z_\sigma\} \simeq \bigcup Z_\sigma$$

Note that the Twin-Nerve Theorem establishes only sufficient conditions because it hinges on the assumption that all intersection of tolerant zones are contractible. This is not the case, for example, in Figure 5, where $\bigcup Z_\sigma$ is clearly homotopy equivalent to the rectangle's boundary, but $Z_{e_2} \cap Z_{e_3}$ includes two connected components.

It is instructive to compare our conclusions with the topological consistency conditions proposed by others. Segal[12] requires that the tolerant zones intersect if and only if the corresponding abstract cells are incident on each other in the boundary representation of a polyhedron. This amounts to requiring an isomorphism between the two nerves, $\mathcal{N}K \cong \mathcal{N}\{Z_\sigma\}$, which is also sufficient but unnecessarily too strong. He makes no mention of contractibility, but many (not all) of tolerant zones are convex sets by construction. By contrast, Jackson [13] makes no assumptions on the shape of the zones and explicitly requires that *connected components* of zone intersections are contractible. He indicates that vertex, edge, and face zones *may* intersect if the corresponding abstract cells are incident, implying that the two nerves do not have to be isomorphic, but without further elaboration. It should be clear from the examples in Figure 4 that these conditions are neither necessary nor sufficient. Additional metric conditions are imposed (such as vertex tolerance should be greater than edge tolerance, etc.), but these are not enough to ensure the required homotopy equivalence in general. Below we analyze a number of special cases, including those proposed in [13], and explain how they can be derived systematically from the Twin-Nerve Theorem.

4.2   Collapsed Intersections

The main utility of the Twin-Nerve Theorem is that it provides a starting point for systematic analysis and derivation of sufficient conditions under which a union of tolerant zones $\bigcup Z_\sigma$ can be used as a tolerant replacement for the exact embedding $|K|$.

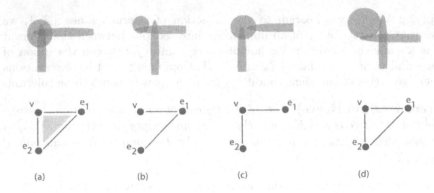

Fig. 7. Collapses in the nerve $\mathcal{N}\{Z_\sigma\}$ preserve its homotopy type. The nerve in (a) can be collapsed to the nerve in (b) or (c), but not to (d). The nerve in (d) cannot be collapsed.

A nerve of a collection of tolerant zones $\{Z_\sigma\}$ is simply a means to impose an ordering on their intersections. The Twin-Nerve Theorem requires this ordering to be "compatible" with the incidence of the corresponding cells $\sigma \in K$. One way to assure this compatibility is to require that the two nerves $\mathcal{N}K$ and $\mathcal{N}\{Z_\sigma\}$ are isomorphic, i.e. are identical under relabeling. But this requirement would rule out many practical solutions, including those proposed in [13], where the zones of some incident cells do not intersect, but their union preserves the homotopy type of K.

Consider, for example, possible intersections between the tolerant zones in the corner of rectangle's boundary representation shown in Figure 7. If all three incident zones of the two edges and a vertex intersect, the nerves $\mathcal{N}\{Z_v, Z_{e_1}, Z_{e_2}\}$ and $\mathcal{N}\{v, e_1, e_2\}$ are simplicial complexes of a single triangle (one 2-simplex, three 1-simplices, and one 0-simplex), as shown in Figure 7(a). It is obvious that the unions of the three zones shown in Figure 7(b) and (c) also preserve the homotopy type of the corner, and this is reflected by the fact that the $\mathcal{N}\{Z_v, Z_{e_1}, Z_{e_2}\}$ is an elementary *collapse*[6] of the nerve $\mathcal{N}\{v, e_1, e_2\}$. In this case, the collapse is achieved by removing the 2-simplex $\langle Z_v, Z_{e_1}, Z_{e_2}\rangle$ and one of its bounding 1-simplices – either $\langle Z_v, Z_{e_1}\rangle$ or $\langle Z_{e_1}, Z_{e_2}\rangle$. Geometrically, the collapse corresponds to the fact that $Z_v \cap Z_{e_1} \cap Z_{e_2} = \emptyset$, and either $Z_v \cap Z_{e_1} = \emptyset$ or $Z_{e_1} \cap Z_{e_2} = \emptyset$. In contrast, all pairwise intersections of zones shown in Figure 7(d) are contractible, but the nerve $\mathcal{N}\{Z_v, Z_{e_1}, Z_{e_2}\}$ is not a collapse of $\mathcal{N}\{v, e_1, e_2\}$, and the union of the three zones does not preserve the homotopy type of K.

4.3   Included Intersections

In order to apply the Twin-Nerve Theorem, we had to assume that all non-empty intersections of tolerant zones were contractible (to a point). This is a common

---

[6] An elementary collapse of a complex is obtained by simultaneously removing a k-simplex σ and one of its free faces τ (i.e. τ is not also a face of another simplex)[30]. More general collapses can be defined as sequences of elementary collapses.

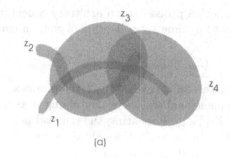

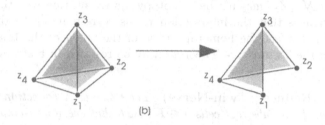

Fig. 8. Reduction of $\mathcal{N}\{Z_\sigma\}$ based on inclusion relationships preserves the homotopy type of the union of zones $\bigcup Z_\sigma$

and a reasonable assumption in many practical situations, for example, when all zones and their intersections are (locally) convex and therefore are contractible. This argument applies to zones constructed as offsets of convex embeddings and to control polygons of Bezier curves, among others. However, contractibility of intersections cannot be assumed in general, and we have two options: either to disallow such intersections altogether and declare the union of zones $\bigcup Z_\sigma$ to be non-tolerant, or to identify the conditions under which such intersections are permitted.

Consider the union of four zones in Figure 8. The union is contractible, but this cannot be deduced from the Nerve Theorem, because the intersection $Z_1 \cap Z_2$ is a disconnected set that is not contractible. The nerve $\mathcal{N}\{Z_i\}$ accounts for *all* intersections and does not take into account the fact that $Z_1 \cap Z_2$ is included (contained) in Z_3. Thus, the topological properties of $Z_1 \cap Z_2$ are irrelevant to the homotopy type of the union $Z_1 \cup Z_2 \cup Z_3 \cup Z_4$. This suggest that use of the full nerve leads to overly conservative results. Instead we should look at a *reduced* nerve that accounts only for those intersections that are not included in other zones. In the example of Figure 8, the full nerve is a tetrahedron. We already observed that $\langle Z_1, Z_2 \rangle$ has no bearing and can be removed, as can be all higher-dimensional simplices that contain $\langle Z_1, Z_2 \rangle$ as a face. In this case, the simplices $\langle Z_1, Z_2, Z_3 \rangle$, $\langle Z_1, Z_2, Z_4 \rangle$, and $\langle Z_1, Z_2, Z_3, Z_4 \rangle$ do not affect the homotopy type of the union. The resulting simplicial complex is the reduced nerve $\mathcal{N}_R\{Z_1, Z_2, Z_3, Z_4\}$.

Generalizing the reduction process to an arbitrary collection of tolerant zones $\{Z_\sigma\}$, let us assume that for some subcollection of zones a containment relationship holds:

$$(Z_{\sigma_1} \cap Z_{\sigma_2} \cap \ldots \cap Z_{\sigma_k}) \subseteq Z_\gamma$$

Then, we remove from the full nerve $\mathcal{N}\{Z_\sigma\}$ the simplex $\langle Z_{\sigma_1}, Z_{\sigma_2}, \ldots, Z_{\sigma_k} \rangle$, and all other higher dimensional simplices containing it as a face, including the simplex $\langle Z_{\sigma_1}, Z_{\sigma_2}, \ldots, Z_{\sigma_k}, Z_\gamma \rangle$. Repeating the removal process for every known containment relationship (in any order), we obtain a new simplicial complex that we will call the **inclusion-reduced nerve** $\mathcal{N}_R\{Z_i\}$.

It is easy to show that the reduction process does not necessarily preserve the homotopy type of the nerve, so that the full nerve $\mathcal{N}\{Z_\sigma\}$ and the inclusion-reduced nerve $\mathcal{N}_R\{Z_\sigma\}$ may not be homotopy equivalent. However, the reduction process guarantees that the intersection terms corresponding to the removed simplices do not affect the homotopy type of the union of the tolerant zones $\bigcup Z_\sigma$. Hence, the Twin-Nerve Theorem may be restated in a stronger form as follows.

Theorem 2 (Reduced Twin-Nerve). *Let $\{Z_\sigma\}$ be a collection of tolerant zones associated with abstract cells $\sigma \in K$, such that every non-empty intersection $\bigcap Z_{\sigma_i}$ is either contractible or is contained in another zone $Z_{\sigma_j}, j \neq i$. Then $\bigcup Z_\sigma$ is a tolerant zone Z_K for the complex K if and only if $\mathcal{N}K \simeq \mathcal{N}_R\{Z_\sigma\}$.*

The Reduced Twin-Nerve Theorem allows verifying homotopy type in some common situations where the zone intersections may not be always contractible. For example, it can be applied directly to the example in Figure 5, where a non-contractible edge intersection $Z_{e_2} \cap Z_{e_3}$ is contained within the vertex zone Z_{v_2}. The theorem also provides at least a partial explanation of why solid boundary representations often assume that a tolerance value δ at a vertex should be equal to or greater than that of the incident edge, which in turn should be equal to or greater than the tolerance associated with the incident face [13].

5   Conclusions

5.1   Summary

This paper is an attempt to formulate the notion of topological consistency for geometric queries that must tolerate inherently imprecise geometric data and rely on approximate numerical computations. We argued that the notion of topological consistency depends (explicitly or implicitly) on the notion of tolerant zones and, at the very minimum, on the requirement that the union of tolerant zones is homotopy equivalent to the combinatorial model of a geometric representation. Sufficient conditions for homotopy equivalence may be derived systematically using the Nerve Theorem. We showed that this approach to topological consistency applies in several important and practical situations, formally justifying earlier heuristic algorithms and identifying several flawed and incomplete arguments.

In the context of this paper, the nerve $\mathcal{N}\{Z_\sigma\}$ of a collection of tolerant zones represents an ordering of all possible (and therefore most stringent) intersection conditions. We have also seen that useful subsets of these intersections may be identified by homotopy preserving transformations of the full nerve (collapses in particular) or of the union $\bigcup Z_\sigma$ of zones themselves (based on known containment relationships). Additional homotopy preserving transformations may apply to other practical situations that are yet to be studied. These observations become particularly important in higher dimensions, because the number of all contractible intersection conditions required by the Nerve Theorem grows exponentially in the total number of incident cells. A typical corner in a boundary representation of a three-dimensional solid includes at least seven different incident cells: one vertex, three edges, and three faces. But based on our analysis in section 4, a much smaller number of intersection conditions may be sufficient to establish the homotopy between the union of the tolerant zones and the corner in a typical solid boundary representation.

The proposed formulation can be applied to a number of practical problems in geometric data translation, validation, and repair. All four hypothetical situations identified in Figure 2 commonly arise in solid modeling, and our formulation suggests a systematic approach for validating any proposed solutions to dealing with imprecision. In practical terms, this requires identifying the tolerance zones implied by known geometric errors, approximations, or algorithms, enumerating all relevant intersection conditions, and verifying that all relevant intersections are contractible — either algorithmically or based on additional *a priori* information. Beyond solid modeling, the proposed homotopy conditions appear to be applicable to a broad range of computational topological consistency problems, from computational geometry [25,31] and geographic information systems and spatial databases[32,33], and mechanical design[34].

5.2   Extensions and Open Issues

Enforcement of homotopy equivalence between the exact and imprecise embeddings in a combinatorial representation is a relatively weak condition, and is only an initial step towards solving a variety of tolerant modeling problem. Additional requirements on the collection of tolerant zones are often needed for specific applications. For example, we would like to refer to the collection $\{Z_\sigma\}$ as a "zone complex," but without additional assumptions this is only a wishful thinking. We would also need to define a boundary operator, and perhaps make sure that the boundary of a tolerant zone is a union of other tolerant zones. However, based on the analysis in this paper, we already know how to make sure that this union is homotopy equivalent to the corresponding boundary in the abstract cell complex.

This research was originally motivated by the need to define a notion of solid boundary representation that can tolerate geometric errors and limited accuracy of algorithms. A set theoretic model for such tolerant solids is described in [6,35]. It is clear that the union of tolerant zones associated with vertices, edges, and faces in the boundary representation must be homotopy equivalent to the intended exact embedding (which may not be known), and this paper shows how this condition may be checked and enforced. However, the homotopy condition alone is not sufficient. A valid boundary representation is also an orientable 2-cycle that separates the Euclidean space into the interior and the exterior of a solid, following the Jordan-Brouwer Theorem. It is not immediately clear how to extend these notions to the union of tolerant zones associated with the boundary representation, but it is likely that the notion of separation used to establish conditions for curve and surface isotopy [36,23,17] will be useful.

The separation property as used in the above references requires that the tolerant zone also contains the exact sets it represents. This condition would have to hold for every cell $\sigma \in K$, $|\sigma| \subseteq Z_\sigma$, and for the whole cell complex $|K| \subseteq \bigcup Z_\sigma$. Satisfaction of the latter condition is not guaranteed even if the homotopy condition $|K| \simeq \bigcup Z_\sigma$ is satisfied. For example, the union of zones in Figure 7(b) is homotopy equivalent to the corner of a quadrilateral's boundary. But notice that $Z_{e_2} \cap Z_v = \emptyset$, which means that the union $Z_{e_2} \cup Z_v$ does not contain the union of exact embeddings $|e_1| \cup |v|$. Adding the containment requirement to the homotopy condition studied in this paper would imply a much more stringent demand that every exact embedding must be a deformation retract of the corresponding tolerant zone. This condition seems to be implied but not explicitly enforced in [13], and no formal arguments either for or against it have been put forward as of now.

Acknowledgements

This research is supported in part by the National Science Foundation grants DMI-0500380 and DMI-0323514. The author is grateful to Prem Mansukhani and Tom Peters for numerous suggestions on improving the presentation in the paper. Responsibility for errors and omissions lies solely with the author.

References

1. Requicha, A.A.G.: Representations for rigid solids: Theory, methods and systems. ACM Computing Surveys 12(4), 437–464 (1980)
2. Shapiro, V.: Solid modeling. In: Farin, G., Hoschek, J., Kim, M.S. (eds.) Handbook of Computer Aided Geometric Design, pp. 473–518. Elsevier Science Publishers, Amsterdam (2002)
3. Shapiro, V.: Maintenance of geometric representations through space decompositions. International Journal of Computational Geometry and Applications 7(4), 383–418 (1997)
4. Qi, J., Shapiro, V., Stewart, N.F.: Single-set and class-of-sets semantics for geometric models. Technical Report 2005-1, Spatial Automation Laborotary, University of Wisconsin - Madison (2005)
5. Tilove, R.B.: Set membership classification: A unified approach to geometric intersection problems. IEEE Transactions on Computer C-29(10) (1980)
6. Qi, J., Shapiro, V.: Epsilon-solidity in geometric data translation. Technical report, SAL 2004-2, Spatial Automation Laboratory, University of Wisconsin-Madison (2004)

7. Kettner, L., Pion, K.M., Schirra, S., Yap, S.: Classroom examples of robustness problems in geometric computations. In: Albers, S., Radzik, T. (eds.) ESA 2004. LNCS, vol. 3221, pp. 702–713. Springer, Heidelberg (2004)

8. Patrikalakis, N.M., Maekawa, T.: Shape Interrogation for Computer Aided Design and Manufacturing. Springer, Heidelberg (2002)

9. Sabin, M.: Subdivision surfaces. In: Farin, G., Hoschek, J., Kim, M.S. (eds.) Handbook of Computer Aided Geometric Design, pp. 309–341. Elsevier Science Publishers, Amsterdam (2002)

10. Peters, J., Wu, X.: SLEVEs for planar spline curves. Computer Aided Geometric Design 21, 615–635 (2004)

11. Segal, M., Sequin, C.: Consistent calculations for solids modeling. In: SCG 1985: Proceedings of the first annual symposium on Computational geometry, pp. 29–38. ACM Press, New York (1985)

12. Segal, M.: Using tolerances to guarantee valid polyhedral modeling results. In: Computer Graphics (Proceedings of ACM SIGGRAPH 1990), pp. 105–114 (1990)

13. Jackson, D.J.: Boundary representation modelling with local tolerancing. In: Proceedings of the 3rd ACM Symposium on Solid Modeling and Applications, Salt Lake City, Utah, pp. 247–253 (1995)

14. Fang, S., Bruderlin, B., Zhu, X.: Robustness in solid modeling: A tolerance-based intuitionistic approach. Computer-Aided Design 25(9), 567–576 (1993)

15. Guibas, L., Salesin, D., Stolfi, J.: Epsilon geometry: Building robust algorithms from imprecise computations. In: Proceedings of the fifth ACM Symposium on Computational Geometry, Saarbruchen, West, Germany, pp. 208–217 (1989)

16. Shen, G., Sakkalis, T., Patrikalakis, N.M.: Analysis of boundary representation model rectification. In: Proceedings of the 6th ACM Symposium on Solid Modeling and Applications, Ann Arbor, Michigan, pp. 149–158 (2001)

17. Chazal, F., Cohen-Steiner, D.: A condition for isotopic approximation. In: Proceedings of the 2004 ACM Symposium on Solid Modeling and Applications, Genova, Italy (2004)

18. Andersson, L.E., Stewart, N.F., Zidani, M.: Error analysis for operations in solid modeling (2004), www.iro.umontreal.ca/~stewart

19. Hoffmann, C.M., Stewart, N.F.: Accuracy and semantics in shape-interrogation applications. Graphical Models (to appear, 2005)

20. Song, X., Sederberg, T.W., Zheng, J., Farouki, R.T., Hass, J.: Linear perturbation methods for topologically consistent representations of free-form surface intersections. Computer Aided Geometric Design 21(3), 303–319 (2004)

21. O'Connor, M.A., Rossignac, J.R.: SGC: A dimension independent model for pointsets with internal structures and incomplete boundaries. In: IFIP/NSF Workshop on Geometric Modeling, Rensselaerville, NY, 1988, North-Holland, Amsterdam (1990)

22. Grandine, T.A., Frederick, W., Klein, I.: A new approach to the surface intersection problem. Comput. Aided Geom. Des. 14(2), 111–134 (1997)

23. Sakkalis, T., Peters, T.J., Bisceglio, J.: Isotopic approximations and interval solids. Computer-Aided Design 36, 1089–1100 (2004)

24. Yap, C.: Robust geometric computation. In: Goodman, J.E., O'Rourke, J. (eds.) Handbook of Discrete and Computational Geometry, CRC Press, Boca Raton (1997)

25. Sugihara, K., Iri, M., Inagaki, H., Imai, T.: Topology-oriented implementation - an approach to robust geometric algorithms. Algorithmica 27(1), 5–20 (2000)

26. Bjorner, A.: Topological methods. In: Graham, R., Grotschel, M., Lovacz, L. (eds.) Handbook of Combinatorics, pp. 1819–1872. Elsevier Science B.V, Amsterdam (1995)
27. Alexandrov, P.: Gestalt u. lage abgeschlossener menge. The Annals of Mathematics 30, 101–187 (1928)
28. Hocking, J.G., Young, G.S.: Topology. Dover Publications, New York (1961)
29. Dey, T., Edelsbrunner, H., Guha, S.: Computational topology. In: Chazelle, B., Goodman, J.E., Pollack, R. (eds.) Advances in Discrete and Computational Geometry (Contemporary mathematics 223), pp. 109–143. American Mathematical Society (1999)
30. Maunder, C.R.F.: Algebraic Topology. Dover Publications, New York (1996)
31. Mehlhorn, K., Yap, C.: Robust Geometric Computation (tentative). Book draft, under preparation (2004), http://www.cs.nyu.edu/~yap/book/egc/
32. Egenhofer, M., Clementini, E., di Felice, P.: Evaluating inconsistencies among multiple representations. In: Sixth International Symposium on Spatial Data Handling, Edinburgh, Scotland, pp. 901–920 (1994)
33. Kang, H.K., Kim, T.W., Li, K.J.: Topological consistency for collapse operation in multi-scale databases. In: Wang, S., Tanaka, K., Zhou, S., Ling, T.-W., Guan, J., Yang, D.-q., Grandi, F., Mangina, E.E., Song, I.-Y., Mayr, H.C. (eds.) ER Workshops 2004. LNCS, vol. 3289, pp. 91–102. Springer, Heidelberg (2004)
34. Armstrong, C.G.: Integrating analysis and design - thoughts for the future. In: State of the Art in CAD/FE Integration - NAFEMS Awareness Seminar, Chester, UK (2002), http://sog1.me.qub.ac.uk/Resources/publications/publications.php
35. Qi, J., Shapiro, V.: Epsilon-topological formulation of tolerant solid modeling. Computer-Aided Design 38(4), 367–377 (2006)
36. Sakkalis, T., Shen, G., Patrikalakis, N.M.: Topological and geometric properties of interval solid models. Graphical Models 63(3), 163–175 (2001)

Transfinite Interpolation for Well-Definition in Error Analysis in Solid Modelling

Neil F. Stewart* and Malika Zidani

Département IRO, Université de Montréal,
C.P. 6128, Succ. CentreVille, Montréal, Qc, H3C 3J7, Canada
{stewart, zidanima}@iro.umontreal.ca

Abstract. An overall approach to the problem of error analysis in the context of solid modelling, analogous to the standard forward/backward error analysis of Numerical Analysis, was described in a recent paper by Hoffmann and Stewart. An important subproblem within this overall approach is the well-definition of the sets specified by inconsistent data. These inconsistencies may come from the use of finite-precision real-number arithmetic, from the use of low-degree curves to approximate boundaries, or from terminating an infinite convergent (subdivision) process after only a finite number of steps.

An earlier paper, by Andersson and the present authors, showed how to resolve this problem of well-definition, in the context of standard trimmed-NURBS representations, by using the Whitney Extension Theorem. In this paper we will show how an analogous approach can be used in the context of trimmed surfaces based on combined-subdivision representations, such as those proposed by Litke, Levin and Schröder.

A further component of the problem of well-definition is ensuring that adjacent patches in a representation do not have extraneous intersections. (Here, 'extraneous intersections' refers to intersections, between two patches forming part of the boundary, other than prescribed intersections along a common edge or at a common vertex.) The paper also describes the derivation of a bound for normal vectors that can be used for this purpose. This bound is relevant both in the case of trimmed-NURBS representations, and in the case of combined subdivision with trimming.

1 Introduction

One of the fundamental problems in proving rigorous theorems in the area of robustness of numerical methods, in the field of solid modelling, is that the data normally provided to the algorithm is not only in error, it may be fundamentally inconsistent. These inconsistencies, in the data purportedly defining a set to be manipulated by a solid-modelling algorithm, come from the use of finite-precision real-number arithmetic, from the use of low-degree curves to approximate boundaries, or from terminating an infinite convergent (subdivision)

---

* The research of the first author was supported in part by a grant from the Natural Sciences and Engineering Research Council of Canada.

P. Hertling et al. (Eds.): Real Number Algorithms, LNCS 5045, pp. 181–192, 2008.

process after only a finite number of steps. Representations often have both a topological component and a geometric component; the geometric component may itself be internally inconsistent, and it may be inconsistent with the topological component. One approach to resolving these inconsistencies is to propose a definition of a set, satisfying strong guarantees of proximity to the given inconsistent input data, and to take this as specifying the input set. This approach permits subsequent rigorous proof, of theorems concerning algorithms that manipulate sets, in terms of well-defined input sets.

It should be noted that we are concerned with the case when the input data is uncertain and, as mentioned above, possibly inconsistent.

In [1] it was shown how the Whitney Extension Theorem [2] can be used to perform transfinite interpolation in order to realize the above goal of well-definition of a given input set, in the context of standard trimmed-NURBS representations. In this paper we will show how it can be used in an analogous way for the well-definition of sets defined by combined subdivision surfaces [3]. Then, later in the paper, we will consider a more general setting, including both of the special cases just mentioned, and provide a result for bounding normal vectors. This result can be used to ensure that adjacent patches in a representation do not have extraneous intersections. (Here, 'extraneous intersections' refers to intersections, between two patches forming part of the boundary, other than prescribed intersections along a common edge or at a common vertex.)

Assuring the well-definition of input sets is an important subproblem within the forward/backward error analysis described by Hoffmann and Stewart [4].

2   Transfinite Interpolation

A transfinite interpolant is a surface that matches data on the entire boundary of a two-dimensional domain, rather than just at a finite number of points. Such surfaces can be obtained, for example, as solutions of the Dirichlet problem, minimizing the functional

$$\iint_D (f_x^2 + f_y^2)dxdy \quad / \quad \iint_D \|f\|^2 dxdy$$

under the Dirichlet boundary conditions [5, p. 110], or by finding area-minimizing solutions. The shape of such solutions is illustrated by a soap-film stretched over a wire-frame (see for example [6, frontispiece]).

Transfinite interpolation has been used since the earliest days of geometric and solid modelling. For example, the Coons patch [7] is a C^1-continuous transfinite interpolant; see also [8]. More recently this kind of interpolation has been used by Gross and Farin to generalize Sibson's interpolant to the case of boundary interpolation [9], and by Shapiro and his students [10] in the context of modelling heterogeneous materials on a point-by-point basis. As mentioned in the Introduction, it has also been used in quite a different way, to provide a definition of a well-formed set in the study of robustness [1,4]. In this last-mentioned context, the transfinite interpolant is not actually computed, but is introduced only to permit proof of rigorous theorems about a single well-defined set that can be viewed as the one specified by the inconsistent data provided to a numerical method. This approach will be described in the following section.

3 Whitney Extension to Provide Transfinite Interpolation in the Context of Solid Modelling

We begin this section by stating a version [1] of the Whitney Extension Theorem [2] that is appropriate for our purposes.

Suppose that we have a mapping

$$\boldsymbol{\epsilon} : C \to R^3, \tag{1}$$

where C is a compact subset of R^2, and suppose that each component ϵ of $\boldsymbol{\epsilon}$ satisfies a Lipschitz condition

$$|\epsilon(\boldsymbol{p_1}) - \epsilon(\boldsymbol{p_2})| \le L \cdot \|\boldsymbol{p_1} - \boldsymbol{p_2}\|, \quad \boldsymbol{p_1}, \boldsymbol{p_2} \in C$$

where

$$L = \sup_{\boldsymbol{p_1}, \boldsymbol{p_2} \in C, \ \boldsymbol{p_1} \neq \boldsymbol{p_2}} \frac{|\epsilon(\boldsymbol{p_1}) - \epsilon(\boldsymbol{p_2})|}{\|\boldsymbol{p_1} - \boldsymbol{p_2}\|} \tag{2}$$

is finite. Then [2] we can extend ϵ to a continuous function on all of R^2 that satisfies a Lipschitz condition with the same Lipschitz constant L.

In fact, let

$$l(\boldsymbol{p}) = \sup_{\boldsymbol{q} \in C}\{\epsilon(\boldsymbol{q}) - L \cdot \|\boldsymbol{p} - \boldsymbol{q}\|\}, \quad \boldsymbol{p} \in R^2, \tag{3}$$

$$u(\boldsymbol{p}) = \inf_{\boldsymbol{q} \in C}\{\epsilon(\boldsymbol{q}) + L \cdot \|\boldsymbol{p} - \boldsymbol{q}\|\}, \quad \boldsymbol{p} \in R^2, \tag{4}$$

and

$$\epsilon(\boldsymbol{p}) = \frac{1}{2}[l(\boldsymbol{p}) + u(\boldsymbol{p})], \quad \boldsymbol{p} \in R^2. \tag{5}$$

It is easy to show [11] that any continuous function satisfying the Lipschitz condition on R^2 is bracketed by the *lower function*, $l(\boldsymbol{p})$, and the *upper function*, $u(\boldsymbol{p})$, and [12] that $l(\boldsymbol{p})$ and $u(\boldsymbol{p})$ are themselves solutions to the extension problem. Furthermore [1], $\epsilon(\boldsymbol{p})$ is a solution to the extension problem satisfying

$$|\epsilon(\boldsymbol{p})| \le \sup_{\boldsymbol{q} \in C} |\epsilon(\boldsymbol{q})|, \quad \boldsymbol{p} \in R^2.$$

Theorem 1. (Whitney, 1934) *The mapping $\epsilon(\boldsymbol{p})$ given in (5) is a continuous function on R^2 that coincides with the mapping (1) given on C. Furthermore, the mapping given in (5) satisfies a Lipschitz condition everywhere in R^2, with Lipschitz constant L defined as in (2), and it satisfies the inequality*

$$|\epsilon(\boldsymbol{p})| \le \sup_{\boldsymbol{q} \in C} |\epsilon(\boldsymbol{q})|, \quad \boldsymbol{p} \in R^2.$$

The Lipschitz constants for each of the components of $\boldsymbol{\epsilon}$ can be used to define a Lipschitz constant for the vector-valued function $\boldsymbol{\epsilon}$.

3.1   The Trimmed-NURBS Case

In this subsection we give a summary of the use of Whitney Extension to define well-formed sets in the case of the standard trimmed-NURBS representation [1,4].

The data Δ in the trimmed-NURBS representation of a solid S is in two parts, the *geometric data* and the *topological data*. The geometric data in Δ comprises a finite set of compact oriented 2-manifolds-with-boundary, and corresponding sets of explicit boundary curves and corner vertices. The underlying surface for each 2-manifold is represented by a spline function over a parametric domain D_0, with range in R^3. The spline function is then restricted to a subset D of D_0, as delineated by certain curves in the parametric domain, yielding a *trimmed NURBS patch*. (Such trimming may arise, for example, as the result of the intersection of two surfaces for representing the boundary of the solid.) Thus, each 2-manifold-with-boundary is represented by a trimmed NURBS patch.

Figure 1 illustrates two trimmed patches which join (approximately) along the intersection of two surfaces F and F', restricted respectively to D and D'. The part of the representation of the solid corresponding to the intended intersection

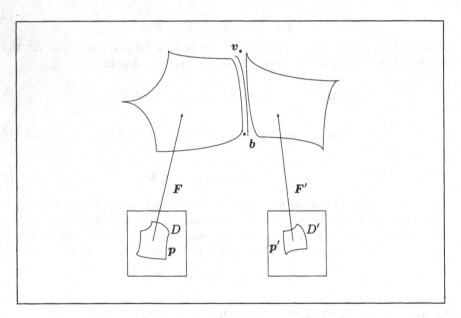

Fig. 1. Two adjoining trimmed NURBS patches

comprises two (almost certainly inconsistent) pre-images, defined by *p-curves*, which are parametric curves p and p' with ranges in the respective parametric domains. In addition, there is often a third representation, usually inconsistent with the other two, which is a parametric curve b, with range in R^3, and which follows closely the images of the p-curves p and p'. Finally, there may also be an explicit representation of each endpoint $v \in R^3$ of the parametric curve $b = b(t)$, as illustrated in Figure 1. Here, however, we consider the more usual case where v always coincides with an endpoint of b.

The representation Δ also contains symbolic information, or *topological data*, describing how the faces, edges and vertices of the cellular decomposition of the boundary ∂S of S fit together. This data defines a topological 2-cycle. Ideally, the geometric and topological data are consistent: for example, corresponding to each 2-cell in the topological data is a trimmed surface patch in R^3 (two of these, $F[D] = \{F(u,v) : (u,v) \in D \subseteq D_0\}$ and $F'[D'] = \{F'(u,v) : (u,v) \in D' \subseteq D_0\}$, are shown in Figure 1).

Unfortunately, as illustrated in Figure 1, the curve $b(t)$ does not usually coincide exactly with the corresponding edge of $F[D]$, nor with the corresponding edge of $F'[D']$, and the question therefore arises, which subset of R^3 should be considered to be represented by the given inconsistent data?

In [1], the set considered to actually be represented by Δ, *i.e.*, the *realization* S of Δ, is defined by its boundary ∂S. This boundary is made up of slightly perturbed trimmed-NURBS patches from Δ, where the perturbation is defined by the Whitney Extension Theorem. The slightly perturbed patches are not necessarily NURBS patches, but they are mutually consistent with the explicit boundary curves $b(t)$, and they all fit together in a way that is exactly consistent with the topological data. This is done by taking $C = \partial D$ and

$$\epsilon(p) = b^k([p^k]^{-1}(p)) - F(p), \quad p \in \partial D$$

in Theorem 1, where $[p^k]^{-1}$ is the inverse of the particular p-curve p^k for which $p^k(t) = p$ for some $t \in [0,1]$. Thus, ϵ is the difference between the edge of the trimmed patch $F[D]$, and the boundary curves b^k, viewed as a function of $p \in \partial D$. (It is assumed that each b^k and each p^k is injective, and that distinct p-curves do not intersect, except at appropriate endpoints.) Then, the Whitney theorem can be used to extend the patch to all of D.

The meaning of the words 'slightly perturbed', in the previous paragraph, is thus quite satisfactory. The perturbation of the trimmed patch, denoted $\epsilon(p)$, is continuous, and the magnitude of the perturbation nowhere exceeds the magnitude of the largest discrepancy between the edge of the given patch and the given neighbouring explicit boundary curves $b(t)$. (In other words, the perturbation is nowhere larger than the largest discrepancy already in the given data, along the edges of the given patch.) In addition, the perturbation satisfies a Lipschitz condition throughout the patch, with a Lipschitz constant for each component ϵ of ϵ equal to

$$L = \sup_{q_1, q_2 \in \partial D, \ q_1 \neq q_2} \frac{|\epsilon(q_1) - \epsilon(q_2)|}{\|q_1 - q_2\|}.$$

(In other words, the perturbation satisfies a Lipschitz condition over the *entire* patch, with a constant equal to the constant in the Lipschitz condition corresponding to the discrepancy present, along the edges of the given patch, in the given data.) This Lipschitz condition is important, because it allows us to bound the change in the normal vector of the perturbed patch, relative to the normal vector n of the given trimmed patch. Such a bound will be necessary when we want to preclude the possibility that adjacent patches have extraneous intersections. This question will be discussed in Section 4, below.

3.2   Combined-Subdivision-with-Trimming Case

A representation permitting trimming of subdivision surfaces was proposed in [3], as an alternative to the classical NURBS representation. The representation is based on the combined subdivision schemes proposed by Levin [13], which permit the construction of subdivision surfaces having arbitrary boundary curves (any piecewise-smooth parametric curve possessing an evaluation procedure). Such schemes modify the subdivision stencils (for example, the Loop or Catmull-Clark stencils) near the boundary, using data from a boundary curve $c = c(u)$ supplied as part of the input [3].

We will refer to the representation of [3] as Combined Subdivision with Trimming (CST). Comparing it with the trimmed-NURBS representation of Subsection 3.1, the supplied boundary curve $c : [0, 1] \to R^3$ corresponds to the combined closed curve made up from the m given boundary segments $b^1(t), \dots, b^m(t)$, joined end to end to form a simple closed curve embedded in R^3. Since c is required only to be piecewise smooth, it is possible to have corner points like those occurring at a join such as $b^{l-1}(1) = b^l(0)$ in the trimmed-NURBS representation. Beyond this, however, the representations are quite different. The surface patch in the CST representation is the limit of a modified combined-subdivision scheme which approximates an initially given input surface, and which, in the absence of roundoff error, interpolates the boundary curve $c = c(u)$ exactly. Note that the transfinite interpolation implied by this limiting process forms part of the actual representation. This is in contrast to the Whitney-Extension transfinite interpolation used in Subsection 3.1 for the purpose of defining a theoretical set determined by inconsistent data in a representation. We will, however, *also* apply Whitney-Extension transfinite interpolation in the case of CST representations.

Following [3], a general subdivision-surface control point i, at level j in the subdivision hierarchy, is denoted p_i^j, and the breakpoint values in the parametric domain of c are denoted u_i^j. A surface called the *original surface* [3], denoted here by Σ, is defined by the (given) control points p_i^0 [3], along with a given subdivision scheme such as the Loop scheme [3,13]. The subsequent trimming algorithm, which produces the actual trimmed-surface Σ_T, involves, first, a local remeshing of the control polyhedron to accommodate the trim curves, together with a sampling of Σ to choose control points for the trimmed surface. This information is used in an approximation stage that modifies the surface shape near the trim curve, in order to ensure proximity to the surface Σ.

The local remeshing will normally cause the assigned initial parameter values u_i^0 to be modified. Again following [3], we will continue to denote these modified values by u_i^0. The surface Σ is then sampled to find points corresponding to the control points of the original mesh.

The approximation stage of the trimming algorithm fits the trimmed surface Σ_T to Σ. This operation is only required near the trim curve, where the control mesh was generated by the remeshing algorithm. Away from this region, the trimmed surface is identical to Σ; within the region, the approximation of Σ is generated as a hierarchy of *detail coefficients* d_i^j, which are (additive) modifications of the control points p_i^j. The depth of the subdivision hierarchy is limited by the introduction of a finite convergence threshold [3, Sec. 3.3].

The final trimmed-surface representation is therefore defined by a hierarchy of u-domain control points u_i^j, which depends on (the possibly modified versions of) the original breakpoints u_i^0, and a hierarchy of control points p_i^j with associated detail vectors d_i^j, which depends on the original p_i^0. As observed in [3], the above trimming algorithm guarantees exact transfinite interpolation of the desired trimmed curve. This statement assumes that the subdivision process is continued until convergence, and without roundoff error. In practice, however, the actual representation will be the result of only a finite number of steps of the combined-subdivision process, and it will be the result of calculations using finite-precision floating-point arithmetic.

It follows from this last remark that if we place ourselves in the context of Subsection 3.1, and seek to prove rigorous mathematical theorems stating that the representation defines a well-formed subset of R^3 which has a boundary that is in some sense close to a given collection of input surfaces $\{\Sigma^k\}_{k=1,...}$, then we must agree on the definition of this subset. One possibility is to define the set in terms of the exactly defined limiting surfaces specifying its boundary patches. To do this, however, it would be necessary to preclude extraneous intersections, as defined at the end of Section 1: intersections between two boundary patches other than prescribed intersections along a common edge or at a common vertex. This, in turn, would require bounds on the normals of the limit surface Σ_T, which is defined in terms of the (rather complicated) process described in [3]. It is not clear whether this approach is feasible.

In this paper we propose an alternative approach, analogous to Subsection 3.1. The situation is illustrated in Figure 2, where two CST patches are intended to meet along a common edge. We assume, of course, that $c^k(u)$ and $c^{k'}(u)$ coincide along this common edge, but in practice, as already mentioned, the trimming algorithm will terminate after a finite number of steps and, furthermore, $c^k(u)$ will be evaluated using finite-precision floating-point arithmetic; consequently, the edge of Σ_*^k, the computed approximation of Σ_T^k, will not coincide exactly with $c^k(u)$. On the other hand (and this is implicit in [3]), it will be possible to obtain very satisfactory bounds for the difference, which we denote by $\epsilon(p)$, where p is a point on the boundary of the final control polyhedron produced by the combined subdivision process. Assuming such bounds have been found, the Whitney Extension Theorem can be applied exactly as in Subsection 3.1 (the set C in Theorem 1 is now taken to be the boundary of the final control polyhedron) to provide a definition of the boundary of the actual subset of R^3 specified by the inconsistent data that specified each of Σ_*^k and $\Sigma_*^{k'}$ separately.

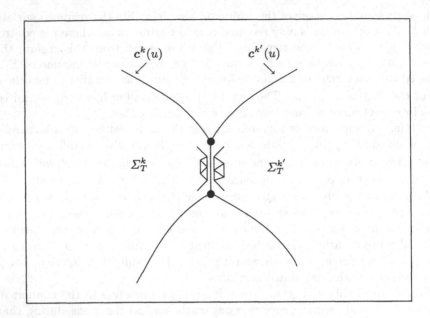

Fig. 2. Two adjoining CST patches

The only remaining issue is the possibility of self-intersection of the combined patches: the patches provided by Whitney Extension guarantee that adjoining patches will meet properly at a common edge, but we must exclude the possibility that there are other intersections (see Figure 3, where the CST patches are shown in cross-section). In the case of CST patches based on Loop or Catmull-Clark subdivision, it has been shown [14] that Σ_*^k (respectively $\Sigma_*^{k'}$) and its derivatives can be evaluated directly, which means that the normal vector n^k (respectively $n^{k'}$) can be estimated, and, say, the method of Volino-Thalmann [15,16,17] applied. This is the subject of Section 4, where an analysis will be presented in a setting that includes both the trimmed-NURBS and CST cases.

4   Bounds on Normal Vectors

The fundamental step required, in order to apply the Volino-Thalmann method for detection of self-intersection of the boundary of the set defined by Whitney Extension, is to bound the normal vector of the perturbed surface in terms of the normal of the original surface. The well-formed set that we are attempting to define is specified in terms of the perturbed surface provided by Whitney Extension, but it is only the normal vector of the original surface that is available to us.

The following analysis is an extension of that found in [1, Sec. 5].

Denote the original patch by F, and the perturbed patch defined by Whitney Extension by $G = F + \epsilon$. (In the trimmed-NURBS case, the notation F here

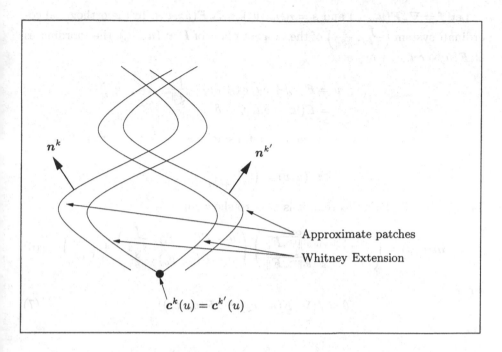

$$c^k(u) = c^{k'}(u)$$

Fig. 3. Extraneous intersection of adjoining CST patches

refers to $F[D]$, the restriction of the NURBS mapping $F(u, v)$ to the trimming domain D, as in Subsection 3.1. In the CST case, Subsection 3.2, F refers to a CST patch such as Σ_T^k.) Also, denote by n_F and n_G the normal vectors at an arbitrary parameter point (U, V). (In the trimmed-NURBS case, see [18]; in the Loop or Catmull-Clark case, see [14].) In order to use a variational argument, we will write $U = u_o + \delta u$, $V = v_o + \delta v$, and consider the case when $U \to u_o$, $V \to v_o$. In the limit, the projection of $F(u_o + \delta u, v_o + \delta v)$ on the tangent plane of F at (u_o, v_o) is equal to $F(u_o, v_o)$, and the normal $n_F(u_o + \delta u, v_o + \delta v)$ is equal to $n(u_o, v_o)$.

We will now give a bound for the angle ϕ, between the two normals n_F and n_G, expressed in terms of the coefficients

$$E = \nabla_u F \cdot \nabla_u F$$
$$F = \nabla_u F \cdot \nabla_v F$$
$$G = \nabla_v F \cdot \nabla_v F$$

of the first fundamental form:

$$I(du, dv) = E du^2 + 2F du dv + G dv^2$$

(see for example [19]).

Let $f = \nabla_u F(u_o, v_o)$ and $s = n(u_o, v_o) \times \nabla_u F(u_o, v_o)$. In the orthogonal coordinate system $\left(\frac{f}{\|f\|}, \frac{s}{\|s\|}\right)$ of the tangent plane of F at (u_o, v_o), the coordinates of $F(u_o + \delta u, v_o + \delta v)$ are:

$$u = F(u_o + \delta u, v_o + \delta v) \cdot \frac{f}{\|f\|}$$
$$v = F(u_o + \delta u, v_o + \delta v) \cdot \frac{s}{\|s\|}.$$

We take u and v as new parameters, and, using the Implicit Function Theorem, define

$$m(u, v) = \begin{pmatrix} u_o + \delta u \\ v_o + \delta v \end{pmatrix}.$$

Then, using Taylor's theorem, it is easy to show that

$$m(u, v) = \begin{pmatrix} \frac{1}{\|\nabla_u F\|} & \frac{-\cos\theta}{\sin\theta \cdot \|\nabla_u F\|} \\ 0 & \frac{1}{\sin\theta \cdot \|\nabla_v F\|} \end{pmatrix} \begin{pmatrix} u - F(u_o, v_o) \cdot \frac{f}{\|f\|} \\ v - F(u_o, v_o) \cdot \frac{s}{\|s\|} \end{pmatrix} + \begin{pmatrix} u_o \\ v_o \end{pmatrix} \quad (6)$$

where

$$\theta = \angle(\nabla_u F(u_o, v_o), \nabla_v F(u_o, v_o)) \quad (7)$$

and

$$G(u, v) = (u, v, 0)^T + (\epsilon(m(u, v)) \cdot \frac{f}{\|f\|}, \epsilon(m(u, v)) \cdot \frac{s}{\|s\|}, \epsilon(m(u, v)) \cdot \frac{n}{\|n\|})^T. \quad (8)$$

The second term on the right is the perturbation vector expressed in the local coordinate system $\left(\frac{f}{\|f\|}, \frac{s}{\|s\|}, \frac{n}{\|n\|}\right)$.

To simplify the notation (and to emphasize the correspondence with the derivation in [1]), let $(e_1, e_2, e_3)^T$ denote the components of ϵ in the parametric domain (u, v). Then, from (8),

$$G(u, v) = \begin{pmatrix} u + e_1 \\ v + e_2 \\ e_3 \end{pmatrix},$$

and in the limit as $\delta u, \delta v \to 0$,

$$n_F(u, v) = (0, 0, 1)^T.$$

It follows that if ϕ is the angle between n_F and n_G, then in the limit

$$\cos\phi = \frac{(G_u \times G_v) \cdot (0, 0, 1)^T}{\|G_u \times G_v\|}$$

where

$$G_u \times G_v = \begin{pmatrix} 1 + e_{1u} \\ e_{2u} \\ e_{3u} \end{pmatrix} \times \begin{pmatrix} e_{1v} \\ 1 + e_{2v} \\ e_{3v} \end{pmatrix} = \begin{pmatrix} e_{2u}e_{3v} - e_{3u}e_{2v} - e_{3u} \\ e_{3u}e_{1v} - e_{3v}e_{1u} - e_{3v} \\ 1 + e_{1u} + e_{2v} + e_{1u}e_{2v} - e_{2u}e_{1v} \end{pmatrix}.$$

It can then be shown, using the same analysis as that in [1], that

$$\cos\phi \cong 1 - \frac{(e_{3u})^2}{2} - \frac{(e_{3v})^2}{2}. \tag{9}$$

In order to bound the derivatives e_{3u} and e_{3v}, we use (6) and the fact that from the Whitney Extension Theorem, if ϵ satisfies a Lipschitz condition with constant L on the boundary of the patch, then it satisfies the same Lipschitz condition everywhere, and consequently

$$\left| \frac{\partial e_3}{\partial m_i} \right| \leq L, \quad i = 1, 2.$$

Using this inequality, and (7) and (9), it follows (after some algebra) that

$$\cos\phi \geq 1 - \frac{L^2}{2} \left[\frac{E + 2F + G}{EG - F^2} \right]$$

where E, F and G are the coefficients of the first fundamental form, given above.

Note that if $\theta = \frac{\pi}{2}$, then $F = 0$, and this lower bound becomes $1 - \frac{L^2}{2} \left[\frac{E+G}{EG} \right]$. If in addition we have $\|\nabla_u F\|^2 = \|\nabla_v F\|^2 = 1$, so that $E = G = 1$, then the lower bound becomes $1 - L^2$. This was the special case treated in [1, Sec. 5].

It would be worthwhile, also, to study the variation of the normals (between the surfaces Σ and Σ_T) induced by the CST scheme, and to modify the scheme, if necessary, to reduce this variation. The additional variation introduced by Whitney Extension will normally be small. Indeed, the value of L used for Whitney Extension will be small provided that the curvature of c is bounded below, and provided that the spacing u_i^j is fine enough, relative to this lower bound. (We are assuming here, however, that the spacing of the u_i^j is still coarse enough so that roundoff error does not play a major role in the variation of ϵ.)

5 Conclusion

In this paper we have shown how Whitney Extension can be used in the well-definition of objects defined by approximate surface patches, in conjunction with the bounds derived here on the variation of normal vectors. These bounds depend on the Lipschitz constant defined by the error in the approximate surface patch along its boundary, and by the (first-normal-form) parameters of the surface.

Acknowledgment

The authors are grateful to Ian Stewart, and to an anonymous referee, for several useful comments. Responsibility for errors or omissions rests solely with the authors.

References

1. Andersson, L.-E., Stewart, N.F., Zidani, M.: Error analysis for operations in solid modeling in the presence of uncertainty. SIAM J. Scientific Computing 29(2), 811–826 (2007)
2. Whitney, H.: Analytic extensions of differentiable functions defined in closed sets. Trans. Amer. Math. Soc. (36), 63–89 (1934)
3. Litke, N., Levin, A., Schröder, P.: Trimming for subdivision surfaces. Computer Aided Geometric Design (18), 463–481 (2001)
4. Hoffmann, C.M., Stewart, N.F.: Accuracy and semantics in shape-interrogation applications. Graphical Models 67(5), 373–389 (2005)
5. Gould, S.H.: Variational Methods for Eigenvalue Problems. University of Toronto Press (1957)
6. Sewell, M.J.: Maximum and Minimum Principles, Cambridge (1987)
7. Coons, S.A.: Surfaces for Computer Aided Design of Space Forms. MIT Project Mac, TR-41, MIT, Cambridge, MA (June 1967)
8. Nielson, G.M.: A transfinite, visually continuous, triangular interpolant. In: Farin, G.E. (ed.) Geometric Modeling: Algorithms and New Trends. SIAM, Philadelphia (1987)
9. Gross, L., Farin, G.: A transfinite form of Sibson's interpolant. Discrete Applied Mathematics (93), 33–50 (1999)
10. Biswas, A., Shapiro, V., Tsukanov, I.: Heterogeneous material modeling with distance fields. Computer Aided Geometric Design (21), 215–242 (2004)
11. Aronsson, G.: Extension of functions satisfying Lipschitz conditions. Arkiv för Matematik, Stockholm (6)(28) (1967)
12. McShane, E.J.: Extension of range of functions. Bull. Amer. Math. Soc. (40), 837–842 (1934)
13. Levin, A.: Combined subdivision schemes for the design of surfaces satisfying boundary conditions. Computer Aided Geometric Design (16), 345–354 (1999)
14. Stam, J.: Exact evaluation of Catmull-Clark subdivision surfaces at arbitrary parameter values. In: Proc. ACM SIGGRAPH, pp. 395–404 (1998)
15. Volino, P., Thalmann, N.M.: Efficient self-collision detection on smoothly discretized surface animations using geometrical shape regularity. In: Daehlen, M., Kjelldahl, K. (eds.) Eurographics 1994, (13)(3), pp. 155–164. Blackwell Publishers, Malden (1994)
16. Grinspun, E., Schröder, P.: Normal bounds for subdivision-surface interference detection. In: Proceedings of the IEEE Conference on Visualization, pp. 333–340 (2001)
17. Andersson, L.-E., Stewart, N.F., Zidani, M.: Proof of a non-selfintersection conjecture. Computer Aided Geometric Design 23, 599–611 (2006)
18. Piegl, L., Tiller, W.: The NURBS Book. Springer, Heidelberg (1997)
19. Pressley, H.: Elementary Differential Geometry. Springer, Heidelberg (2001)

Theory of Real Computation According to EGC*

Chee Yap

Courant Institute of Mathematical Sciences
Department of Computer Science
New York University

Abstract. The Exact Geometric Computation (EGC) mode of computation has been developed over the last decade in response to the widespread problem of numerical non-robustness in geometric algorithms. Its technology has been encoded in libraries such as LEDA, CGAL and Core Library. The key feature of EGC is the necessity to decide zero in its computation. This paper addresses the problem of providing a foundation for the EGC mode of computation. This requires a theory of real computation that properly addresses the Zero Problem. The two current approaches to real computation are represented by the analytic school and algebraic school. We propose a variant of the analytic approach based on **real approximation**.

- To capture the issues of representation, we begin with a reworking of van der Waerden's idea of explicit rings and fields. We introduce explicit sets and explicit algebraic structures.
- Explicit rings serve as the foundation for real approximation: our starting point here is not $\mathbb{R}$, but $\mathbb{F} \subseteq \mathbb{R}$, an explicit ordered ring extension of $\mathbb{Z}$ that is dense in $\mathbb{R}$. We develop the approximability of real functions within standard Turing machine computability, and show its connection to the analytic approach.
- Current discussions of real computation fail to address issues at the intersection of continuous and discrete computation. An appropriate computational model for this purpose is obtained by extending Schönhage's pointer machines to support both algebraic and numerical computation.
- Finally, we propose a synthesis wherein both the algebraic and the analytic models coexist to play complementary roles. Many fundamental questions can now be posed in this setting, including transfer theorems connecting algebraic computability with approximability.

1 Introduction

Software breaks down due to numerical errors. We know that such breakdown has a numerical origin because when you tweak the input numbers, the problem

---

* Expansion of a talk by the same title at Dagstuhl Seminar on "Reliable Implementation of Real Number Algorithms: Theory and Practice", Jan 7-11, 2006. This work is supported by NSF Grant No. 043086.

P. Hertling et al. (Eds.): Real Number Algorithms, LNCS 5045, pp. 193–237, 2008.

goes away. Such breakdown may take the dramatic form of crashing or loop-ing. But more insidiously, it may silently produce *qualitatively* wrong results. Such qualitative errors are costly to catch further down the data stream. The economic consequences of general software errors have been documented[1] in a US government study [32]. Such problems of geometric nonrobustness are also well-known to practitioners, and to users of geometric software. See [22] for the anatomy of such breakdowns in simple geometric algorithms.

In the last decade, an approach called **Exact Geometric Computation** (EGC) has been shown to be highly effective in eliminating nonrobustness in a large class of basic geometric problems. The fundamental analysis and prescrip-tion of EGC may be succinctly stated as follows:

> *"Geometry is concerned with relations among geometric objects. Basic geometric objects (e.g., points, half-spaces) are parametrized by numbers. Geometric algorithms (a) construct geometric objects and (b) determine geometric relations. In real geometry, these relations are determined by evaluating the signs of real functions, typically polynomials. Algorithms use these signs to branch into different computation paths. Each path cor-responds to a particular output geometry. So the EGC prescription says that, in order to compute the correct geometry, it is sufficient to ensure that the correct path is taken. This reduces to error-free sign computa-tions in algorithms."*

How do algorithms determine the sign of a real quantity x? Typically, we compute approximations $\tilde{x}$ with increasing precision until the error $|x - \tilde{x}|$ is known to be less than $|\tilde{x}|$; then we conclude $\mathbf{sign}(x) = \mathbf{sign}(\tilde{x})$. Note that this requires interval arithmetic, to bound the error $|x - \tilde{x}|$. But in case $x = 0$, interval arithmetic does not help – we may reduce the error as much as we like, but the stopping condition (i.e., $|x - \tilde{x}| < |\tilde{x}|$) will never show up. This goes to the heart of EGC computation: how to decide if $x = 0$ [43]. This **zero problem** has been extensively studied by Richardson [33,34]. Numerical computation in this **EGC mode**[2] has far reaching implications for software and algorithms. Currently software such as LEDA [11,27], CGAL [17] and the Core Library [21] supports this EGC mode. We note that EGC assumes that the numerical input is exact (see discussion in [43]).

In this paper, we address the problem of providing a computability theory for the EGC mode of computation. Clearly, EGC requires arbitrary precision computation and falls under the scope of the theory of real computation. While the theory of computation for discrete domains (natural numbers $\mathbb{N}$ or strings Σ^*) has a widely accepted foundation, and possesses a highly developed com-plexity theory, the same cannot be said for computation over an uncountable

---

[1] A large part of the report focused on the aerospace and automobile industries. Both industries are major users of geometric software such as CAD modelers and simula-tion systems. The numerical errors in such software are well-known and so we infer that part of the cost comes from the kind of error of interest to us.

[2] Likewise, one may speak of the "numerical analysis mode", the "computer algebra mode", or the "interval arithmetic mode" of computing. See [42].

and continuous domain such as $\mathbb{R}$. Currently, there are two distinct approaches to the theory of real computation. We will call them the **analytic school** and the **algebraic school** respectively.

The analytic school goes back to Turing (1936), Grzegorczyk (1955) and Lacombe (1955) (see [40]). Modern proponents of this theory include Weihrauch [40], Ko [23], Pour-El and Richards [31] and others. In fact, there are at least six equivalent versions of analytic computability, depending on one's preferred starting point (metric spaces, domain-theory, etc) [20, p. 330]. In addition, there are complementary logical or descriptive approaches, based on algebraic specifications or equations. (e.g., [20]). But approaches based on Turing machines are most congenial to our interest in complexity theory. Here, real numbers are represented by rapidly converging Cauchy sequences. But an essential extension of Turing machines is needed to handle such inputs. In Weihrauch's TTE approach [40], Turing machines are allowed to compute forever to handle infinite input/output sequences. For the purposes of defining complexity, we prefer Ko's variant [23], using oracle Turing machines that compute in finite time. There is an important branch of the analytic school, sometimes known[3] as the Russian Approach [40, Chap. 9]. Below, we will note the close connections between the Russian Approach (Kolmogorov, Uspenskiĭ, Mal'cev) and our work.

The algebraic school goes back to the study of algebraic complexity [7,10]. The original computational model here is non-uniform (e.g., straightline programs). The uniform version of this theory has been advocated as a theory of real computation by Blum, Shub and Smale [6]. Note that this uniform algebraic model is also the *de facto* computational model of theoretical computer science and algorithms. Here, the preferred model is the Real RAM [2], which is clearly equivalent to the BSS model. In the algebraic school, real numbers are directly represented as atomic objects. The model allows the primitive operations to be carried out in a single step, without error. We can also compare real numbers without error. Although the BSS computational model [6] emphasizes ring operations $(+, -, \times)$, in the following, we will allow our algebraic model to use any chosen set Ω of real operations.

We have noted that the zero problem is central to EGC: sign determination implies zero determination. Conversely, if we can determine zero, we can also determine sign under mild assumptions (though the complexity can be very different). A natural hierarchy of zero problems can be posed [43]. This hierarchy can be used to give a natural complexity classification of geometric problems. But these distinctions are lost in the analytic and algebraic approaches: the zero problem is undecidable in analytic approach ([23, Theorem 2.5, p. 44]); it is trivial in the algebraic computational model. We need a theory where the complexity of the zero problems is more subtly portrayed, consistent with practical experience in EGC computation.

In numerical analysis, it is almost axiomatic that input numbers are inexact. In particular, this justifies the **backward analysis** formulation of solving problems. On the other hand, the EGC approach and the zero problem makes sense only if

---

[3] We might say the main branch practices the Polish Approach.

numerical inputs are exact. Hence, we must reject the oft-suggested notion that *all real inputs are inherently inexact*. In support of this notion, it is correctly noted that all physical constants are inherently inexact. Nevertheless, there are many applications of numerical computing for which such a view is untenable, in areas such as computer algebra, computational number theory, geometric theorem proving, computational geometry, and geometric modeling. The exact input assumption is also the dominant view in the field of algorithmics [13]. We also refute a related suggestion, that *the backward analysis solution is inevitable for real input problems*. A common argument to support this view says that "even if the input is exact, computers can only compute real quantities with limited precision". But the success of EGC, and existence of software such as LEDA, CGAL and Core Library, provides a wealth of counter examples. Nevertheless, the backward analysis problem formulation has an important advantage over EGC: it does not require a model of computation that must be able to decide zero. But the point of this paper is to shed some light on those computational problems for which we must decide zero.

The goal of this paper is three-fold: First, we propose a variant of the analytic approach that is suitable for studying the zero problem and for modeling contemporary practice in computation. Second, we propose a computational model suitable for "semi-numeric problems" – these are problems such as in computational geometry or topology where the input and output comprise a combination of numeric and discrete data. This is necessary to formalize what we mean by computation in the EGC mode. Finally, we propose a framework whereby the algebraic approach can be discussed in its natural relation to the analytic approach. Basic questions such as "transfer theorems" can be asked here.

2   Explicit Set Theory

Any theory of computation ought to address the representation of the underlying mathematical domains, especially with a view to their computation on machines. Thus Turing's 1936 analysis of the concept of computability began with a discussion of representation of real numbers. But the issue of representation usually is de-emphasized in discrete computability theory because we usually define our problems directly[4] on a canonical universe such as strings (Σ^* in Turing computability) or natural numbers ($\mathbb{N}$ in recursive function theory). Other domains (e.g., finite graphs) must be suitably encoded in this canonical universe. The encoding of such discrete domains does not offer conceptual difficulties, although it may have significant complexity theoretic implications. But for continuous or uncountable domains such as $\mathbb{R}$, choice of representation may be critical (e.g., representing reals by its binary expansion is limiting [40]). See Weihrauch and Kreitz [24,40] for topological investigations of representations (or "naming systems") for such sets.

---

[4] I.e., we simply say that our problems are functions or relations on this canonical universe, thereby skipping the translation from the canonical universe to our intended domain of application.

The concept of representation is implicit (sic) in van der Waerden's notion of "explicit rings and fields" (see Fröhlich and Shepherdson [18]). To bring representation into algebraic structures, we must first begin with the more basic concept of "explicit sets". These ideas were extensively developed by the Russian school of computability. Especially relevant is Mal'cev's theory of **numbered algebraic systems** [26, Chapters 18,25,27]. A **numbering** of an arbitrary set S is an onto, total function $\alpha : D \to S$ where $D \subseteq \mathbb{N}$. The pair (S, α) is called a **numbered set**. A numbered algebraic system, then, is an algebraic system (see below) whose carrier set S has a numbering α, and whose functions and predicates are represented by functions on $\mathbb{N}$ that are compatible with α in the natural way.

Intuitively, an explicit set is one in which we can recognize the identity of its elements through its names or representation elements. This need for recognizing elements of a set is well-motivated by EGC's need to decide zero "explicitly", and to perform exact operations. Explicitness is a form of effectivity. Traditionally, one investigates effectivity of subsets of $\mathbb{N}$, and such sets can be studied via Kleene or Gödel numberings. But the issues are more subtle when we treat arbitrary sets. Unlike $\mathbb{N}$, which has a canonical representation, general sets which arise in algebra or analysis may be specified in a highly abstract (prescriptive) manner that gives no hint as to their representation. A central motivation of Mal'cev [26, p. 287] is to study numberings of arbitrary sets. His theorems are inevitably about numbered sets, not just the underlying sets (cf. [26, Chapter 25]). Our point of departure is the desire to have a numbering-independent notion of explicitness. We expect that in a suitable formulation of explicit sets, certain basic axioms in naive set theory [19] will become theorems in explicit set theory. E.g., the well-ordering principle for sets.

Partial functions. If S, T are sets, we shall denote[5] a partial function from T to S by $f : T \dashrightarrow S$ (dashed arrow). If $f(x)$ is undefined, we write $f(x) =\uparrow$; otherwise we write $f(x) =\downarrow$. Call T the **nominal domain** of f; the **proper domain** of f is $\mathrm{domain}(f) := \{x \in T : f(x) =\downarrow\}$. Relative to f, we may refer to $x \in T$ as **proper** if $f(x) =\downarrow$; otherwise x is **improper**. If $\mathrm{domain}(f) = T$, then f is **total**, and we indicate total functions in the usual way, $f : T \to S$. Similarly, the sets S and $\mathrm{range}(f) := \{f(s) : s \in \mathrm{domain}(f)\}$ are (resp.) the **nominal** and **proper ranges** of f. We say f is **onto** if $\mathrm{range}(f) = S$, and f is **1-1** if $f(x) = f(y) \neq\uparrow$ implies $x = y$.

In composing partial functions, we use the standard rule that a function value is undefined if any input argument is undefined. We often encounter predicates such as "$f(x) = g(y)$". We interpret such equalities in the "strong sense", meaning that the left side $f(x)$ is defined iff the right side $g(y)$ is defined. To indicate this **strong equality**, we write "$f(x) \equiv g(y)$".

Partial recursive partial functions. Partial functions on strings, $f : \Sigma^* \dashrightarrow \Sigma^*$ have a special status in computability theory: we say f is **partial recursive**

---

[5] This notation is used, e.g., by Weihrauch [24] and Mueller [29]. We thank Norbert Mueller for his contribution of this partial function symbol. An alternative notation is $f :\subseteq T \to S$ [40].

if there is a (deterministic) Turing machine that, on input $w \in \Sigma^*$, halts with the string $f(w)$ in the output tape if $f(w) = \downarrow$, and does not halt (i.e., loops) if $f(w) = \uparrow$. If f is a total function, and f is partial recursive, then we say that f is **total recursive**. Both these definitions are standard. We need an intermediate version of these two concepts: assume that our Turing machines have two special states, $q_\downarrow$ (**proper state**) and $q_\uparrow$ (**improper state**). A Turing machine is **halting** if it halts on all input strings. We say[6] that f is **recursive** if there is a halting Turing machine M that, for all $w \in \Sigma^*$, if $f(w) = \uparrow$, M halts in the improper state $q_\uparrow$; if $f(w) = \downarrow$, then M halts in the proper state $q_\downarrow$ with the output tape containing $f(w)$. If $TOT\text{-}REC$ (resp., $PART\text{-}REC$, REC) denote the set of total recursive (resp., partial recursive, recursive) functions, then we have $TOT\text{-}REC \subseteq REC \subseteq PART\text{-}REC$. These inclusions are all strict.

Let $S \subseteq \Sigma^*$. For our purposes, we define the **characteristic function** of S to be $\chi_S : \Sigma^* \dashrightarrow \{1\}$ where $\chi_S(w) = 1$ if $w \in S$, and $\chi_S(w) = \uparrow$ otherwise. A closely related function is the **partial identity function**, $\iota_S : \Sigma^* \dashrightarrow S$ where $\iota_S(w) = w$ iff $w \in S$ and $\iota_S(w) = \uparrow$ otherwise. We say[7] S is (partial) **recursive** if χ_S (equivalently ι_S) is (partial) recursive.

Notes on Terminology. In conventional computability, it would be redundant to say "partial recursive *partial* function". Likewise, it would be an oxymoron to call a partial function "recursive". Note that a recursive function f in our sense is equivalent to "f is partial recursive with a recursive domain" in standard terminology. Our notion of recursive function anticipates its use in various concepts of "explicitness". In our applications, we are not directly computing over Σ^*, but are computing over some abstract domain S that is represented through Σ^*. Thus, our Turing machine computation over Σ^* is interpreted via this representation. We use the explicitness terminology in association with such concepts of "interpreted computations" – explicitness is a form of "interpreted effectivity". In recursive function theory, we know that the treatment of partial functions is an essential feature. In algebraic computations, we have a different but equally important reason for treating of partial functions: algebraic operations (e.g., division, squareroot, logarithm) are generally partial functions. The standard definitions would classify division as "partial recursive" (more accurately, partially explicit). But all common understanding of recursiveness dictates that division be regarded as "recursive" (more accurately, explicit). Thus, we see that our terminology is more natural.

In standard computability theory (e.g., [35]), the terms "computable" and "recursive" are often interchangeable. In this paper, the terms "recursive" and "partial recursive" are only applied to computing over strings (Σ^*), "explicit" refer to computing over some abstract domain (S). Further, we reserve the term

---

[6] In the literature, "total recursive" is normally shortened to "recursive"; so our definition of "recursive" forbids such an abbreviation.

[7] It is more common to say that S is recursively enumerable if χ_S is partial recursive. But note that Mal'cev [26, p.164–5] also calls S partial recursive when χ_S is partial recursive.

"computability" for real numbers and real functions (see Section 4), in the sense used by the analytic school [40,23].

Representation of sets. We now consider arbitrary sets S, T. Call $\rho : T \dashrightarrow S$ a **representation** of S (with **representing set** T) if ρ is an onto function. If $\rho(t) = s$, we call t a **representation element** of s. More precisely, t is called a ρ-**name** of s. Relative to ρ, we say t, t' are **equivalent**, denoted $t \equiv t'$ if $\rho(t) \equiv \rho(t')$ (recall $\equiv$ means equality is in the strong sense). In case $T = \Sigma^*$ for some alphabet Σ, we call ρ a **notation**. It is often convenient to identify Σ^* with $\mathbb{N}$, under some bijection that takes $n \in \mathbb{N}$ to $\underline{n} \in \Sigma^*$. Hence a representation of the form $\rho : \mathbb{N} \dashrightarrow S$ is also called a notation.

We generally use 'ν' instead of 'ρ' for notations. For computational purposes (according to Turing), we need notations. Our concept of notation ν and Mal'cev's numbering α are closely related: the fact that $\mathtt{domain}(\alpha) \subseteq \mathbb{N}$ while $\mathtt{domain}(\nu) \subseteq \Sigma^*$ is not consequential since we may identify $\mathbb{N}$ with Σ^*. But it is significant that ν is partial, while α is total. Unless otherwise noted, we may (wlog) assume $\Sigma = \{0, 1\}$. Note that if a set has a notation then it is countable.

A notation ν is **recursive** if the set[8] $E_\nu := \{(w, w') \in (\Sigma^*)^2 : \nu(w) \equiv \nu(w')\}$ is recursive. In this case, we say S is ν-**recursive**. If S is ν-recursive then the set $D_\nu := \{w \in \Sigma^* : \nu(w) = \downarrow\}$ $(= \mathtt{domain}(\nu))$ is recursive: to see this, note that $w \in D_\nu$ iff $(w, w_\uparrow) \notin E_\nu$ where $w_\uparrow$ is any word such that $\nu(w_\uparrow) = \uparrow$.

It is important to note that "recursiveness of ν" does not say that the function ν is a recursive function. Indeed, such a definition would not make sense unless S is a set of strings. The difficulty of defining explicit sets amounts to providing a substitute for defining "recursiveness of ν" as a function. Tentatively, suppose we say S is "explicit" (in quotes) if there "exists" a recursive notation ν for S. Clearly, the "existence" here cannot have the standard understanding – otherwise we have the trivial consequence that a set S is "explicit" iff it is countable. One possibility is to understand it in some intuitionistic sense of "explicit existence" (e.g., [4, p. 5]). But we prefer to proceed classically.

To illustrate some issues, consider the case where $S \subseteq \mathbb{N}$. There are two natural notations for S: the **canonical notation** of S is $\nu_S : \mathbb{N} \dashrightarrow S$ where $\nu_S(n) = n$ if $n \in S$, and otherwise $\nu_S(n) = \uparrow$. The **ordered notation** of S is $\nu'_S : \mathbb{N} \dashrightarrow S$ where $\nu'_S(n) = i$ if $i \in S$ and the set $\{j \in S : j < i\}$ has n elements. Let S be the halting set $K \subseteq \mathbb{N}$ where $n \in K$ iff the nth Turing machine on input string $\underline{n}$ halts. The canonical notation ν_K is not "explicit" since E_{ν_K} is not recursive. But the ordered notation ν'_K is "explicit" since $E_{\nu'_K}$ is the trivial diagonal set $\{(n, n) : n \in \mathbb{N}\}$. On the other hand, ν'_K does not seem to be a "legitimate" way of specifying notations (for instance, it is even not a computable function). The problem we face is to distinguish between notations such as ν_K and ν'_K.

Our first task is to distinguish a legitimate set of notations. We consider three natural ways to construct notations: let $\nu_i : \Sigma_i^* \dashrightarrow S_i$ $(i = 1, 2)$ be notations and $\#$ is a symbol not in $\Sigma_1 \cup \Sigma_2$.

---

[8] The alphabet for the set E_ν may be taken to be $\Sigma \cup \{\#\}$ where $\#$ is a new symbol, and we write "(w, w')" as a conventional rendering of the string $w \# w'$.

1. (Cartesian product) The notation $\nu_1 \times \nu_2$ for the set $S_1 \times S_2$ is given by:

$$\nu_1 \times \nu_2 : (\Sigma_1^* \cup \Sigma_2^* \cup \{\#\})^* \dashrightarrow S_1 \times S_2$$

where $(\nu_1 \times \nu_2)(w_1 \# w_2) = (\nu_1(w_1), \nu_2(w_2))$ provided $\nu_i(w_i) =\downarrow$ $(i = 1, 2)$; for all other w, we have $(\nu_1 \times \nu_2)(w) =\uparrow$.

2. (Kleene star) The notation ν_1^* for finite strings over S_1 is given by:

$$\nu_1^* : (\Sigma_1^* \cup \{\#\})^* \dashrightarrow S_1^*$$

where $\nu_1^*(w_1 \# w_2 \# \cdots \# w_n) = \nu_1(w_1)\nu_1(w_2) \cdots \nu_1(w_n)$ provided $\nu_1(w_j) =\downarrow$ for all j; for all other w, we have $(\nu_1^*)(w) =\uparrow$.

3. (Restriction) For an arbitrary function $f : \Sigma_1^* \dashrightarrow \Sigma_2^*$, we obtain the following notation

$$\nu_2|_f : \Sigma_1^* \dashrightarrow T$$

where $\nu_2|_f(w) = \nu_2(f(w))$ and $T = \mathbf{range}(\nu_2 \circ f) \subseteq S_2$. Thus $\nu_2|_f$ is essentially the function composition, $\nu_2 \circ f$, except that the nominal range of $\nu_2 \circ f$ is S_2 instead of T. If f is a recursive function, then we call this operation **recursive restriction**.

We now define "explicitness" by induction: a notation $\nu : \Sigma^* \dashrightarrow S$ where ν is $1 - 1$ and S is finite is called a **base notation**. a notation ν is **explicit** if ν is a base notation or (inductively) there exist explicit notations ν_1, ν_2 such that ν is one of the notations

$$\nu_1 \times \nu_2, \quad \nu_1^*, \quad \nu_1|_f$$

where f is recursive. A set S is **explicit** if there exists an explicit notation for S.

Informally, an explicit set is obtained by repeated application of Cartesian product, Kleene star and recursive restriction, starting from base notations. Note that Cartesian product, Kleene star and restriction are analogues (respectively) of the Axiom of pairing, Axiom of powers and Axiom of Specification in standard set theory ([19, pp. 9,6,19]). Let us note some applications of recursive restriction: suppose $\nu : \Sigma^* \dashrightarrow S$ is an explicit notation.

- (Change of alphabet) Notations can be based on any alphabet Γ: we can find a suitable recursive function f such that $\nu|_f : \Gamma^* \dashrightarrow S$ is an explicit notation. We can further make $\nu|_f$ a 1-1 function.
- (Identity) The identity function $\nu : \Sigma^* \to \Sigma^*$ is explicit: to see this, begin with $\nu_0 : \Sigma^* \dashrightarrow \Sigma$ where $\nu_0(a) = a$ if $a \in \Sigma$ and $\nu_0(w) =\uparrow$ otherwise. Then ν can be obtained as a recursive restriction of ν_0^*. Thus, Σ^* is an explicit set.
- (Subset) Let T be a subset of S such that $D = \{w : \nu(w) \in T\}$ is recursive. If ι_D is the partial identity function of D, then $\nu|_{\iota_D}$ is an explicit notation for T. We denote $\nu|_{\iota_D}$ more simply by $\nu|_T$.
- (Quotient) Let $\sim$ be an equivalence relation on S, we want a notation for $S/\sim$ (the set of equivalence classes of $\sim$). Consider the set $E = \{(w, w') : \nu(w) \sim \nu(w') \text{ or } \nu(w) = \nu(w') =\uparrow\}$. We say $\sim$ is **recursive relative to** ν if E is a recursive set. Define the function $f : \Sigma^* \to \Sigma^*$ via $f(w) = \min\{w' : (w, w') \in E\}$(where min is based on any lexicographic order $\leq_{\text{LEX}}$ on Σ^*). If

E is recursive then f is clearly a recursive function. Then the notation $\nu|_f$, which we denote by

$$\nu/\sim \tag{1}$$

can be viewed as a notation for $S/\sim$, *provided* we identify $S/\sim$ with a subset of S (namely, each equivalence class of $S/\sim$ is identified with a representative from the class). This identification device will be often used below.

We introduce a normal form for explicit notations. Define a **simple notation** to be one obtained by applications of the Cartesian product and Kleene-star operator to a base case (i.e., to a notation $\nu : \Sigma^* \dashrightarrow S$ that is $1-1$ and S is finite). A **simple set** is one with a simple notation. In other words, simple sets do not need recursive restriction for their definition. A **normal form notation** ν_S for a set S is one obtained as the recursive restriction of a simple notation: $\nu_S = \nu|_f$ for some simple notation ν and recursive function f.

Lemma 1 (Normal form). *If S is explicit, then it has a normal form notation ν_S.*

Proof. Let $\nu_0 : \Sigma^* \dashrightarrow S$ be an explicit notation for S.
(0) If S is a finite set, then the result is trivial.
(1) If $\nu_0 = \nu_1 \times \nu_2$ then inductively assume the normal form notations $\nu_i = \nu_i'|_{f_i'}$ ($i = 1, 2$). Let $\nu = \nu_1' \times \nu_2'$ and for $w_i \in \text{domain}(\nu_i')$, define f by $f(w_1\#w_2) = f_1(w_1)\#f_2(w_2) \in S$. Clearly f is recursive and $\nu|_f$ is an explicit notation for S.
(2) If $\nu_0 = \nu_1^*$ then inductively assume a normal form notation $\nu_1 = \nu_1'|_{f_1'}$. Let $\nu = (\nu_1')^*$ and for all $w_j \in \text{domain}(\nu_1')$, define $f(w_1\#w_2\#\cdots\#w_n) = f_1(w_1)\#f_1(w_2)\#\cdots\#f_1(w_n) \in S$. So $\nu|_f$ is an explicit notation for S.
(3) If $\nu_0 = \nu_1|_{f_1}$, then inductively assume the normal form notation $\nu_1 = \nu_1'|_{f_1'}$. Let $\nu = \nu_1$ and $f = f_1' \circ f_1$. Clearly, $\nu|_f$ is an explicit notation for S.   **Q.E.D.**

The following is easily shown using normal form notations:

Lemma 2. *If ν is explicit, then the sets E_ν and D_ν are recursive.*

In the special case where $S \subseteq \mathbb{N}$ or $S \subseteq \Sigma^*$, we obtain:

Lemma 3. *A subset of $S \subseteq \mathbb{N}$ is explicit iff S is partial recursive (i.e., recursively enumerable).*

Proof. If S is explicit, then any normal form notation $\nu : \Sigma^* \dashrightarrow S$ has the property that ν is a recursive function, and hence S is recursively enumerable. Conversely, if S is recursively enumerable, it is well-known that there is a total recursive function $f : \mathbb{N} \to \mathbb{N}$ whose range is S. We can use f as our explicit notation for S.   **Q.E.D.**

We thus have the interesting conclusion that the halting set K is an explicit set, but *not* by virtue of the canonical (ν_K) or ordered (ν_K') notations discussed above. Moreover, the complement of K is not an explicit set, confirming that our concept of explicitness is non-trivial (i.e., not every countable set is explicit).

Our lemma suggests that explicit sets are analogues of recursively enumerable sets. We could similarly obtain analogues of recursive sets, and a complexity theory of explicit sets can be developed.

The following data structure will be useful for describing computational structures. Let L be any set of symbols. To motivate the definition, think of L as a set of labels for numerical expressions. E.g., $L = \mathbb{Z} \cup \{+, -, \times\}$, and we want to define arithmetic expressions labeled by L.

An **ordered L-digraph** $G = (V, E; \lambda)$ is a directed graph (V, E) with vertex set $V = \{1, \ldots, n\}$ (for some $n \in \mathbb{N}$) and edge set $E \subseteq V \times V$, and a labeling function $\lambda : V \to L$ such that the set of outgoing edges from any vertex $v \in V$ is totally ordered. Such a graph may be represented by a set $\{L_v : v \in V\}$ where each L_v is the **adjacency list** for v, having the form $L_v = (v, \lambda(v); u_1, \ldots, u_k)$ where k is the outdegree of v, and each (v, u_i) is an edge. The total order on the set of outgoing edges from v is specified by this adjacency list. We deem two such graphs $G = (V, E; \lambda)$ and $G' = (V', E'; \lambda')$ to be the same if, up to a renaming of the vertices, they have the same set of adjacency lists (so the identity of vertices is unimportant, but their labels are). Let $\mathcal{DG}(L)$ be the set of all ordered L-digraphs.

Convention for Representation Elements. In normal discourse, we prefer to focus on a set S rather than on its representing set T (under some representation $\rho : T \dashrightarrow S$). We introduce an "underbar-convention" to facilitate this. If $x \in S$, we shall write $\underline{x}$ (x-underbar) to denote *some* representing element for x (so $\underline{x} \in T$ is an "underlying representation" of x). This convention makes sense when the representation ρ is understood or fixed. Writing "$\underline{x}$" allows us to acknowledge that it is the representation of x in an unobtrusive way. The fact that "$\underline{x}$" is under-specified (non-unique) is usually harmless. We use this convention in the proof of the next lemma:

Lemma 4. *Let S, T be explicit sets. Then the following sets are explicit:*
(i) Disjoint union $S \uplus T$,
(ii) Finite power set $\widehat{2^S}$ (set of finite subsets of S),
(iii) The set of ordered S-digraphs $\mathcal{DG}(S)$.

Proof. Let $\nu_1 : \Sigma^* \dashrightarrow S$ and $\nu_2 : \Sigma^* \dashrightarrow T$ be explicit notations. We use the above convention that $\underline{s}$ denotes a representation element for $s \in S$ (i.e., $\nu_1(\underline{s}) = s$).
(i) Let ν_0 be a base notation for the set $\{0, 1\}$, and fix $s_0 \in S$ and $t_0 \in T$ (we may assume S, T to be non-empty). So the Cartesian product $\nu_0 \times \nu_1 \times \nu_2$ is an explicit notation for $\{0, 1\} \times S \times T$. Consider the restriction f for $\nu_0 \times \nu_1 \times \nu_2$ where

$$f(b\#\underline{s}\#\underline{t}) = \begin{cases} b\#\underline{s_0}\#\underline{t} & \text{if } b = \underline{0} \\ b\#\underline{s}\#\underline{t_0} & \text{if } b = \underline{1} \end{cases}.$$

Also f is undefined in all other cases. We can regard $(\nu_0 \times \nu_1 \times \nu_2)|_f$ as an explicit notation for $S \uplus T$. (ii) For finite power set $\widehat{2^S}$, we use the fact that S

is well-ordered by an ν_1-explicit total ordering $<_{\nu_1}$ (see Lemma 10 below). We apply recursive restriction to the notation ν^* for S^*: define $f(x_1 \# \cdots \# x_n) = x_{\pi(1)} \# \cdots \# x_{\pi(m)}$ where $x_{\pi(1)} <_{\nu_1} \cdots <_{\nu_1} x_{\pi(m)}$, assuming that $\{x_1, \ldots, x_n\} = \{x_{\pi(1)}, \ldots, x_{\pi(m)}\} \subseteq S$. Also f is undefined in other cases. Then $\nu_1^*|_f$ is a notation for $\widehat{2}^S$, assuming that we identify $\widehat{2}^S$ with a suitable subset of S^*.

(iii) Recall the representation of an ordered S-digraph above, as a set of adjacency lists, $\{L_v : v \in V\}$ and $V = \{1, \ldots, n\}$. Vertices $v \in \mathbb{N}$ are represented by binary numbers. The set of all adjacency lists L_v can be represented via Cartesian product, Kleene star and recursive restriction. Finite sets of adjacency lists can be represented by the finite power set method of (ii). We further need restriction on the sets of adjacency lists, to ensure that each set has the form $\{L_v : v \in V\}$ and each vertex in list L_v belongs to V. **Q.E.D.**

Example: Dyadic numbers. Let $\mathbb{D} := \mathbb{Z}[\frac{1}{2}] = \{m2^n : m, n \in \mathbb{Z}\}$ denote the set of **dyadic numbers** (or **bigfloats**, in programming contexts). Let $\Sigma_2 = \{0, 1, \bullet, +, -\}$. A string

$$w = \sigma b_{n-1} b_{n-2} \cdots b_k \bullet b_{k-1} \cdots b_0 \tag{2}$$

in Σ_2^* is **proper** if $\sigma \in \{+, -\}$, $n \geq 0$, $k = 0, \ldots, n$ and each $b_j \in \{0, 1\}$. The **dyadic notation**

$$\nu_2 : \Sigma_2^* \dashrightarrow \mathbb{D} \tag{3}$$

takes the proper string w in (2) to the number $\nu_2(w) = \sigma 2^{-k} \sum_{i=0}^{n-1} b_i 2^i \in \mathbb{D}$; otherwise $\nu_2(w) = \uparrow$. In order to consider ν_2 as an explicit notation, we will identify $\mathbb{D}$ with a suitable subset $\mathbb{D} \subseteq \Sigma_2^*$. $\mathbb{D}$ comprises the proper strings (2) with additional properties: $(n > k \Rightarrow b_{n-1} = 1)$, $(k \geq 1 \Rightarrow b_0 = 1)$ and $(n = k = 0 \Rightarrow \sigma = +)$. Thus the strings $+\bullet$, $-1\bullet$, $+10\bullet$, $-11\bullet$, $+ \bullet 1$, $-1 \bullet 01$, etc. are identified with the numbers $0, -1, 2, -3, 0.5, -1.25$, etc. We can identify $\mathbb{N}$ and $\mathbb{Z}$ as suitable subsets of $\mathbb{D}$, and thus obtain notations for $\mathbb{N}, \mathbb{Z}$ by recursive restrictions of ν_2. All these are explicit notations. We also extend ν_2 to a notation for $\mathbb{Q}$: consider the representation of rational numbers by pairs of integers, $\rho_{\mathbb{Q}} : \mathbb{Z}^2 \dashrightarrow \mathbb{Q}$ where $\rho_{\mathbb{Q}}(m, n) = m/n$ if $n \neq 0$, and $\rho_{\mathbb{Q}}(m, 0) = \uparrow$. This induces an equivalence relation $\sim$ on $\mathbb{Z}^2$ where $(m, n) \sim (m', n')$ iff $\rho_{\mathbb{Q}}(m, n) \equiv \rho_{\mathbb{Q}}(m', n')$. We then obtain a notation for $\mathbb{Q}$ by the composition $\rho_{\mathbb{Q}} \circ (\nu_2 \times \nu_2)$. This example illustrates the usual way in which representations ρ of mathematical domains are built up by successive composition of representations, $\rho = \rho_1 \circ \rho_2 \circ \cdots \circ \rho_k$ $(k \geq 1)$. If ρ_k is a notation, then ρ is also a notation. Although ρ may not be an explicit notation, it can be converted into an explicit notation by a natural device. E.g., $\rho_{\mathbb{Q}} \circ (\nu_2 \times \nu_2)$ is not an explicit notation for $\mathbb{Q}$. To obtain an explicit notation for $\mathbb{Q}$, we use the quotient notation $(\nu_2 \times \nu_2)/\sim$, as described in (1). This required an identification of $\mathbb{Q}$ with a subset of $\mathbb{Z}^2$. In fact, we typically identify $\mathbb{Q}$ with the subset of $\mathbb{Z}^2$ comprising relatively prime pairs $(p, q) \in \mathbb{Z}^2$ where $q > 0$. We next formalize this procedure.

Suppose we want to show the explicitness of a set S, and we have a "natural" representation $\rho : T \dashrightarrow S$. For instance, $\rho_{\mathbb{Q}}$ is surely a "natural" representation of $\mathbb{Q}$. We proceed by choosing an explicit notation $\nu : \Sigma^* \dashrightarrow T$ for T. Then

$$\rho \circ \nu$$

is a notation for S, but not necessarily explicit. Relative to ρ, we say that the set S is **naturally identified** in T if we identify each $x \in S$ as some element of the set $\rho^{-1}(x)$. Under this identification, we have $S \subseteq T$. Moreover, the representation ρ is the identity function when its nominal domain is restricted to S. The following lemma then provides the condition for concluding that S is an explicit set.

Lemma 5. Let $\rho : T \dashrightarrow S$ be a representation of S. Suppose $\nu : \Sigma^* \dashrightarrow T$ is an explicit notation such that $\rho \circ \nu : \Sigma^* \dashrightarrow S$ is a recursive notation. Then the set S is an explicit set under some natural identification of S in T. In fact, there is recursive function $f : \Sigma^* \dashrightarrow \Sigma^*$ such that $\nu|_f : \Sigma^* \dashrightarrow S$ is an explicit notation.

Explicit Subsets. So far, set intersection $S \cap T$ and union $S \cup T$ are not among the operations we considered. To discuss these operations, we need a "universal set", taken to be another explicit set U.

Let U be a ν-explicit set where $\nu : \Sigma^* \dashrightarrow U$. We call[9] $S \subseteq U$ a **(partially)** ν-**explicit subset** of U if the set $\{w \in \Sigma^* : \nu(w) \in S\}$ is (partially) recursive. By definition, S is an explicit subset of U iff U is an **explicit superset** of S.

Lemma 6. Let U be ν-explicit, and S and T are ν-explicit subsets of U.
(i) The sets S, $U \setminus S$, $S \cup T$ and $S \cap T$ are all explicit sets.
(ii) The set of ν-explicit subsets of U is closed under the Boolean operations (union, intersection, complement).
(iii) The set of partially ν-explicit subsets of U is closed under union.
(iv) There exists U and a partially ν-explicit subset $S \subseteq U$ such that $U \setminus S$ is not an explicit set.

Proof. To show (iv), we let $U = \Sigma^*$, where ν is the identity function, and let S be the halting set.   **Q.E.D.**

Abstract and Concrete Sets. Before concluding this section, we make two remarks on the interplay between abstract and concrete sets. For this discussion, let us agree to call a set **concrete** if it is a subset of some Σ^*; other kinds of set are said to be **abstract**. Following Turing, algorithms can only treat concrete sets; but in mathematics, we normally treat abstract sets like $\mathbb{Q}, \mathbb{R}, Hom(A, B), SO(3)$, etc. It is seen from the above development that if we want explicit notations for abstract sets, we must ultimately identify them with a suitable subset of a simple set (cf. (1)). Thus, the set of all subsets of strings is our "canonical universe" for

---

[9] In this paper, any appearance of the parenthetical "(partially)" conveys two parallel statements: one with all occurrences of "(partially)" removed, and another with "partially" inserted.

abstract sets. In practice, making such identifications of abstract sets is not an arbitrary process (it is not enough to obtain just any bijection). Abstract sets in reality have considerable structure and relation with other abstract sets, and these must be preserved and made available for computation. Our explicit sets can encode such information, but this is is not formalized. In other words, our sets are conceptually "flat". Exploring structured sets is a topic for future work. Without formalizing these requirements, we must exercise judgment in making such identifications on a case by case basis, but typically they are non-issues as in, e.g., $\mathbb{Q}$ (above), ideals $\mathcal{ID}_n$ (Lemma 9) and real closure $\overline{F}$ (Theorem 13). The second remark concerns the manipulation of an abstract set S through its notation ν. Strictly speaking, all our development could have been carried out by referring only to the equivalence relation E_ν, and never mentioning S. But this approach would be tedious and unnatural for normal human comprehension. Being able to talk directly about S is more efficient. For this practical[10] reason we prefer to talk about S, and (like the underbar-convention) only allude to the contingent E_ν.

3 Explicit Algebraic Structures

In this section, we extend explicit sets to explicit algebraic structures such as rings and fields. To do this, we need to introduce explicit functions.

Function representations. To represent functions, we assume a representation of its underlying sets. Let $\rho : T \dashrightarrow S$ be a representation. If $f : S^2 \dashrightarrow S$ is a function, then the function $\underline{f} : T \times T \dashrightarrow T$ is called a ρ-**representation** of f if for all $x, y \in T$,

$$\rho(\underline{f}(x,y)) \equiv f(\rho(x), \rho(y)).$$

In case ρ is a notation, we call $\underline{f}$ a ρ-**notation** for f.

We say $f : S^2 \dashrightarrow S$ is ν-**explicit** (or simply "explicit") if S is ν-explicit and there exists a recursive ν-notation $\underline{f}$ for f. Although we never need to discuss "partially explicit sets", we will need "partially explicit functions": we say f is **partially ν-explicit** (or simply "partially explicit") if S is ν-explicit and there is a ν-notation $\underline{f}$ for f which is partial recursive.

These concepts are naturally extended to general k-ary functions, and to functions whose range and domain involve different sets. E.g., if $f : S \dashrightarrow T$ where S is ν-explicit and T is ν'-explicit, we can talk about f being (ν, ν')-explicit. These concepts also apply to the special case of predicates: we define a **predicate on S** to be any partial function $R : S \dashrightarrow V$ where V is any finite set. Usually $|V| = 2$, but geometric predicates often have $|V| = 3$ (e.g., $V = \{IN, OUT, ON\}$).

An **(algebraic) structure** is a pair (S, Ω) where S is a set (the carrier set) and Ω is a set of predicates and algebraic operations on S. By an **(algebraic)**

---

[10] If not for some deeper epistemological reason.

operation on S we mean any partial function $f : S^k \dashrightarrow S$, for some $k \geq 0$ (called the **arity** of f).

As an example, and one that is used extensively below, an **ordered ring** S is an algebraic structure with $\Omega = \{+, -, \times, 0, 1, \leq\}$ such that $(S, +, -, \times, 0, 1)$ is a ring[11] and $\leq$ is a total order on S with the following properties: the ordering defines a **positive set** $P = \{x \in R : x > 0\}$ that is closed under $+$ and $\times$, and for all $x \in S$, exactly one of the following conditions is true: $x = 0$ or $x \in P$ or $-x \in P$. See [41, p. 129].

Ordered rings are closely related to another concept: a ring R is **formally real** if $0 \notin R^{(2)}$ where $R^{(2)}$ is the set of all sums of non-zero squares. When the rings are fields or domains, etc, we may speak of ordered fields, formally real domains, etc. Note that ordered rings are formally real, and formally real rings must be domains and can be extended into formally real fields. Conversely, formally real fields can be extended into an ordered field (its real constructible closure) [41, Theorem 5.6, p. 134]. Clearly, formally real fields have characteristic 0.

The key definition is this: an algebraic structure (S, Ω) is ν-**explicit** if its carrier set is ν-explicit, and each $f \in \Omega$ is ν-explicit. Thus we may speak of explicit rings, explicit ordered fields, etc.

We must discuss substructures. By a **substructure** of (S, Ω) we mean (S', Ω') such that $S' \subseteq S$, there is a bijection between Ω' and Ω, and each $f' \in \Omega'$ is the restriction to S' of the corresponding operation or predicate $f \in \Omega$. Thus, we may speak of subfields, subrings, etc. If (S, Ω) is ν-explicit, then (S', Ω') is a ν-**explicit substructure** of (S, Ω) if S' is a ν-explicit subset of S and (S', Ω') is a substructure of (S, Ω). Thus, we have explicit subrings, explicit subfields, etc. If (S', Ω') is an explicit substructure of (S, Ω), we call (S, Ω) an **explicit extension** of (S', Ω').

The following shows the explicitness of some standard algebraic constructions:

Lemma 7. *Let ν_2 be the normal form notation for $\mathbb{D} = \mathbb{Z}[\frac{1}{2}]$ in (3).*
(i) $\mathbb{D}$ is a ν_2-explicit ordered ring.
(ii) $\mathbb{N} \subseteq \mathbb{Z} \subseteq \mathbb{D}$ are ν_2-explicit ordered subrings.
(iii-a) If D is an ν-explicit domain, then the quotient field $Q(D)$ is explicit.
(iii-b) Moreover, the divides relation $a|b$ is ν-explicit in D iff D is an explicit subring of $Q(D)$.
(iv) If R is an explicit ring, and W is an explicit set of indeterminates, then the polynomial ring $R[W]$ is an explicit ring extension of R.
(v) If F is an explicit field, then any simple algebraic extension $F(\theta)$ is an explicit field extension of F.

Proof. (i)-(ii) are obvious. The constructions (iii)-(v) are standard algebraic constructions; these constructions can be implemented using operations of explicit sets. Briefly:
(iii-a) The explicit notation $\nu_{Q(D)}$ for $Q(D)$ is a standard generalization of the construction of $\mathbb{Q}$ from $\mathbb{Z}$, illustrated above. In fact, $\nu_{Q(D)}$ is a recursive restriction of $\nu \times \nu$, after an identification of $Q(D)$ with a subset of $D \times D$ (see

---

[11] All our rings are commutative with unit 1.

Lemma 5). To conclude that $Q(D)$ is an explicit field, we verify that the field operations of $Q(D)$ are $\nu_{Q(D)}$-explicit.

(iii-b) Recall that a divides b, written $a|b$, iff there is some $c \in D$ such that $ac = b$. In addition to identifying $Q(D)$ with a subset of $D \times D$, we may assume this subset includes $D \times \{1\}$. If we further identify D with $D \times \{1\}$, we see that D is a subset of $Q(D)$. Now D is an explicit subring of $Q(D)$ iff D is an explicit subset of $Q(D)$. So it suffices to show that $a|b$ is ν-explicit iff we can explicitly decide if any $a/b \in Q(D)$ belongs to D. [Pf: ($\Rightarrow$) Given a/b, we can compute $\underline{a'}, \underline{b'}$ for some $a', b' \in D$ such that $a'/b' = a/b$. Then we conclude that $a/b \in D$ iff $a'|b'$. By assumption, we know how to check $a'|b'$ in D. ($\Leftarrow$) Given $\underline{a}, \underline{b}$, we first compute a/b. Then $a|b$ iff a/b denotes an element of D. By assumption, we know how to check membership in D.]

(iv) First consider the case where $W = \{X\}$ with just one indeterminate. The standard representation of $R[X]$ uses R^* (Kleene star) as representing set: $\rho : R^* \dashrightarrow R[X]$. Since R is explicit, so is R^*, and hence $R[X]$. All the polynomial ring operations are also recursive relative to this notation for $R[X]$. It is now easy to see that the argument extends to any explicit set W of indeterminates.

(v) Assume θ is the root of an irreducible polynomial $p(X) \in F[X]$ of degree n. The elements of $F(\theta)$ can be directly represented by elements of F^n (n-fold Cartesian product): thus, if ν is an explicit notation for F, then ν^n is an explicit notation of $F^n = F(\theta)$. The ring operations of $F(\theta)$ are reduced to polynomial operations modulo $p(X)$, and division is reduced to computing inverses using the Euclidean algorithm. It is easy to check that these operations are ν^n-explicit.

$$\text{Q.E.D.}$$

This lemma is essentially a restatement of corresponding results in [18]. In particular (iii-a, iii-b) may be found in [18, Theorem 3.3]. It is interesting to point out[12] the following counterpart of (iii-b):

Proposition 8 (Volker Bosserhoff). *There is an ν-explicit domain D such that the divides predicate $a|b$ is not ν-explicit.*

Proof. Define the domains

$$D' := \mathbb{Z}[Y, X_i : i \in \mathbb{N}], \qquad D := \mathbb{Z}[Y, X_j, YX_i : j \in H, i \in \mathbb{N}]$$

where $H \subseteq \mathbb{N}$ is the halting set. So D is a subring of D'. Clearly, D' has a explicit notation ν' in which the function $i \mapsto X_i$ is ν'-explicit. There is a standard recursive function $f : \mathbb{N} \to H$ which is a bijection. Using f, we can construct a $1-1$ recursive function $F : \Sigma^* \to \Sigma^*$ whose range is the set of all ν'-names of elements of D. Thus $\nu := \nu'|_F$ is an explicit notation for D. If $a \in D$, write $\underline{a} \in \Sigma^*$ for the ν-name of a. The ring operations of D are ν-explicit: say we want to ν-compute $a + b$ for $a, b \in D$. Given $\underline{a}, \underline{b}$, we first compute $F(\underline{a}), F(\underline{b})$. These are just the ν'-names of a, b. Since D' is ν'-explicit, we can perform the ring operation $a + b$ in D' using their ν'-names $F(\underline{a})$ and $F\underline{b})$. The result is the

---

[12] See also [39, Example 4.3.9].

ν'-name z of $a + b$. By exhaustive search, we find the w such that $F(w) = z$. We output w since $w = \underline{a + b}$. This establishes D' as a ν-explicit domain.

Now suppose the predicate $a|b$ is ν-explicit. We derive a contradiction by showing how to decide the halting problem: given $i \in \mathbb{N}$, we want to decide if $i \in H$. From i, we can compute the ν'-name of X_i, as noted above. Using a fixed ν'-name of Y, we can compute a ν'-name of $X_i Y$. If the ν'-name is u_i, we can use exhaustive search to find the ν-name of $X_i Y$, i.e., the w_i such that $F(w_i) = u_i$. As usual, we write $\underline{X_i Y}$ for w_i. Since the divides predicate in D is ν-explicit, we can decide whether $Y|X_i Y$ in D by applying a ν-algorithm for the divides predicate to $\underline{X_i Y}$ and a ν-name $\underline{Y}$ of Y. But $Y|X_i Y$ iff $X_i \in D$ iff $i \in H$.

Q.E.D.

It follows from Lemma 7 that standard algebraic structures such as the structure $\mathbb{Q}[X_1, \ldots, X_n]$ or algebraic number fields are explicit. Clearly, many more standard constructions can be shown explicit (e.g., matrix rings). The next lemma uses constructions whose explicitness are less obvious: let $\mathcal{ID}_n(F)$ denote the set of all ideals of $F[X_1, \ldots, X_n]$ where F is a field. For ideals $I, J \in \mathcal{ID}_n(F)$, we have the ideal operations of sum $I + J$, product IJ, intersection $I \cap J$, quotient $I : J$, and radical $\sqrt{I}$ [41, p. 25]. These operations are all effective, for instance, using Gröbner basis algorithms [41, chap. 12].

Lemma 9. *Let F be an explicit field. Then the set $\mathcal{ID}_n(F)$ of ideals has an explicit notation ν, and the ideal operations of sum, product, intersection, quotient and radical are ν-explicit.*

Proof. From Lemma 7(iv), $F[X_1, \ldots, X_n]$ is explicit. From Lemma 4(ii), the set S of finite subsets of $F[X_1, \ldots, X_n]$ is explicit. Consider the map $\rho : S \to \mathcal{ID}_n(F)$ where $\rho(\{g_1, \ldots, g_m\})$ is the ideal generated by $g_1, \ldots, g_m \in F[X_1, \ldots, X_n]$. By Hilbert's basis theorem [41, p. 302], ρ is an onto function, and hence a representation. If $\nu : \Sigma^* \dashrightarrow S$ is an explicit notation for S, then $\rho \circ \nu$ is a notation for $\mathcal{ID}_n(F)$. To show that this notation is explicit, it is enough to show that the equivalence relation E_ρ is decidable (cf. (1)). This amounts to checking if two finite sets $\{f_1, \ldots, f_\ell\}$ and $\{g_1, \ldots, g_m\}$ of polynomials generate the same ideal. This can be done by computing their Gröbner bases (since such operations are all rational and thus effective in an explicit field), and seeing each reduces the other set of polynomials to 0. Let S/E_ρ denote the equivalence classes of S; by identifying S/E_ρ with the set $\mathcal{ID}_n(F)$, we obtain an explicit notation for $\mathcal{ID}_n(F)$, $\nu/E_\rho : \Sigma^* \dashrightarrow S/E_\rho$. The (ν/E_ρ)-explicitness of the various ideal operations now follows from known algorithms, using the notation (ν/E_ρ).

Q.E.D.

Well-ordered sets. Many algebraic constructions (e.g., [38, chap. 10]) are transfinite constructions, e.g., the algebraic closure of fields. The usual approach for showing closure properties of such constructions depends on the well-ordering of sets (Zermelo's theorem), which in turn depends on the Axiom of Choice (e.g., [19] or [38, chap. 9]). Recall that a strict total ordering $<$ of a set S is a well-ordering if every non-empty subset of S has a least element. In explicit set theory, we can replace such axioms by theorems, and replace non-constructive constructions by explicit ones.

Lemma 10. *A ν-explicit set is well-ordered. This well-ordering is ν-explicit.*

Proof. Let $\nu : \Sigma^* \dashrightarrow S$ be an explicit notation for S. Now Σ^* is well-ordered by any lexicographical order $\leq_{\mathrm{LEX}}$ on strings. This induces a well-ordering $\leq_\nu$ on the elements $x, y \in S$ as follows: let $w_x := \min\{w \in \Sigma^* : \nu(w) = x\}$. Define $x \leq_\nu y$ if $w_x \leq_{\mathrm{LEX}} w_y$. The predicate $\leq_\nu$ is clearly ν-explicit. Moreover it is a well-ordering.   **Q.E.D.**

The proof of Theorem 13 below depends on such a well-ordering.

Expressions. Expressions are basically "universal objects" in the representation of algebraic constructions.

Let $\widehat{\Omega}$ be a (possibly infinite) set of symbols for algebraic operations, and $k : \widehat{\Omega} \to \mathbb{N}$ assigns an "arity" to each symbol in $\widehat{\Omega}$. The pair $(\widehat{\Omega}, k)$ is also called a signature. Suppose Ω is a set of operations defined on a set S. To prove the closure of S under the operations in Ω, we consider "expressions" over $\widehat{\Omega}$, where each $\widehat{g} \in \widehat{\Omega}$ is interpreted by a corresponding $g \in \Omega$ and $k(\widehat{g})$ is the arity of g. To construct the closure of S under Ω, we will use "expressions over $\widehat{\Omega}$" as the representing set for this closure.

Let $\widehat{\Omega}^{(k)}$ denote the subset of $\widehat{\Omega}$ comprising symbols with arity k. Recall the definition of the set $\mathcal{DG}(\widehat{\Omega})$ of ordered $\widehat{\Omega}$-digraphs. An **expression over** $\widehat{\Omega}$ is a digraph $G \in \mathcal{DG}(\widehat{\Omega})$ with the property that (i) the underlying graph is acyclic and has a unique source node (the root), and (ii) the outdegree of a node v is equal to the arity of its label $\lambda(v) \in \widehat{\Omega}$. Let $Expr(\widehat{\Omega}, k)$ (or simply, $Expr(\widehat{\Omega})$) denote the set of expressions over $\widehat{\Omega}$.

Lemma 11. *Suppose $\widehat{\Omega}$ is a ν-explicit set and the function $k : \widehat{\Omega} \to \mathbb{N}$ is[13] ν-explicit. Then the set $Expr(\widehat{\Omega})$ of expressions is an explicit subset of $\mathcal{DG}(\widehat{\Omega})$.*

Proof. The set $\mathcal{DG}(\widehat{\Omega})$ is explicit by Lemma 4(iii). Given a digraph $G = (V, E; \lambda) \in \mathcal{DG}(\widehat{\Omega})$, it is easy to algorithmically check properties (i) and (ii) above in our definition of expressions.   **Q.E.D.**

Universal Real Construction. A fundamental result of field theory is Steinitz's theorem on the existence and uniqueness of algebraic closures of a field F [38, chap. 10]. In standard proofs, we only need the well-ordering principle. To obtain the "explicit version" of Steinitz's theorem, it is clear that we also need F to be explicit. But van der Waerden pointed out that this may be insufficient: in general, we need another explicitness assumption, namely the ability to factor over $F[X]$ (see [18]). Factorization in an explicit UFD (unique factorization domain) such as $F[X]$ is equivalent to checking irreducibility [18, Theorem 4.2].

If F is a formally real field, then a **real algebraic closure** $\overline{F}$ of F is an algebraic extension of F that is formally real, and such that no proper algebraic

---

[13] Strictly speaking, k is (ν, ν')-explicit where ν' is the notation for $\mathbb{N}$.

extension is formally real. Again $\overline{F}$ exists [41, chap. 5], and is unique up to isomorphism. Our goal here is to give the explicit analogue of Steinitz's theorem for real algebraic closure.

If $p, q \in F[X]$ are polynomials, then we consider the operations of computing their remainder $p \bmod q$, their quotient $p \operatorname{\mathbf{quo}} q$, their gcd $\operatorname{GCD}(p, q)$, their resultant $\operatorname{\mathbf{resultant}}(p, q)$, the derivative $\frac{dp}{dX}$ of p, the square-free part $\operatorname{\mathbf{sqfree}}(p)$ of p, and the Sturm sequence $\operatorname{Sturm}(p)$ of p. Thus $\operatorname{Sturm}(p)$ is the sequence $(p_0, p_1, \ldots, p_k)$ where $p_0 = p$, $p_1 = \frac{dp}{dX}$, and $p_{i+1} = p_{i-1} \bmod p_i$ $(i = 1, \ldots, k)$, and $p_{k+1} = 0$. These are all explicit in an explicit field:

Lemma 12. *If F is a ν-explicit field, and $p(X), q(X) \in F[X]$, then the following operations are*[14] *ν-explicit:*

$$p \bmod q, \quad p \operatorname{\mathbf{quo}} q, \quad \operatorname{GCD}(p, q), \quad \frac{dp}{dX}, \quad \operatorname{\mathbf{resultant}}(p, q), \quad \operatorname{\mathbf{sqfree}}(p), \quad \operatorname{Sturm}(p).$$

Proof. Let $prem(p, q)$ and $pquo(p, q)$ denote the pseudo-remainder and pseudo-quotient of $p(X), q(X)$ [41, Lemmas 3.5, 3.8]. Both are polynomials whose coefficients are determinants in the coefficients of $p(X)$ and $q(X)$. Hence $prem(p, q)$ and $pquo(p, q)$ are explicit operations. The leading coefficients of $prem(p, q)$ and $pquo(p, q)$ can be detected in an explicit field. Dividing out by the leading coefficient, we can obtain $p \bmod q$ and $p \operatorname{\mathbf{quo}} q$ from their pseudo-analogues. Similarly, $\operatorname{GCD}(p, q)$ and $\operatorname{\mathbf{resultant}}(p, q)$ can be obtained via subresultant computations [41, p. 90ff]. Clearly, differentiation $\frac{dp}{dX}$ is a ν-explicit operation. We can compute $\operatorname{\mathbf{sqfree}}(p)$ as $p/\operatorname{GCD}(p, dp/dX)$. Finally, we can compute the Sturm sequence $\operatorname{Sturm}(p)$ because we can differentiate and compute $p \bmod q$, and can test when a polynomial is zero.   **Q.E.D.**

Some common predicates are easily derived from these operations, and they are therefore also explicit predicates: (a) $p|q$ (p divides q) iff $p \bmod q = 0$. (b) p is squarefree iff $\operatorname{\mathbf{sqfree}}(p) = p$.

Let F be an ordered field. Given $p \in F[X]$, an interval I is an **isolating interval** of p in one of the following two cases: (i) $I = [a, a]$ and $p(a) = 0$ for some $a \in F$, (ii) $I = (a, b)$ where $a, b \in F$, $a < b$, $p(a)p(b) < 0$, and the Sturm sequence of p evaluated at a has one more sign variation than the Sturm sequence of p evaluated at b. It is clear that an isolating interval uniquely identifies a root α in the real algebraic closure of F. Such an α is called a **real root** of p. In case p is square-free, we call the pair (p, I) an **isolating interval representation** for α. We may now define the operation $\operatorname{Root}_k(a_0, \ldots, a_n)$ $(k \geq 1, a_i \in F)$ that extracts the kth largest real root of the polynomial $p(X) = \sum_{i=0}^{n} a_i X^i$. This operation is undefined in case $p(X)$ has less than k real roots. For the purposes of this paper, we shall define the **real algebraic closure** of F, denoted $\overline{F}$, to be the smallest ordered field that is an algebraic extension of F and that is closed under the operation $\operatorname{Root}_k(a_0, \ldots, a_n)$ for all $a_0, \ldots, a_n \in \overline{F}$ and $k \in \mathbb{N}$. For other characterizations of real algebraic closures, see e.g., [41, Theorem 5.11].

---

[14] Technically, these operations are ν'-explicit where $F[X]$ is an ν'-explicit set, and ν' is derived from ν using the above standard operators.

Theorem 13. *Let F be an explicit ordered field. Then the real algebraic closure $\overline{F}$ of F is explicit. This field is unique up to F-isomorphism (isomorphism that leaves F fixed).*

Unlike Steinitz's theorem [38, chap. 10], this result does not need the Axiom of Choice; and unlike the explicit version of Steinitz's theorem [18], it does do not need factorization in $F[X]$. But the ordering in F must be explicit.

Proof. For simplicity in this proof, we will assume the existence of $\overline{F}$ (see [41]). So our goal is to show its explicitness, i.e., we must show an explicit notation ν for $\overline{F}$, and show that the field operations as well as $\text{Root}_k(a_0, \ldots, a_n)$ are ν-explicit. Consider the set

$$\widehat{\Omega} := F \cup \{+, -, \times, \div\} \cup \{\text{Root}_k : n \in \mathbb{N}, 1 \leq k \leq n\}$$

of operation symbols. The arity of these operations are defined as follows: the arity of $x \in F$ is 0, arity of $g \in \{+, -, \times, \div\}$ is 2, and arity of Root_k is $n + 1$. It is easy to see that $\widehat{\Omega}$ is explicit, and hence $Expr(\widehat{\Omega})$ is explicit. Define a natural evaluation function,

$$\text{Eval} : Expr(\widehat{\Omega}) \dashrightarrow \overline{F}. \tag{4}$$

Let e be an expression. We assign $\text{val}(u) \in \overline{F}$ to each node u of the underlying DAG of e, in bottom-up fashion. Then $\text{Eval}(e)$ is just the value of the root. If any node has an undefined value, then $\text{Eval}(e) = \uparrow$. The leaves are assigned constants $\text{val}(u) \in F$. At an internal node u labeled by a field operation $(+, -, \times, \div)$, we obtain $\text{val}(u)$ as the result of the corresponding field operation on elements of $\overline{F}$. Note that a division by 0 results in $\text{val}(u) = \uparrow$. Similarly, if the label of u is Root_k and its children are $u_0, \ldots, u_n$ (in this order), then $\text{val}(u)$ is equal to the kth largest real root (if defined) of the polynomial $\sum_{j=0}^n \text{val}(u_j) X^j$. We notice that the evaluation function (4) is onto, and hence is a representation of the real algebraic closure $\overline{F}$. Since $Expr(\widehat{\Omega})$ is an explicit set, Eval is a notation of $\overline{F}$. It remains to show that Eval is an explicit notation.

To conclude that $\overline{F}$ is an explicit set via the notation (4), we must be able to decide if two expressions e, e' represent the same value. By forming the expression $e - e'$, this is reduced to deciding if the value of a proper expression e is 0. To do this, assume that for each expression e, we can either determine that it is improper or else we can compute an isolating interval representation $(P_e(X), I_e)$ for its value $\text{val}(e)$. To determine if $\text{val}(e) = 0$, we first compute the isolating interval representation (P_e, I_e). If $P_e(X) = \sum_{i=0}^n a_i X^i$, and if $\text{val}(e) \neq 0$, then Cauchy's lower bound [41, Lem. 6.7] holds: $|\text{val}(e)| > |a_0|/(|a_0| + \max\{|a_i| : i = 1, \ldots, n\})$. Let $w(I) = b - a$ denote the width of an interval $I = (a, b)$ or $I = [a, b]$. Therefore, if

$$w(I_e) < B_0 := \frac{|a_0|}{|a_0| + \max\{|a_i| : i = 1, \ldots, n\}}, \tag{5}$$

we can decide whether $\text{val}(e) = 0$ as follows: $\text{val}(e) = 0$ iff $0 \in I_e$. To ensure (5), we first compute B_0 and then we use binary search to narrow the width of

I_e: each binary step on the current interval $I' = (a', b')$ amounts to computing $m = (a' + b')/2$ and testing if $P_e(m) = 0$ (which is effective in F). If so, $\mathtt{val}(e) = m$ and we can tell if $m = 0$. Otherwise, we can replace I' by (a', m) or (m, b'). Specifically, we choose (a', m) if $P_e(a')P(m) < 0$ and otherwise choose (m, b'). Thus the width of the interval is halved. We repeat this until the width is less than B_0.

Given $P(X) \in F[X]$, we need to compute a complete set of isolating intervals for all the real roots of $P(X)$ using Sturm sequences. This is well known if $F \subseteq \mathbb{R}$, but it is not hard to see that everything extends to explicit ordered fields. Briefly, three ingredients are needed: (i) The Sturm sequence $\mathrm{Sturm}_P(X)$ of $P(X)$ is $\mathrm{Sturm}_P(X) := (P_0, P_1, \ldots, P_h)$ where $P_0 = P(X)$, $P_1 = dP(X)/dX$, and $P_{i+1} = P_{i-1} \bmod P_i$ $(i = 1, \ldots, h)$, and $P_{h+1} = 0$. To compute $\mathrm{Sturm}_P(X)$, we need to compute polynomial remainders for polynomials in $F[X]$, and be able to detect zero coefficients (so that we know the leading coefficients). (ii) We need an upper bound B_1 on the magnitude of all roots. The Cauchy bound may be used: choose $B_1 = 1 + (\max_{i=0}^{n-1} |a_i|)/|a_n|$ where $a_0, \ldots, a_n$ are coefficients of the polynomial [41, p. 148]. (iii) Let $V_P(a)$ denote the number of sign variations of the Sturm sequence of P evaluated at a. The usual theory assumes that $P(a)P(b) \neq 0$. In this case, the number of distinct real roots of $P(X)$ in an interval (a, b) is given by $V_P(a) - V_P(b)$. But in case $P(a) = 0$ and/or $P(b) = 0$, we need to compute $V_P(a^+)$ and/or $V_P(b^-)$. As noted in [16], if P is square-free, then for all $a, b \in F$, we have $V_P(a^+) = V_P(a)$ and $V_P(b^-) = \delta(P(b)) + V_P(b)$ where $\delta(x) = 1$ if $x = 0$ and $\delta(x) = 0$ otherwise. This computation uses only ring operations and sign determination.

We now return to our main problem, which is to compute, for given expression e, an isolating interval representation of $\mathtt{val}(e)$ or determine that e is improper. The algorithm imitates the preceding bottom-up assignment of values $\mathtt{val}(u)$ to each node u, except that we now compute an isolating interval representation of $\mathtt{val}(u)$ at each u. Let this isolating interval representation be $(P_u(X), I_u)$. If u is a leaf, $P_u = X - \mathtt{val}(u)$, and $I_u = [\mathtt{val}(u), \mathtt{val}(u)]$. Inductively, we have two possibilities:

(I) Suppose $\lambda(u)$ is a field operation $\diamond \in \{+, -, \times, \div\}$ and the children of u are v, w. Recursively, assume $\mathtt{val}(v), \mathtt{val}(w)$ are defined and we have computed the isolating interval representations (P_v, I_v) and (P_w, I_w). In case $\diamond$ is division, we next check if $\mathtt{val}(w) = 0$. If so, $\mathtt{val}(u)$ is undefined. If not, then $P_u(X)$ may be given by the square-free part of the following resultants (respectively):

$$v \pm w : \ \mathbf{res}_Y(P_v(Y), P_w(X \mp Y))$$
$$v \times w : \ \mathbf{res}_Y(P_v(Y), Y^{\deg(P_w)} P_w(X/Y))$$
$$v/w : \ \mathbf{res}_Y(P_v(Y), X^{\deg(P_w)} P_w(Y/X)).$$

See [41, p. 158]; the last case can be deduced from the reciprocal formula there. If the operation is division, we first use the binary search procedure above to narrow the width of I_w until either $0 \notin I_w$ or we detect that $\mathtt{val}(w) = 0$ i.e., $w(I_w) < B_0$ and $0 \in I_w$. In the latter case, v/w is not defined. Next, tentatively set $I_u = I_v \diamond I_w$ (using interval arithmetic) and test if it is an isolating interval of $P_u(X)$. This is done by using Sturm sequences: by evaluating the Sturm sequence at the endpoints of I_u, we can tell if I_u is isolating. If not, we will half the intervals I_v and I_w using bisection search as above, and repeat the test. It is clear that this process terminates.

(II) Suppose $\lambda(u)$ is $\mathtt{Root}_i(u_0, \ldots, u_n)$. To compute $P_u(X) \in F[X]$, we recall that by the theory of elementary symmetric functions, there is a polynomial $P(X) \in F[X]$ of degree at most $D = \prod_{j=0}^n \deg(P_{u_j})$ such that $Q(X)|P(X)$ and $Q(X) = \sum_{j=0}^n \mathtt{val}(u_j)X^j$ [41, Proof of Theorem 6.24]. We can construct an expression over the ring $\overline{F}[X]$ to represent $Q(X)$ (since the coefficient of $Q(X)$ are elements of $\overline{F}$). Moreover, $F[X]$ is an explicit polynomial ring, so we can systematically search for a $P(X) \in F[X] \subseteq \overline{F}[X]$ that is divisible by $Q(X)$. We must show that the predicate $Q(X)|P(X)$ is effective. This is equivalent to checking if $P(X) \bmod Q(X) = 0$. As noted in the proof of Lemma 12, $R(X) := P(X) \bmod Q(X)$ is a polynomial whose coefficients are determinants in the coefficients of $P(X), Q(X)$. Using the resultant techniques in (I), we can therefore compute isolating interval representations of the coefficients of $R(X)$. Thus $R(X) = 0$ iff all the coefficients of $R(X)$ vanishes, a test that is effective. (Note: this test is a kind of "bootstrap" since we are using isolating interval representations to determine the isolating interval representation of u.) Finally, from $P(X)$, we compute its square-free part P_u.

Next, we must compute an isolating interval I_u of P_u for the ith largest root of the polynomial $Q(X)$. Although we can isolate the real roots of P_u, we must somehow identify the root that corresponds to the ith largest root of $Q(X)$. The problem is that $Q(X) \in \overline{F}[X] \setminus F[X]$, and so we cannot directly use the Sturm method of (I) above. Nevertheless, it is possible to use the Sturm method in a bootstrap manner, by exploiting the isolating interval representation (similar to checking if $Q(X)|P(X)$ above). Indeed, all the operations in Lemma 12 can be implemented in this bootstrap manner. Once we have isolated these roots of $Q(X)$, we can continue to narrow the intervals as much as we like. In particular, we must narrow the ith largest root of $Q(X)$ until it is also an isolating interval for $P_u(X)$. This completes our proof.   **Q.E.D.**

REMARK: An alternative method to isolate all the real roots of $Q(X) \in \overline{F}[X]$ may be based on a recent result of Eigenwillig et al [1]. They showed an algorithm to isolate the real roots of square-free polynomials with real coefficients where the coefficients are given as *potentially infinite bit streams*. To apply this algorithm we must first make $Q(X)$ square-free, but this can be achieved using the bootstrap method.

4   Real Approximation

We now address the problem of real computation. Our approach is a further elaboration of [42].

All realizable computations over abstract mathematical domains S are ultimately reduced to manipulation of their representations. If we accept Turings' general analysis of computation, then these representations of S are necessarily notations. But it is impossible to provide notations if S is an uncountable set such as $\mathbb{R}$. There are two fundamental positions to take with respect to this dilemma.

(A) One position insists that a theory of real computation must be able to compute with all of $\mathbb{R}$ from the start. Both the (Polish) analytic school and the algebraic school adopt this line. The analytic school proceeds to generalize the concept of computation to handle representations (of real number, real functions, real operators, etc). So we must generalize Turing machines to handle such non-notation representations: two examples of such generalizations are oracle machines [23] and TTE machines [40]. The algebraic school chooses an abstract computational model that directly manipulates the real numbers, in effect ignoring all representation issues. The decision to embrace all of $\mathbb{R}$ from the start exacts a heavy toll. We believe that the standard criticisms of both schools stem from this fundamental decision: the theory is either unrealizable (algebraic school) or too weak (analytic school, which treats only continuous functions). The resulting complexity theory is also highly distorted unless it is restricted in some strong way. In the algebraic school, one approach is to focus on the "Boolean part" (e.g., [3]). Another is to analyze complexity as a function of condition number [6]. In the analytic school, one focuses[15] mainly on "precision complexity", which is essentially local complexity or complexity at a point.

(B) The alternative position is to accept that there will be real numbers that are simply "inaccessible" for computation. This phenomenon seems inevitable. Once we accept this principle, there is no reason to abandon or generalize Turing's fundamental analysis – for the "accessible reals", we can stick to computations over notations (using standard Turing machines). The Russian branch of the analytic school [40, Chap. 9] takes this approach, by identifying the accessible reals with the computable reals (see below; also Spreen [37]). A real number is now represented by its Gödel numbers (names of programs for computing its Cauchy sequences). Alas, this view also swallows too much in one gulp, again leading to a distorted complexity theory. Our approach [42] takes position (B), but adopts a more constructive view about which real numbers ought to be admitted from the start. Nevertheless, the set of reals that can be studied by our approach is not fixed in advance (see below).

---

[15] That is, the only useful parameter in complexity functions is the precision parameter p. E.g., for $f : \mathbb{R} \to \mathbb{R}$, the main complexity function we can associate with f is $T(p)$, giving the worst number of steps of an oracle Turing machine for approximating p-bits of $f(x)$ for all $x \in \mathbb{R}$. There is no natural way to use the real value x (or $|x|$) as a complexity in computing $f(x)$. Thus Ko [23, p. 57] defines the complexity function $T(x, p)$ as the time to compute $f(x)$ to p-bits of absolute accuracy. But the x parameter is instantly factored out by considering uniform time complexity, and never used in actual complexity results. The real parameter x fails to behave properly as a complexity parameter: the function $T(x, p)$ is not monotonic in x (for $x > 0$). Even if x is rational, the monotonicity property fails. To be concrete, suppose $T(x, p)$ be the time to compute a p-bit approximation to $\sqrt{x}$. The inequality $T(x, p) \leq T(2, p)$ fails as badly as we like, by choosing $x = (n + 1)/n$ as $n \to \infty$.

Base Reals. We begin with a set of "base reals" that is suitable for approximating other real numbers. Using the theory of explicit algebraic structures, we can now give a succinct definition (cf. [42]): a subset $\mathbb{F} \subseteq \mathbb{R}$ is called a **ring of base reals** if $\mathbb{F}$ is an explicit ordered ring extension of the integers $\mathbb{Z}$, such that $\mathbb{F}$ is dense in $\mathbb{R}$. Elements of $\mathbb{F}$ are called **base reals**.

The rational numbers $\mathbb{Q}$, or the dyadic (or bigfloat) numbers $\mathbb{D} = \mathbb{Z}[\frac{1}{2}]$, or even the real algebraic numbers, can serve as the ring of base reals. Since $\mathbb{F}$ is an explicit ring, we can perform all the ring operations and decide if two base reals are equal. Being dense, we can use $\mathbb{F}$ to approximate any real number to any desired precision. We insist that all inputs and outputs of our algorithms are base reals. This approach reflects very well the actual world of computing: in computing systems, floating point numbers are often called "reals". Basic foundation for this form of real computation goes back to Brent [8,9]. It is also clear that all practical development of real computation (e.g., [28,30,25]), as in our work in EGC, also ultimately depend on approximations via base reals. The choice of $\mathbb{D}$ as the base reals is the simplest: assuming that $\mathbb{F}$ is closed under the map $x \mapsto x/2$, then $\mathbb{D} \subseteq \mathbb{F}$. *In the following, we shall assume this property.* Then we can do standard binary searches (divide by 2), work with dyadic notations, and all the results in [42] extends to our new setting.

Error Notation. We consider both absolute and relative errors: given $x, \widetilde{x}, p \in \mathbb{R}$, we say that $\widetilde{x}$ is an **absolute p-bit approximation** of x if $|\widetilde{x} - x| \le 2^{-p}$. We say $\widetilde{x}$ is a **relative p-bit approximation** of x if $|\widetilde{x} - x| \le 2^{-p}|x|$.

The inequality $|\widetilde{x} - x| \le 2^{-p}$ is equivalent to $\widetilde{x} = x + \theta 2^{-p}$ where $|\theta| \le 1$. To avoid introducing an explicit variable θ, we will write this in the suggestive form "$\widetilde{x} = x \pm 2^{-p}$". More generally, whenever we use the symbol '$\pm$' in a numerical expression, the symbol $\pm$ in the expression should be replaced by the sequence "$+\theta$" where θ is a real variable satisfying $|\theta| \le 1$. Like the big-Oh notations, we think of the $\pm$-convention as a variable hiding device. As further example, the expression "$x(1 \pm 2^{-p})$" denotes a relative p-bit approximation of x. Also, write $(x \pm \varepsilon)$ and $[x \pm \varepsilon]$ (resp.) for the intervals $(x - \varepsilon, x + \varepsilon)$ and $[x - \varepsilon, x + \varepsilon]$.

Absolute and Relative Approximation. The ring $\mathbb{F}$ of base reals is used for approximation purposes. So all approximation concepts will depend on this choice. If $f : S \subseteq \mathbb{R}^n \dashrightarrow \mathbb{R}$ is[16] a real function, we call a function

$$\widetilde{f} : (S \cap \mathbb{F}^n) \times \mathbb{F} \dashrightarrow \mathbb{F} \tag{6}$$

an **absolute approximation** of f if for all $\mathbf{d} \in S \cap \mathbb{F}^n$ and $p \in \mathbb{F}$, we have $\widetilde{f}(\mathbf{d}, p) =\downarrow$ iff $f(\mathbf{d}) =\downarrow$. Furthermore, when $f(\mathbf{d}) =\downarrow$ then $\widetilde{f}(\mathbf{d}, p) = f(\mathbf{d}) \pm 2^{-p}$. We can similarly define what it means for $\widetilde{f}$ to be a **relative approximation** of f. Let

$$\mathcal{A}_f, \qquad \mathcal{R}_f$$

---

[16] This is just a short hand for "$f : S \dashrightarrow \mathbb{R}$ and $S \subseteq \mathbb{R}^n$". Similarly, $f : S \subseteq \mathbb{R}^n \dashrightarrow T \subseteq \mathbb{R}$ is shorthand for $f : S \dashrightarrow T$ with the indicated containments for S and T.

denote the set of all absolute, respectively relative, approximations of f. If $\tilde{f} \in \mathcal{A}_f \cup \mathcal{R}_f$, we also write "$\tilde{f}(\mathbf{d})[p]$" instead of $\tilde{f}(\mathbf{d}, p)$ to distinguish the **precision parameter** p. We remark that this parameter p could also be restricted to $\mathbb{N}$ for our purposes; we often use this below.

We say f is **absolutely approximable** (or $\mathcal{A}$-approximable) if some $\tilde{f} \in \mathcal{A}_f$ is explicit. Likewise, f is **partially absolutely approximable** (or partially $\mathcal{A}$-approximable) if some $\tilde{f} \in \mathcal{A}_f$ is partially explicit. Analogous definitions hold for f being **relatively approximable** (or $\mathcal{R}$-approximable) and **partially relatively approximable** (or partially $\mathcal{R}$-approximable). The concept of approximability (in the four variants here) is the starting point of our approach to real computation. Notice that "real approximation" amounts to "explicit computation on the base reals".

Remark on nominal domains of partial functions. It may appear redundant to consider a function f that is a partial function *and* whose nominal domain S is a proper subset of $\mathbb{R}^n$. In other words, by specifying $S = \mathbb{R}^n$ or $S = \text{domain}(f)$, we can either avoid partial functions, or avoid $S \neq \mathbb{R}^n$. This attitude is implicit in recursive function theory, for instance. It is clear that the choice of S affects the computability of f since S determines the input to be fed to our computing devices. In the next section, the generic function $f(x) = \sqrt{x}$ is used to illustrate this fact. Intuitively, the definability of f at any point x is intrinsic to the function f, but its points of undefinability is only incidental to f (an artifact of the choice of S). Unfortunately, this intuition can be wrong: the points of undefinability of f can tell us much about the global nature of f. To see this, consider the fact that the choice of S is less flexible in algebra than in analysis. In algebra, we are not free to turn the division operation in a field into a total function, by defining it only over non-zero elements. In analysis, it is common to choose S so that f behaves nicely: e.g., f has no singularity, f is convergent under Newton iteration, etc. But even here, this choice is often not the best and may hide some essential difficulties. So in general, we do not have the option of specifying $S = \mathbb{R}^n$ or $S = \text{domain}(f)$ for a given problem.

Much of what we say in this and the next section are echos of themes found in [23,40]. Our two main goals are (i) to develop the computability of f in the setting of a general nominal domain S, and (ii) to expose the connection between computability of f with its approximability. A practical theory of real computability in our view should be largely about approximability.

Regular Functions. Let $f : S \subseteq \mathbb{R} \dashrightarrow \mathbb{R}$. In [23] and [40], the real functions are usually restricted to $S = [a, b], (a, b)$ or $S = \mathbb{R}$; this choice is often essential to the computability of f. To admit S which goes beyond the standard choices, we run into pathological examples such as $S = \mathbb{R} \setminus \mathbb{F}$. This example suggests that we need an ample supply of base reals in S. We say that a set $S \subseteq \mathbb{R}$ is **regular** if for all $x \in S$ and $n \in \mathbb{N}$, there exists $y \in S \cap \mathbb{F}$ such that $y = x \pm 2^{-n}$. Thus, S contains base reals arbitrarily close to any member. We say f is **regular** if $\text{domain}(f)$ is regular. Note that regularity, like all our approximability concepts, is defined relative to $\mathbb{F}$.

Cauchy Functions. The case $n = 0$ in (6) is rather special: in this case, f is regarded as a constant function, representing some real number $x \in \mathbb{R}$. An absolute approximation of x is any function $\widetilde{f} : \mathbb{F} \to \mathbb{F}$ where $\widetilde{f}(p) = x \pm 2^{-p}$ for all $p \in \mathbb{F}$. We call $\widetilde{f}$ a **Cauchy function** for x. The sequence $(\widetilde{f}(0), \widetilde{f}(1), \widetilde{f}(2), \ldots)$ is sometimes called a **rapidly converging Cauchy sequence** for x; relative to $\widetilde{f}$, the p-th **Cauchy convergent** of x is $\widetilde{f}(p)$.

Extending the above notation, we may write $\mathcal{A}_x$ for the set of all Cauchy functions for x. But note that $\widetilde{f}$ is not just an approximation of x, but it uniquely identifies x. Thus $\widetilde{f}$ is a representation of x. So by our underbar convention, we prefer to write "$\underline{x}$" for any Cauchy function of x. Also write "$\underline{x}[p]$" (instead of $\underline{x}(p)$) for the pth convergent of $\underline{x}$.

We can also let $\mathcal{R}_x$ denote the set of relative approximations of x. If some $\underline{x} \in \mathcal{A}_x$ ($\underline{x} \in \mathcal{R}_x$) is explicit, we say x is $\mathcal{A}$-**approximable** ($\mathcal{R}$-**approximable**). Below we show that x is $\mathcal{R}$-approximable iff x is $\mathcal{A}$-approximable. Hence we may simply speak of "approximable reals" without specifying whether we are concerned with absolute or relative errors.

Among the Cauchy functions in $\mathcal{A}_x$, we identify one with nice monotonicity properties: every real number x can be written as

$$n + 0.b_1 b_2 \cdots$$

where $n \in \mathbb{Z}$ and $b_i \in \{0, 1\}$. The b_i's are uniquely determined by x when $x \notin \mathbb{D}$. Otherwise, all b_i's are eventually 0 or eventually 1. For uniqueness, we require all b_i's to be eventually 0. Using this unique sequence, we define the **standard Cauchy function** of x via

$$\beta_x[p] = n + \sum_{i=1}^{p} b_i 2^{-i}.$$

For instance, $-5/3$ is written $-2 + 0.01010101\cdots$. This defines the Cauchy function $\beta_x[p]$ for all $p \in \mathbb{N}$. Technically, we need to define $\beta_x[p]$ for all $p \in \mathbb{F}$: when $p < 0$, we simply let $\beta_x[p] = \beta_x[0]$; when $p > 0$ and is not an integer, we let $\beta_x[p] = \beta_x[\lceil p \rceil]$. We note some useful facts about this standard function:

Lemma 14. *Let $x \in \mathbb{R}$ and $p \in \mathbb{N}$.*
(i) $\beta_x[p] \leq \beta_x[p+1] \leq x$.
(ii) $x - \beta_x[p] < 2^{-p}$.
(iii) If $y = \beta_x[p] \pm 2^{-p}$, then for all $n \leq p$, we also have $y = \beta_x[n] \pm 2^{-n}$. In particular, there exists $\underline{y} \in \mathcal{A}_y$ such $\underline{y}[n] = \beta_x[n]$ for all $n \leq p$.
(iv) There is a recursive procedure $B : \mathbb{F} \times \mathbb{N} \to \mathbb{F}$ such that for all $x \in \mathbb{F}, p \in \mathbb{N}$, $B(x, p) = \beta_x[p]$. In particular, for each $x \in \mathbb{F}$, the standard Cauchy function of x is recursive.

To see (iii), it is sufficient to verify that if $y = \beta_x[p] \pm 2^{-p}$ then $y = \beta_x[p-1] \pm 2^{1-p}$. Now $\beta_x[p] = \beta_x[p-1] + \delta 2^{-p}$ where $\delta = 0$ or 1. Hence $y = \beta_x[p] \pm 2^{-p} = (\beta_x[p-1] \pm 2^{-p}) \pm 2^{-p} = \beta_x[p-1] \pm 2^{1-p}$.

Explicit computation with one real transcendental. In general, it is not known how to carry out explicit computations (e.g., decide zero) in transcendental extensions of $\mathbb{Q}$ (but see [12] for a recent positive result). However, consider the field $F(\alpha)$ where α is transcendental over F. If F is ordered, then the field $F(\alpha)$ can also be ordered using an ordering where $a <' \alpha$ for all $a \in F$. Further, if F is explicit, then $F(\alpha)$ is also an explicit ordered field with this ordering $<'$. But the ordering $<'$ is clearly non-Archimedean (i.e., there are elements $a, x \in F(\alpha)$ such that for all $n \in \mathbb{N}$, $n|a| < |x|$). Now suppose $F \subseteq \mathbb{R}$ and $\alpha \in \mathbb{R}$ (for instance, $F = \mathbb{Q}$ and $\alpha = \pi$). Then $F(\alpha) \subseteq \mathbb{R}$ can be given the standard (Archimedean) ordering $<$ of the reals.

Theorem 15. *If $F \subseteq \mathbb{R}$ is an explicit ordered field, and $\alpha \in \mathbb{R}$ is an approximable real that is transcendental over F, then the field $F(\alpha)$ with the Archimedean order $<$ is an explicit ordered field.*

Proof. The field $F(\alpha)$ is isomorphic to the quotient field of $F[X]$, and this field is explicit by Lemma 7(iii,iv). It remains to show that the Archimedean order $<$ is explicit. Let $P(\alpha)/Q(\alpha) \in F(\alpha)$ where $P(X), Q(X) \in F[X]$ and $Q(X) \neq 0$. It is enough to show that we can recognize the set of positive elements of $F(\alpha)$. Now $P(\alpha)/Q(\alpha) > 0$ iff $P(\alpha)Q(\alpha) > 0$. So it is enough to recognize whether $P(\alpha) > 0$ for any $P(\alpha) \in F[\alpha]$. First, we can verify that $P(\alpha) \neq 0$ (this is true iff some coefficient of $P(\alpha)$ is nonzero). Next, since α is approximable, we find increasingly better approximations $\underline{\alpha}[p] \in \mathbb{F}$ of α, and evaluate $P(\underline{\alpha}[p])$ for $p = 0, 1, 2, \ldots$. To estimate the error, we derive from Taylor's expansion the bound $P(\alpha) = P(\underline{\alpha}[p]) \pm \delta_p$ where $\delta_p = \sum_{i \geq 1} 2^{-ip}|P^{(i)}(\underline{\alpha}[p])|$. We can easily compute an upper bound $\beta_p \geq |\delta_p|$, and stop when $|P(\underline{\alpha}[p])| > \beta_p$. Since $\delta_p \to 0$ as $p \to \infty$, we can also ensure that $\beta_p \to 0$. Hence termination is assured. Upon termination, we know that $P(\alpha)$ has the sign of $P(\underline{\alpha}[p])$.   **Q.E.D.**

In particular, this implies that $\mathbb{D}(\pi)$ or $\mathbb{Q}(e)$ can serve as the set $\mathbb{F}$ of base reals. The choice $\mathbb{D}(\pi)$ may be appropriate in computations involving trigonometric functions, as it allows exact representation of the zeros of such functions, and thus the possibility to investigate the neighborhoods of such zeros computationally. Moreover, we can extend the above technique to any number of transcendentals, provided they are algebraically independent. For instance, π and $\Gamma(1/3) = 2.678938\ldots$ are algebraically independent and so $\mathbb{D}(\pi, \Gamma(1/3))$ would be an explicit ordered field.

Real predicates. Given $f : S \subseteq \mathbb{R} \dashrightarrow \mathbb{R}$, define the predicate $\text{Sign}_f : S \dashrightarrow \{-1, 0, 1\}$ given by

$$\text{Sign}_f(x) = \begin{cases} 0 & \text{if } f(x) = 0, \\ +1 & \text{if } f(x) > 0, \\ -1 & \text{if } f(x) < 0, \\ \uparrow & \text{else.} \end{cases}$$

Define the related predicate $\text{Zero}_f : S \dashrightarrow \{0, 1\}$ where $\text{Zero}_f(x) \equiv |\text{Sign}_f(x)|$ (so $\text{range}(\text{Zero}_f) \subseteq \{0, 1\}$). By the fundamental analysis of EGC (see Introduction), Sign_f is the critical predicate for geometric algorithms. We usually prefer to focus on the simpler $Zero_f$ predicate because the approximability of these two predicates are easily seen to be equivalent in our setting of base reals (cf. [42]).

In general, a **real predicate** is a function $P : S \subseteq \mathbb{R}^n \dashrightarrow \mathbb{R}$ where $\mathbf{range}(P)$ is a finite set. The approximation of real predicates is somewhat simpler than that of general real functions.

To treat the next result, we need some new definitions. Let $S \subseteq D \subseteq \Sigma^*$. We say S is **recursive modulo** D if there is a Turing machine that, on input x taken from the set D, halts in the state $q_\downarrow$ if $x \in S$, and in the state $q_\uparrow$ if $x \notin S$. Similarly, S is **partial recursive modulo** D if there is a Turing machine that, on input x taken from D, halts iff $x \in S$. Let $S \subseteq D \subseteq U$ where U is a ν-explicit set and $\nu : \Sigma^* \dashrightarrow U$. We say S is a (partially) ν-**explicit subset of** U **modulo** D if the set $\{w \in \Sigma^* : \nu(w) \in S\}$ is (partial) recursive modulo $\{w \in \Sigma^* : \nu(w) \in D\}$. Also, denote by $\mathbf{range}_\mathbb{F}(f) := \{f(x) : x \in \mathbb{F} \cap S\}$, the range of f when its domain is restricted to base real inputs.

Lemma 16. *For a real predicate* $P : S \subseteq \mathbb{R} \dashrightarrow \mathbb{R}$, *the following are equivalent:*
(i) P is partially $\mathcal{R}$-approximable
(ii) P is partially $\mathcal{A}$-approximable
(iii) Each $a \in \mathbf{range}_\mathbb{F}(P)$ is a computable real number and the set $P^{-1}(a) \cap \mathbb{F}$ is partially explicit modulo S.

Proof. (i) implies (ii): Let $\widehat{P} \in \mathcal{R}_P$ be a partially explicit function. Our goal is to compute some $\widetilde{P} \in \mathcal{A}_P$. Let the input for $\widetilde{P}$ be $(x, p) \in S \times \mathbb{N}$. First, compute $y = \widehat{P}(x)[1]$. If $y =\uparrow$, then we do not halt. So let $y =\downarrow$. We know that $P(x) = 0$ iff $\widehat{P}(x)[1] = 0$ ([42]). Hence we can define $\widetilde{P}(x)[p] = 0$ when $y = 0$. Now assume $y \neq 0$. Then $|y| \geq |P(x)|/2$, and we may compute and output $z := \widehat{P}(x)[p + 1 + \lceil \log_2 |y| \rceil]$. This output is correct since $z = P(x)(1 \pm 2^{-p-1-\lceil \log_2 |y| \rceil}) = P(x) \pm 2^{-p}$.

(ii) implies (iii): Let $\widetilde{P} \in \mathcal{A}_P$ be a partially explicit function. Fix any $a \in \mathbf{range}_\mathbb{F}(P)$. To see that a is a computable real, for all p, we can compute $\underline{a}[p]$ as $\widetilde{P}(x)[p]$ (for some fixed $x \in P^{-1}(a) \cap \mathbb{F}$). To show that $P^{-1}(a) \cap \mathbb{F}$ is a partially explicit subset of $\mathbb{F}$ modulo S, note that there is a p' such that for all $b, b' \in \mathbf{range}_\mathbb{F}(P)$, $b \neq b'$ implies $|b - b'| \geq 2^{-p'}$. We then choose an $\widetilde{a} \in \mathbb{F}$ such that $\widetilde{a}$ is a $(p' + 3)$-bit absolute approximation of a. Then we verify that $P^{-1}(a) \cap \mathbb{F}$ is equal to

$$\{x \in S : |\widetilde{P}(x)[p' + 3] - \widetilde{a}| \leq 2^{-p'-1}\}.$$

Thus, given $x \in S$, we can partially decide if $x \in P^{-1}(a)$ by first computing $\widetilde{P}(x)[p' + 3]$. If this computation halts, then $x \in P^{-1}(a)$ iff $|\widetilde{a} - \widetilde{P}(x)[p' + 3]| \leq 2^{-p'-1}$. Thus $P^{-1}(a) \cap \mathbb{F}$ is a partially explicit subset of $\mathbb{F}$ modulo S.

(iii) implies (i): Given $x \in S$ and $p \in \mathbb{N}$, we want to compute some $z = P(x)(1 \pm 2^{-p})$. We first determine the $a \in \mathbf{range}_\mathbb{F}(P)$ such that $P(x) = a$. We can effectively find a by enumerating the elements of $P^{-1}(b) \cap \mathbb{F} \cap S$ for each $b \in \mathbf{range}_\mathbb{F}(P)$ until x appears (this process does not halt iff $P(x) =\uparrow$). Assume a is found. If $a = 0$, then we simply output 0. Otherwise, we compute $\underline{a}[i]$ for

$i = 0, 1, 2, \ldots$ until $|\underline{a}[i]| > 2^{-i}$. Let i_0 be the index when this happens. Then we have $|a| > b := |\underline{a}[i_0]| - 2^{-i_0}$. Set $q := p - \lfloor \log_2 b \rfloor$ and output $z := \underline{a}[q]$. As for correctness, note that $z = a \pm 2^{-q} = a \pm 2^{-p}b = a(1 \pm 2^{-p})$.   **Q.E.D.**

There is an analogous result where we remove the "partially" qualifications in the statement of this lemma. However in (iii), we need to add the requirement that the set $S \setminus \text{domain}(f)$ must be explicit relative to S. In view of this lemma, we can simply say that a real predicate is "(partially) approximable" instead of (partially) $\mathcal{A}$-approximable or $\mathcal{R}$-approximable. This lemma could be extended to "generalized predicates" $P : S \subseteq \mathbb{R}^n \dashrightarrow \mathbb{R}$ whose range is **discrete** in the sense that for some $\varepsilon > 0$, for all $x, y \in \mathbb{R}^n$, $P(x) = P(y) \pm \varepsilon$ implies $P(x) = P(y)$.

On relative versus absolute approximability. In [42], we proved that a partial function $f : \mathbb{R} \dashrightarrow \mathbb{R}$ is $\mathcal{R}$-approximable iff it is $\mathcal{A}$-approximable and Zero_f is explicit. The proof extends to:

Theorem 17. *Let $f : S \subseteq \mathbb{R} \dashrightarrow \mathbb{R}$. Then f is (partially) $\mathcal{R}$-approximable iff f is (partially) $\mathcal{A}$-approximable, and Zero_f is (partially) approximable.*

Relative approximation is dominant in numerical analysis: machine floating systems are all based on relative precision (for example, the IEEE Standard). See Demmel et al [15,14] for recent work in this connection. Yet the analytic school exclusively discusses absolute approximations. This theorem shows why: relative approximation requires solving the zero problem, which is undecidable in the analytic approach.

5   Computable Real Functions

We now study computable real functions following the analytic school. Our main goal is to show the exact relationship between the approximation approach and the analytic school.

Let $f : S \subseteq \mathbb{R} \dashrightarrow \mathbb{R}$ be a partial real function. Following Ko [23], we will use **oracle Turing machines** (OTM) as devices for computing f. A real input x is represented by any Cauchy function $\underline{x} \in \mathcal{A}_x$. An OTM M has, in addition to the usual tape(s) of a standard Turing machine, two special tapes, called the **oracle tape** and the **precision tape**. It also has two special states,

$$q_?, q_! \tag{7}$$

called the **query state** and the **answer state**. We view M as computing a function (still denoted) $M : \mathbb{R} \times \mathbb{F} \to \mathbb{R}$. A real input x is represented by an arbitrary $\underline{x} \in \mathcal{A}_x$, which serves as an oracle. The input $p \in \mathbb{F}$ is placed on the precision tape. Whenever the computation of M enters the query state $q_?$, we require the oracle tape to contain a binary number $\underline{k}$. In the next instant, M will enter the answer state $q_!$, and simultaneously the string $\underline{k}$ is replaced by a representation of the kth convergent $\underline{x}[k]$. Then M continues computing, using this oracle answer. Eventually, there are two possible outcomes: either M loops and we write $M(\underline{x}, p) = \uparrow$, or it halts and its output tape holds a representation $\underline{d}$ for some $d \in \mathbb{F}$, and this defines the output, $M(\underline{x}, p) = d$. Equivalently, we write[17] "$M^{\underline{x}}[p] = d$". We say M **computes** a function $f : S \subseteq \mathbb{R} \dashrightarrow \mathbb{R}$, if for all $x \in S$ and $p \in \mathbb{F}$, we have that for all $\underline{x} \in A_x$,

$$M^{\underline{x}}[p] = \begin{cases} \uparrow & \text{if } f(x) = \uparrow \\ f(x) \pm 2^{-p} & \text{if } f(x) = \downarrow . \end{cases}$$

This definition extends to computation of multivariate partial functions of the form $f : S \subseteq \mathbb{R}^n \dashrightarrow \mathbb{R}$. Then the OTM M takes as input n oracles (corresponding to the n real arguments). Each query on the oracle tape (when we enter state $q_?$) consists of a pair $(\underline{i}, \underline{k})$ indicating that we want the kth convergent of the ith oracle. Since such an extension will be trivial to do in most of our results, we will generally stay with the univariate case.

Our main definition is this: a real function $f : S \subseteq \mathbb{R}^n \dashrightarrow \mathbb{R}$ is **computable** if there is a OTM that computes f. When f is total on S, we also say that it is **total computable**. In case $n = 0$, f denotes a real number x; the corresponding OTM is actually an ordinary Turing machine since we never use the oracle tape. In the analytic approach, x is known as a **computable real**. But it is easy to see that it is equivalent to x being an approximable real.

Example: consider $f(x) = \sqrt{x}$ where $f(x) = \downarrow$ iff $x \geq 0$. If $S = \mathbb{R}$, then it is easy to show that f is not computable by oracle Turing machines (the singularity at $x = 0$ cannot be decided in finite time). If we choose $S = [0, \infty)$, then f can be shown to be computable by oracle Turing machines even though the singularity at $x = 0$ remains. But we might prefer $S = [1, 4)$ in implementations since $\sqrt{x}$ can be reduced to $\sqrt{y}$ where $y \in [1, 4)$, by first computing $y = x \cdot 4^n$, $n \in \mathbb{Z}$.

Notice that our definition of "computable functions" includes[18] partial functions. In this case, improper inputs for OTM's are handled in the conventional way of computability theory, with the machine looping. This follows Ko [23, p. 62]. We do not require OTM's to be halting because it would result in there being only trivial examples of computable real functions that are not total.

Lemma 18. *If* $f : S \subseteq \mathbb{R}^n \dashrightarrow \mathbb{R}$ *is computable then* f *is partially $\mathcal{A}$-approximable.*

Proof. There is nothing to show if $n = 0$. To simplify notations, assume $n = 1$ (the case $n > 1$ is similar). Let M be an OTM that computes f. We define a function $\widetilde{f} \in \mathcal{A}_f$ where $\widetilde{f} : \mathbb{F}^2 \dashrightarrow \mathbb{F}$. To compute $\widetilde{f}$, we use an ordinary Turing machine N that, on input $d, p \in \mathbb{F}$, simulates M on the oracle β_d (the standard Cauchy function of d) and p. Note that by Lemma 14(iv), we can compute $\beta_d[k]$ for any $k \in \mathbb{N}$. If $M^{\beta_d}[p]$ outputs z, N will output z. Clearly, the choice $\widetilde{f}(d)[p] = z$ is correct. If $d \notin \text{domain}(f)$, the computation loops.   **Q.E.D.**

---

[17] This follows the convention of putting the oracle argument in the superscript position, and our convention of putting the precision parameter in square brackets.

[18] So alternatively, we could say that "f is *partial* computable" instead of "f is computable".

Recursively open sets. Let $\phi : \mathbb{N} \to \mathbb{F}$. This function defines a sequence $(I_0, I_1, I_2, \ldots)$ of open intervals where $I_n = (\phi(2n), \phi(2n+1))$. We say ϕ is an **interval representation** of the open set $S = \bigcup_{n \geq 0} I_n$. A set $S \subseteq \mathbb{R}$ is **recursively open** if S has an interval representation ϕ that is explicit. We say S is **recursively closed** if its complement $\mathbb{R} \setminus S$ is recursively open.

Note that if $\phi(2n) \geq \phi(2n+1)$ then the open interval $(\phi(2n), \phi(2n+1))$ is empty. In particular, the empty set $S = \emptyset$ is recursively open. So is $S = \mathbb{R}$.

The following result is from Ko [23, Theorem 2.31].

Proposition 19. *A set $T \subseteq \mathbb{R}$ is recursively open iff there is a computable function $f : \mathbb{R} \dashrightarrow \mathbb{R}$ with $T = \mathrm{domain}(f)$.*

Modulus of continuity. Consider a partial function $f : S \subseteq \mathbb{R} \dashrightarrow \mathbb{R}$. We say f is **continuous** if for all $x \in \mathrm{domain}(f)$ and $\delta > 0$, there is an $\epsilon = \epsilon(x, \delta) > 0$ such that if $y \in \mathrm{domain}(f)$ and $y = x \pm \epsilon$ then $f(y) = f(x) \pm \delta$. We say f is **uniformly continuous** if for all $\delta > 0$, there is an $\epsilon = \epsilon(\delta) > 0$ such that for all $x, y \in \mathrm{domain}(f)$, $y = x \pm \epsilon$ implies $f(y) = f(x) \pm \delta$.

Let $m : S \times \mathbb{N} \dashrightarrow \mathbb{N}$. We call m a **modulus function** if for all $x \in S, p \in \mathbb{N}$, we have $m(x, p) = \uparrow$ iff $m(x, 0) = \uparrow$. We define $\mathrm{domain}(m) := \{x \in S : m(x, 0) = \downarrow\}$. Such a function m is called a **modulus of continuity** (or simply, modulus function) for f if $\mathrm{domain}(f) = \mathrm{domain}(m)$ and for all $x, y \in \mathrm{domain}(f), p \in \mathbb{N}$, if $y = x \pm 2^{-m(x,p)}$ then $f(y) = f(x) \pm 2^{-p}$.

Call a function $m : \mathbb{N} \to \mathbb{N}$ a **uniform modulus of continuity** (or simply, a uniform modulus function) **for** f if for all $x, y \in \mathrm{domain}(f), p \in \mathbb{N}$, if $y = x \pm 2^{-m(p)}$ then $f(y) = f(x) \pm 2^{-p}$. To emphasize the distinction between uniform modulus function and the non-uniform version, we might describe the latter as "local modulus functions". The following is immediate:

Lemma 20. *Let $f : S \subseteq \mathbb{R} \dashrightarrow \mathbb{R}$. Then f is continuous iff it has a modulus of continuity. Then f is uniformly continuous iff it has a uniform modulus of continuity.*

Ko [23] uses uniform[19] continuity to characterize computable total real functions of the form $f : [a, b] \to \mathbb{R}$. Our goal is to generalize his characterization to capture real functions with non-compact domains, as well as those that are partial. Our results will characterize computable real functions $f : S \dashrightarrow \mathbb{R}$ where $S \subseteq \mathbb{R}^n$ is regular. Use of local continuity simplifies proofs, and avoids any appeal to the Heine-Borel theorem.

Multivalued Modulus function of an OTM. We now show how modulus functions can be computed, but they must be generalized to multivalued functions. Such functions are treated in Weihrauch [40]; in particular, computable modulus functions are multivalued [40, Cor. 6.2.8]. Let us begin with the usual (single-valued) modulus function $m : S \times \mathbb{N} \dashrightarrow \mathbb{N}$. It is computed by OTMs since one

---

[19] Indeed, uniform modulus functions are simply called "modulus functions" in Ko. He has a notion of "generalized modulus function" that is similar to our local modulus functions.

of m's arguments is a real number. If $(\underline{x}, p)$ is the input to an OTM N which computes m, we still write "$N^{\underline{x}}[p]$" for the computation of N on $(\underline{x}, p)$. Notice that, since the output of N comes from the discrete set $\mathbb{N}$, we do not need an extra "precision parameter" to specify the precision of the output (unlike the general situation when the output is a real number).

Let M be an OTM that computes some function $f : S \subseteq \mathbb{R} \dashrightarrow \mathbb{R}$. Consider the Cauchy function $\beta_x^+ \in \mathcal{A}_x$ where $\beta_x^+[n] = \beta_x[n+1]$ for all $n \in \mathbb{N}$. (Thus β_x^+ is a "sped up" form of the standard Cauchy function β_x.) Consider the function $k_M : S \times \mathbb{N} \dashrightarrow \mathbb{N}$ where

$$k(x, p) = k_M(x, p) := \text{the largest } k \text{ such that the} \tag{8}$$
$$\text{computation of } M^{\beta_x^+}[p] \text{ queries } \beta_x^+[k].$$

This definition depends on M using β_x^+ as oracle. If $M^{\beta_x^+}[p] =\uparrow$ then $k(x, p) =\uparrow$; if $M^{\beta_x^+}[p] =\downarrow$ but the oracle was never queried, define $k(x, p) = 0$. Some variant of $k(x, p)$ was used by Ko to serve as a modulus function for f. Unfortunately, we do not know how to compute $k(x, p)$ as that seems to require simulating the oracle β_x^+ using an arbitrary oracle $\underline{x} \in \mathcal{A}_x$. Instead, we proceed as follows. For any $\underline{x} \in \mathcal{A}_x$, let $\underline{x}^+$ denote the oracle in $\mathcal{A}_x$ where $\underline{x}^+[n] = \underline{x}[n+1]$ for all n. Define, in analogy to (8), the function $\underline{k}_M : \mathcal{A}_S \times \mathbb{N} \dashrightarrow \mathbb{N}$ where $\mathcal{A}_S = \cup \{\mathcal{A}_x : x \in S\}$ and

$$\underline{k}(\underline{x}, p) = \underline{k}_M(\underline{x}, p) := \text{the largest } k \text{ such that the} \tag{9}$$
$$\text{computation of } M^{\underline{x}^+}[p] \text{ queries } \underline{x}^+[k].$$

As before, if $M^{\underline{x}^+}[p] =\uparrow$ (resp., if the oracle was never queried), then $\underline{k}(\underline{x}, p) =\uparrow$ (resp., $\underline{k}(\underline{x}, p) = 0$). The first argument of $\underline{k}$ is an oracle, not a real number. For different oracles from the set $\mathcal{A}_x$, we might get different results; such functions $\underline{k}$ are[20] called **intensional functions**. Computability of intensional functions is defined using OTM's, just as for real functions. We can naturally interpret $\underline{k}_M$ as representing a **multivalued function** (see [40, Section 1.4]) which may be denoted[21] $k_M : S \times \mathbb{N} \to 2^{\mathbb{N}}$ with $k_M(x, p) = \{\underline{k}_M(\underline{x}, p) : \underline{x} \in \mathcal{A}_x\}$. Statements about k_M can be suitably interpreted as statements about $\underline{k}_M$ (see below).

The following lemma is key:

Lemma 21. *Let* $f : S \subseteq \mathbb{R} \dashrightarrow \mathbb{R}$ *be computed by an OTM* M*, and* $p \in \mathbb{N}$*.*
(i) If $x \in \text{domain}(f)$ *and* $\underline{x} \in \mathcal{A}_x$*, then* $(x \pm 2^{-\underline{k}_M(\underline{x}, p) - 1}) \cap S \subseteq \text{domain}(f)$*.*
(ii) If, in addition, $y = x \pm 2^{-\underline{k}_M(\underline{x}, p) - 1}$ *and* $y \in S$ *then* $f(y) = f(x) \pm 2^{1-p}$*.*

Proof. Let $y = x \pm 2^{-\underline{k}_M(\underline{x}, p) - 1}$ where $x \in \text{domain}(f)$ and $y \in S$.
(i) We must show that $y \in \text{domain}(f)$. For this, it suffices to show that M halts

---

[20] Intensionality is viewed as the (possible) lack of "extensionality". We say $\underline{k}$ is **extensional** if for all $a, b \in \mathcal{A}_x$ and $p \in \mathbb{N}$, $\underline{k}(a, 0) =\downarrow$ iff $\underline{k}(a, p) =\downarrow$; moreover, $\underline{k}(a, p) \equiv \underline{k}(b, p)$. Extensional functions can be interpreted as (single-valued) partial functions on real arguments.

[21] Or, $k_M : S \times \mathbb{N} \overset{\to}{\to} \mathbb{N}$.

on the input $(\underline{y}, p)$ for some $\underline{y} \in \mathcal{A}_y$. Consider the modified Cauchy function $\underline{y}'$ given by

$$\underline{y}'[n] = \begin{cases} \underline{x}^+[n] & \text{if } n \leq \underline{k}_M(\underline{x}, p) \\ \underline{y}[n] & \text{else.} \end{cases}$$

To see that $\underline{y}' \in \mathcal{A}_y$, we only need to verify that $y = \underline{y}'[n] \pm 2^{-n}$ for $n \leq \underline{k}_M(\underline{x}, p)$. This follows from

$$y = x \pm 2^{-\underline{k}_M(\underline{x}, p) - 1} = (\underline{x}[n+1] \pm 2^{-n-1}) \pm 2^{-\underline{k}_M(\underline{x}, p) - 1} = \underline{x}^+[n] \pm 2^{-n} = \underline{y}'[n] \pm 2^{-n}.$$

Since the computation of $M^{\underline{x}^+}[p]$ does not query $\underline{x}^+[n]$ for $n > \underline{k}_M(\underline{x}, p)$, it follows that this computation is indistinguishable from the computation of $M^{\underline{y}'}[p]$. In particular, both $M^{\underline{x}^+}[p]$ and $M^{\underline{y}'}[p]$ halt. Thus $y \in \text{domain}(f)$.

(ii) We further show

$$|f(y) - f(x)| \leq |f(y) - M^{\underline{y}'}[p]| + |M^{\underline{x}^+}[p] - f(x)|$$
$$\leq 2^{-p} + 2^{-p} = 2^{1-p}.$$

Q.E.D.

Let M be an OTM. Define the intensional function $\underline{m} : \mathcal{A}_S \times \mathbb{N} \dashrightarrow \mathbb{N}$ by

$$\underline{m}(\underline{x}, p) := \underline{k}_M(\underline{x}, p + 1) + 1. \tag{10}$$

We call $\underline{m}$ an (intensional) **modulus of continuity** for $f : S \subseteq \mathbb{R} \dashrightarrow \mathbb{R}$ if[22] for all $x \in S$ and $p \in \mathbb{N}$, we have $\underline{m}(\underline{x}, p) = \downarrow$ iff $x \in \text{domain}(f)$. In addition, for all $x, y \in \text{domain}(f)$, if $y = x \pm 2^{-\underline{m}(\underline{x}, p)}$ then $f(y) = f(x) \pm 2^{-p}$. The multivalued function $m : S \times \mathbb{N} \to 2^{\mathbb{N}}$ corresponding to $\underline{m}$ will be called a **multivalued modulus of continuity** of f. In this case, define $\text{domain}(m) = \text{domain}(\underline{m})$ to be $\text{domain}(f)$. It is easy to see that f has a multivalued modulus of continuity iff it has a (single-valued) modulus of continuity. The next result may be compared to [40, Corollary 6.2.8].

Lemma 22. *If $f : S \subseteq \mathbb{R} \dashrightarrow \mathbb{R}$ is computed by an OTM M then the function $\underline{m}(\underline{x}, p)$ of (10) is a modulus of continuity for f. Moreover, $\underline{m}$ is computable.*

Proof. To show that $\underline{m}$ is a modulus of continuity for f, we may assume $x \in S$. Consider two cases: if $x \notin \text{domain}(f)$ then $\underline{k}(\underline{x}, p + 1)$ is undefined. Hence $\underline{m}(\underline{x}, p)$ is undefined as expected. So assume $x \in \text{domain}(f)$. Suppose $y = x \pm 2^{-\underline{m}(\underline{x}, p)} = x \pm 2^{-\underline{k}_M(\underline{x}, p+1) - 1}$ and $y \in S$. By the previous lemma, we know that $y \in \text{domain}(f)$ and $f(y) = f(x) \pm 2^{-p}$. Thus $\underline{m}(\underline{x}, p)$ is a modulus of continuity for f.

---

[22] In discussing intensional functions, it is convenient to assume that whenever we introduce a quantified real variable x, we simultaneously introduce a corresponding universally-quantified Cauchy function variable $\underline{x} \in \mathcal{A}_x$. These two variables are connected by our under-bar convention. That is, "$(Qx \in S \subseteq \mathbb{R})$" should be translated "$(Qx \in S \subseteq \mathbb{R})(\forall \underline{x} \in \mathcal{A}_x)$" where $Q \in \{\forall, \exists\}$.

To show that $\underline{m}$ is computable, we construct an OTM N which, on input $\underline{x} \in \mathcal{A}_x$ and $p \in \mathbb{F}$, simulates the computation of $M^{\underline{x}^+}[p+1]$. Whenever the machine M queries $\underline{x}[n]$ for some n, the machine N queries $\underline{x}^+[n] = \underline{x}[n+1]$ instead. When the simulation halts, N outputs the largest $k+1$ such that $\underline{x}^+[k]$ was queried (or $k = 0$ if there were no oracle queries).   **Q.E.D.**

The above proof shows that $m(x, p) := k_M(x, p+1)+1$ is a modulus of continuity in the following "strong" sense: a multivalued modulus function m for $f : S \dashrightarrow \mathbb{R}$ is said to be **strong** if $x \in \mathrm{domain}(f)$ implies $[x \pm 2^{-\underline{m}(\underline{x}, p)}] \cap S \subseteq \mathrm{domain}(f)$. Note that if m is strong then S is regular implies $\mathrm{domain}(m)$ is regular. Thus:

Corollary 23. *If $f : S \subseteq \mathbb{R} \dashrightarrow \mathbb{R}$ is computable, then it has a strong multivalued modulus function that is computable. In particular, f is continuous.*

Modulus cover. We introduce an alternative formulation of strong modulus of continuity: let

$$\Box\mathbb{F} := \{(a, b) : a < b, a, b \in \mathbb{F}\}$$

denote the set of open intervals over $\mathbb{F}$. A **modulus cover** refers to any subset $G \subseteq \Box\mathbb{F} \times \mathbb{N}$. For simplicity, the typical element in G is written (a, b, p) instead of the more correct $((a, b), p)$. We call G a **modulus cover of continuity** (or simply, a modulus cover) for $f : S \subseteq \mathbb{R} \dashrightarrow \mathbb{R}$ if the following two conditions hold:

(a) For each $p \in \mathbb{N}$ and $x \in \mathrm{domain}(f)$, there exists $(a, b, p) \in G$ with $x \in (a, b)$.
(b) For all $(a, b, p) \in G$, we have $(a, b) \cap S \subseteq \mathrm{domain}(f)$. Moreover, $x, y \in (a, b) \cap S$ implies $f(x) = f(y) \pm 2^{-p}$.

If the characteristic function $\chi_G : \Box\mathbb{F} \times \mathbb{N} \dashrightarrow \{1\}$ of G is (resp., partially) explicit then we say G is (resp., partially) explicit.

The advantage of using G over a modulus function m is that we avoid multi-valued functions, and the triples of G are parametrized by base reals. Thus we compute the characteristic function of G using ordinary Turing machines while m must be computed by OTMs. We next show that we could interchange the roles of G and m.

Lemma 24. *For $f : S \subseteq \mathbb{R} \dashrightarrow \mathbb{R}$, the following statements are equivalent:*
(i) f has a modulus cover G that is partially explicit.
(ii) f has a strong multivalued modulus function m that is computable.

Proof. (i) implies (ii): If G is available, we can define $\underline{m}(\underline{x}, p)$ via the following dove-tailing process: let the input be a Cauchy function $\underline{x}$. For each $(a, b, p) \in \Box\mathbb{F} \times \mathbb{N}$, we initiate a (dovetailed) computation to do three steps:
(1) Check that $(a, b, p) \in G$.
(2) Find the first $i = 0, 1, \ldots$ such that $[\underline{x}[i] \pm 2^{-i}] \subseteq (a, b)$.
(3) Output $k = -\lfloor \log_2 \min\{\underline{x}[i] - 2^{-i} - a, b - 2^{-i} - \underline{x}[i]\} \rfloor$.
Correctness of this procedure: since G is partially explicit, step (1) will halt if $(a, b, p) \in G$. Step (2) amounts to checking the predicate $a < x < b$. If $x \in \mathrm{domain}(f)$ then steps (1) and (2) will halt for some $(a, b, p) \in G$. The output k in step (3) has the property that if $y = x \pm 2^{-k}$ and $y \in S$ then

$y \in \mathrm{domain}(f)$ and $f(y) = f(x) \pm 2^{-p}$. Thus $\underline{m}$ is a strong modulus of continuity of f, and our procedure shows $\underline{m}$ to be computable.

(ii) implies[23] (i): Suppose f has a modulus function $\underline{m}$ that is computed by the OTM M. A finite sequence $\sigma = (x_0, x_1, \ldots, x_k)$ is called a **Cauchy prefix** if there exists a Cauchy function $\underline{x}$ such that $\underline{x}[i] = x_i$ for $i = 0, \ldots, k$. We say $\underline{x}$ **extends** σ in this case. Call σ a **witness** for a triple $(a, b, p) \in \mathbb{F} \times \mathbb{F} \times \mathbb{N}$ provided the following conditions hold:

(4) $[a, b] \subseteq \bigcap_{i=0}^{k} [x_i \pm 2^{-i}]$.

(5) If $\underline{x}$ extends σ then the computation $M^{\underline{x}}[p]$ halts and does not query the oracle for $\underline{x}[n]$ for any $n > k$.

(6) If $M^{\underline{x}}[p]$ outputs $\ell = \underline{m}(\underline{x}, p)$, then we have $0 < b - a < 2^{-\ell}$.

Let G comprise all (a, b, p) that have a witness. The set G is partially explicit since, on input (a, b, p), we can dovetail through all sequences $\sigma = (x_0, \ldots, x_k)$, checking if σ is a witness for (a, b, p). This amounts to checking conditions (4)-(6). To see that G is a modulus cover for f, we first note that if $(a, b, p) \in G$ then $(a, b) \cap S \subseteq \mathrm{domain}(M) = \mathrm{domain}(f)$. Moreover, for all $p \in \mathbb{N}$ and $x \in \mathrm{domain}(f)$, we claim that there is some $(a, b) \in \Box \mathbb{F}$ where $(a, b, p) \in G$ and $x \in (a, b)$. To see this, consider the computation of $M^{\beta_x}[p]$ where β_x is the standard Cauchy function of x. If the largest query made by this computation to the oracle β_x is k, then consider the sequence $\sigma = (\beta_x[0], \ldots, \beta_x[k])$. Note that x is in the interior of $[\beta_x[k] \pm 2^{-k}] = \bigcap_{i=0}^{k} [\beta_x[i] \pm 2^{-i}]$. If $M^{\beta_x}[p] = \ell$, then we can choose $(a, b) \subseteq (\beta_x[k] \pm 2^{-k})$ such that $b - a \leq 2^{-\ell}$ and $a < x < b$. Also for all $y, y' \in (a, b)$, we have $y' = y \pm 2^{\underline{m}(y, p)}$ and hence $|f(y') - f(y)| \leq 2^{-p}$.   **Q.E.D.**

Corollary 25. *If the function $f : S \dashrightarrow \mathbb{R}$ is computable then it has a partially explicit modulus cover G. Moreover, if S is regular then f is regular.*

Proof. By Corollary 23 and Lemma 24, we see that such a G exists. Let S be regular. To see that f is regular, note that $x \in \mathrm{domain}(f)$ implies that for all $p \in \mathbb{N}$, we have $x \in (a, b)$ for some $(a, b, p) \in G$. So $(x \pm \varepsilon) \subseteq (a, b)$ for sufficiently small $\varepsilon > 0$. Regularity of S implies $(x \pm \varepsilon) \cap S \cap \mathbb{F}$ is non-empty. But $(a, b) \cap S \subseteq \mathrm{domain}(f)$ implies $(x \pm \varepsilon) \cap \mathrm{domain}(f) \cap \mathbb{F} = (x \pm \varepsilon) \cap S \cap \mathbb{F}$. The non-emptiness of $(x \pm \varepsilon) \cap \mathrm{domain}(f) \cap \mathbb{F}$ proves that $\mathrm{domain}(f)$ is regular.
Q.E.D.

Main Result. We now characterize computability of real functions in terms of two explicitness concepts ($\mathcal{A}$-approximability and explicit modulus cover). In one direction, we also need a regularity condition.

Theorem 26 (Characterization of computable functions). *Let $f : S \subseteq \mathbb{R} \dashrightarrow \mathbb{R}$. If f is computable then the following two conditions hold:*
(i) f is partially $\mathcal{A}$-approximable.
(ii) f has a partially explicit modulus cover.
Conversely, if S is regular then (i) and (ii) implies f is computable.

---

[23] This proof is kindly provided by V. Bosserhoff and another referee. My original argument required S to be regular.

Proof. If f is computable, then conditions (i) and (ii) hold by Lemma 18 and Corollary 25 (resp.). Conversely, suppose (i) f is partially $\mathcal{A}$-approximable via a partially explicit $\widetilde{f} \in \mathcal{A}_f$, and (ii) f has a modulus cover $G \subseteq \square\mathbb{F} \times \mathbb{N}$ that is partially explicit. Consider the following OTM M to compute f: given a Cauchy function $\underline{x} \in \mathcal{A}_x$ and precision $p \in \mathbb{N}$:

STEP 1: Perform a dovetailed computation over all $(k, a, b) \in \mathbb{N} \times \square\mathbb{F}$. For each (k, a, b), check if $[\underline{x}[k] \pm 2^{-k}] \subseteq (a, b)$ and $(a, b, p+1) \in G$.

STEP 2: Suppose (k, a, b) passes the test in STEP 1. Perform a dovetailed computation over all $c \in \mathbb{F} \cap (a, b)$: for each c, compute $\widetilde{f}(c, p+1)$. If any such computation halts, output its result.

First we show partial correctness: if M outputs $z = \widetilde{f}(c, p+1)$, then $z = f(c) \pm 2^{-p-1}$. Since $f(c) = f(x) \pm 2^{-p-1}$, we conclude that $z = f(x) \pm 2^{-p}$, as desired. Now we show conditional termination of M: for $x \in S$, we show that $x \in \mathrm{domain}(f)$ iff M halts on $\underline{x}, p$. But $x \in \mathrm{domain}(f)$ iff $x \in (a, b)$ for some $(a, b, p+1) \in G$. Then for k large enough, $[\underline{x}[k] \pm 2^{-k}] \subseteq (a, b)$, and so STEP 1 will halt. Conversely, if $x \notin \mathrm{domain}(f)$ then STEP 1 does not halt. We finally show the halting of STEP 2, assuming $x \in \mathrm{domain}(f)$. The regularity of S and $x \in (a, b)$ implies there exists $c \in (a, b) \cap S \cap \mathbb{F}$. By definition of modulus cover, $c \in (a, b) \cap S \subseteq \mathrm{domain}(f)$. Hence STEP 2 will halt at such a value c.   **Q.E.D.**

This theorem is important because it tells us exactly what we are giving up when we abandon computability for absolute approximability: we give up precisely one thing, continuity. That is exactly the effect we want in EGC, since continuous functions are too restrictive in our applications. We obtain a stronger characterization of f in the important case where $S = [a, b]$ (essentially [23, Corollary 2.14] in Ko). Now we need to invoke the Heine-Borel theorem:

Theorem 27 (Ko). *Let a, b be computable reals. A total function $f : [a, b] \to \mathbb{R}$ is computable iff it is $\mathcal{A}$-approximable and it has an explicit uniform modulus function.*

Proof. By the characterization theorem, computability of f is equivalent to (i) the partial $\mathcal{A}$-approximability of f, and (ii) existence of a partially explicit modulus cover G. Since f is a total function, it is $\mathcal{A}$-approximable. It remains to show that the existence of G is equivalent to f having an explicit uniform modulus function $m : \mathbb{N} \to \mathbb{N}$.

One direction is easy: if $m : \mathbb{N} \to \mathbb{N}$ is an explicit uniform modulus function for f, then we can define G to comprise all (c, d, p) such that $c, d \in \mathbb{F}$, $c < d < c+1$ and $p \in \mathbb{N}$ satisfies $d - c \leq 2^{-m(p)}$. Conversely, suppose G exists. From part(a) in the definition of modulus cover of f, for any $p \in \mathbb{N}$, the set $\{(c, d) : (c, d, p) \in G\}$ of open intervals is a cover for $[a, b]$. By Heine-Borel, there is a finite subcover $C = \{I_j : j = 0, \ldots, k\}$. Wlog, C is a minimal cover. Then the intervals have a uniquely ordering $I_0 < I_1 < \ldots < I_k$ induced by sorting their left (equivalently, right) endpoints. Let $J_i = I_{i-1} \cap I_i$ ($i = 1, \ldots, k$). By minimality of C, we see that the J_i's are pairwise disjoint. Let $w(C) = \min\{w(J_i) : i = 1, \ldots, k\}$ where $w(I) = d - c$ is the width of an interval $I = (c, d)$. It follows that if $x, y \in [a, b]$ and $|x - y| < w(C)$ then $\{x, y\} \subseteq I_j$ for some $j = 0, \ldots, k$. This would imply

$f(x) = f(y) \pm 2^{-p}$. Therefore if we define $m(p) := - \lfloor \log_2 w(C) \rfloor$ then m would be a uniform modulus function for f.

To show that m is explicit, we convert the preceding outline into an effective procedure. To compute $m(p)$, we first search for a cover as follows: we dovetail all the computations to search for triples (c, d, p) in G (for all $c, d \in \mathbb{F}$). We maintain a current minimal set C of intervals, initially C is empty. Inductively, assume $\bigcup C$ is equal to the union of all intervals found so far. For each triple $(c, d, p) \in G$, we discard (c, d) if $(c, d) \subseteq \bigcup C$. Otherwise, we add (c, d) to C, and remove any resulting redundancies in C. Next, we check if $[a, b] \subseteq \bigcup C$, and if so we can compute $m(p) := - \lfloor \log_2 w(C) \rfloor$, as described above. If not, we continue the search for more intervals (c, d). We note that checking if $[a, b] \subseteq \bigcup C$ is possible since a, b are[24] computable. **Q.E.D.**

6   Unified Framework for Algebraic-Numeric Computation

We have seen that our real approximability approach is a modified form of the analytic approach. We now address the algebraic approach. The main proponent of this approach is Blum, Cucker, Shub and Smale [6]. For an articulate statement of their vision, we refer to their manifesto [5], reproduced in [6, Chap. 1]. Many researchers have tried to refine and extend the algebraic model, by considering the Boolean part of its complexity classes, by introducing non-unit cost measures, by injecting an error parameter into complexity functions, etc. These refinements can be viewed as attempting to recover some measure of representational complexity into the algebraic model. We do not propose to refine the algebraic approach. Instead, we believe a different role is reserved for the algebraic approach.

To motivate this role, consider the following highly simplified two-stage scheme of how computational scientists goes about solving a practical numerical problem (e.g., solving a PDE model or a numerical optimization problem).

STEP A: First, we determine the abstract algorithmic **Problem** P to be solved: in the simplest form, this amounts to specifying the input and output in mathematical terms. We then design an **Ideal Algorithm** A. This algorithm assumes certain real operations such as $+, -, \times, \exp()$, etc. Algorithm A may also use standard computational primitives for discrete computation such as found in programming languages, and abstract data types such as queues or heaps. We then show that Algorithm A solves the problem P in this ideal setting.

STEP B: We proceed to implement Algorithm A as a **Numerical Algorithm** B in some actual programming language. Algorithm B must now address concrete representation issues: concrete data structures that implement abstract data types, possible error in the input data, approximation

---

[24] A referee pointed out that it suffices that a be a left-computable, and b a right-computable, real number.

of the operations $+, -, \times, \exp()$, etc. We also define the sense in which an implementation constitutes an acceptable approximation of algorithm A, for example, in the sense of backwards analysis, or in the EGC sense (below). Finally, we prove that Algorithm B satisfies this requirement.

What is the conceptual view of these two steps? Basically, we are asking for computational models for Algorithms A and B. It seems evident that Algorithm A is a program in some algebraic model like the BSS model [6] or the Real RAM. Since Algorithm A uses the primitives $+, -, \times, \exp(), \ldots$, we say that its **computational basis** is the set $\Omega = \{+, -, \times, \exp(), \ldots\}$. What about Algorithm B? If we accept the Church-Turing Thesis, then we could say that Algorithm B belongs to the Turing model. This suggestion is appropriate if we are only interested in computability issues. But for finer complexity distinctions, we want a computational model that better reflects how numerical analysts design algorithms, a "numerical computational model" that is more structured than Turing machines. We outline such a model below.

Suppose we now have two computational models: an algebraic model α for Algorithm A, and a numerical model β for Algorithm B. We ask a critical question: can Algorithm A be implemented by some Program B? In other words, we want "Transfer Theorems" that assure us that every program of α can be successfully implemented as a program in β.

There is much (psychological) validity in this 2-stage scheme: certainly, theoretical computer scientists and computational geometers design algorithms this way, using the Real RAM model. But even numerical analysts proceed in this manner. Indeed, most numerical analysis books take STEP A only, and rarely discuss the issues in taking STEP B. From a purely practical viewpoint, the algebraic model provides a useful level of abstraction that guide the eventual transfer of algorithmic ideas into actual executable programs. We thus see that on the one hand, the algebraic model is widely used, and on the other hand, severe criticisms arise when it is proposed as the computational model of numerical analysis (and indeed of all scientific computation). This tension is resolved through our scheme where the algebraic model takes its proper place.

Pointer Machines. We now face the problem of constructing a computational framework in which the algebraic and numerical worlds can co-exist and complement each other. We wish to ensure from the outset that both discrete combinatorial computation and continuous numerical computation can be naturally expressed in this framework. Following Knuth, we may describe such computation as "semi-numerical". Current theories of computation (algebraic or analytic or standard complexity theories) do not adequately address such problems. For instance, real computation is usually studied as the problem of computing a real function $f : \mathbb{R} \dashrightarrow \mathbb{R}$ even though this is a very special case with no elements of combinatorial computing. In Sections 4 and 5 we followed this tradition. On the other hand, algorithms of computational geometry are invariably semi-numerical [43]. Following [42], we will extend Schönhage's elegant Pointer Machine Model [36] to provide such a framework.

We briefly recall the concept of a pointer machine (or storage modification machine). Let Δ be any set of symbols, which we call **tags**. Pointer machines manipulates graphs whose edges are labeled by tags. More precisely, a **tagged graph** (or Δ-**graph**) is a finite directed graph $G = (V, E)$ with a distinguished node $s \in V$ called the **origin** and a label function that assigns tags to edges such that outgoing edges from any node have distinct tags. We can concisely write $G = (V, s, \tau)$ where $s \in V$ and $\tau : V \times \Delta \dashrightarrow V$ is the **tag function**. Note that τ is a partial function, and it implicitly defines E, the edge set: $(u, v) \in E$ iff $\tau(u, a) = v$ for some $a \in \Delta$. Write $u \xrightarrow{a} v$ if $\tau(u, a) = v$. The edge (u, v) is also called a **pointer**, written $u \to v$, and its tag is a. Each word $w \in \Delta^*$ defines at most one path in G that starts at the origin and follows a sequence of edges whose tags appear in the order specified by w. Let $[w]_G$ (or $[w]$ if G is understood) denote the last node in this path. If no such path exists, then $[w] = \uparrow$. It is also useful to let w^- denote the word where the last tag a in w is removed: so $w^- a = w$. In case $w = \epsilon$ (empty word), then let w^- denote ϵ. Thus, if $w \neq \epsilon$ then the last edge in the path is $[w^-] \to [w]$. This is the edge that will be modified in an assignment statement of the form $w \leftarrow w'$ (see (11) next). A node u is **accessible** if there is a w such that $[w] = u$; otherwise, it is inaccessible. If we prune all inaccessible nodes and edges issuing from them, we get a **reduced tagged graph**. We distinguish tagged graphs only up to **equivalence**, defined as isomorphism on reduced tagged graphs. For any Δ-graph G, let $G|w$ denote the Δ-graph that is identical to G except that the origin of G is replaced by $[w]_G$.

Let $\mathcal{G}_\Delta$ denote the set of all Δ-graphs. Pointer machines manipulate Δ-graphs. Thus, the Δ-graphs play the role of strings in Turing machines, and $\mathcal{G}_\Delta$ is the analogue of Σ^* as the universal representation set. The key operation[25] of pointer machines is the **pointer assignment instruction**: if $w, w' \in \Delta^*$, then the assignment

$$w \leftarrow w' \tag{11}$$

modifies the current Δ-graph G by redirecting or creating a single pointer. This operation is defined iff $[w^-]$ and $[w']$ are both defined. We have two possibilities: if $w = \epsilon$, then this amounts to changing the origin to $[w']$. Else, if $w = w^- a$ then the pointer from $[w^-]$ with tag a will be redirected to point to $[w']$. If there was no previous pointer with tag a, then this operation creates the new pointer $[w^-] \xrightarrow{a} [w']$. If G' denotes the Δ-graph after the assignment, we generally have the effect that $[w]_{G'} = [w']_G$. But this equation fails in general.

E.g., let $w = abaa$ and $w' = ab$ where $[\epsilon]_G, [a]_G, [ab]_G$ are three distinct nodes but $[aba]_G = [\epsilon]_G = [abb]_G$. Then we have $[w]_{G'} \neq [w']_G$.

Assignment, plus three other instructions of the pointer machines, are summarized in rows (i)-(iv) of Table 1. A **pointer machine**, then, is a finite sequence of these four types of pointer instructions, possibly with labels. With suitable conventions for input, output, halting in state $q_\uparrow$ or $q_\downarrow$, etc, which the reader may readily supply (see [42] for details), we now see that each pointer machine computes a partial function $f : \mathcal{G}_\Delta \dashrightarrow \mathcal{G}_\Delta$. It is easy to see that pointer machines can

---

[25] The description here is a generalization of the one in [42], which also made the egregious error of describing the result of $w \leftarrow w'$ by the equation $[w]_{G'} = [w']_G$.

Table 1. Instruction Set of Pointer Models

Type	Name	Instruction	Effect (G is transformed to G')
(i)	Pointer Assignment	$w \leftarrow w'$	$[w']_G$ is new origin of G' if $w = \epsilon$; else $w = w^- a$ and $[w^-]_G \overset{a}{\to} [w']_G$ holds in G'.
(ii)	Node Creation	$w \leftarrow \mathbf{new}$	$[w^-]_G \overset{a}{\to} u$ holds in G' where u is a new node
(iii)	Node Comparison	**if** $w \equiv w'$ **goto** L	Branch to L if $[w]_G = [w']_G$, but $G' = G$
(iv)	Halt and Output	$\mathbf{HALT}(w)$	Output $G' = G|w$
(v)	Value Comparison	**if** $(w \diamond w')$ **goto** L where $\diamond \in \{=, <, \leq\}$	Branch to L if $Val_G(w) \diamond Val_G(w')$ but $G' = G$.
(vi)	Value Assignment	$w := o(w_1, \ldots, w_m)$ where $o \in \Omega$ and $w, w_i \in \Delta^*$	$Val_{G'}(w) = o(Val_G(w_1), \ldots, Val_G(w_n))$

simulate Turing machines. The converse simulation is also possible. The merit of pointer machines lies in their naturalness in modeling combinatorial computing – in particular, it can directly represent graphs, in contrast to Turing machines that must "linearize" graphs into strings.

Semi-numerical Problems and Real Pointer Machines. Real RAM's and BSS-machines have the advantage of being natural for numerical and algebraic computation. We propose to marry these features with the combinatorial elegance of pointer machines.

We extend tagged graphs to support real computation by associating a real **val**$(u) \in \mathbb{R}$ with each node $u \in V$. Thus a **real Δ-graph** is given by $G = (V, s, \tau, \mathbf{val})$. Let $\mathcal{G}_\Delta(\mathbb{R})$ (or simply $\mathcal{G}(\mathbb{R})$) denote the set of such **real Δ-graphs**. For a **real pointer machine** to manipulate such graphs, we augment the instruction set with two instructions, as specified by rows (v)-(vi) in Table 1. Instruction (v) compares the values of two nodes and branches accordingly; instruction (vi) applies an algebraic operation g to the values specified by nodes. The set of algebraic operations g comes from a set Ω of algebraic operations which we call the **computational basis** of the model. The simplest computational basis is $\Omega_0 = \{+, -, \times\} \cup \mathbb{Z}$, resulting in nodes with only integer values. Each real pointer machine computes a partial function

$$f : \mathcal{G}(\mathbb{R}) \dashrightarrow \mathcal{G}(\mathbb{R}) \tag{12}$$

For simplicity, we define[26] a **semi-numerical problem** to also be a partial function of the form (12). The objects of computational geometry can be represented by real tagged graphs (see [43]). Thus, the problems of computational geometry can be regarded as semi-numerical problems. A semi-numerical problem (12) is Ω-**solvable** if there is a halting real pointer machine over the basis Ω

---

[26] That is analogous to defining problems in discrete complexity to be a function $f : \Sigma^* \dashrightarrow \Sigma^*$. So we side-step the issues of representation.

that solves it. Another example of a semi-numerical problem is the **evaluation function** $\mathrm{Eval}_\Omega : Expr(\Omega) \dashrightarrow \mathbb{R}$ (cf. (4)) where the set $Expr(\Omega)$ of expressions is directly represented by tagged graphs.

Real pointer machines constitute our idealized algebraic model for STEP A in our 2-stage scheme. Since real pointer machines are equivalent in power to real RAMs or BSS machines, the true merit of real pointer machines lies in their naturalness for capturing semi-numerical problems. For STEP B, the Turing model is adequate[27] but not natural. For instance, numerical analysts do not think of their algorithms as pushing bits on a tape, but as manipulating higher-level objects such as numbers or matrices with appropriate representations.

To provide a model closer to this view, we introduce **numerical Δ-graphs** which is similar to real Δ-graphs except that the value at each node is a base real from $\mathbb{F}$. The instructions for modifying numerical tagged graphs are specified by rows (i)-(v) in Table 1, plus a modified row (vi). The modification is that each $g \in \Omega$ is replaced by a relative approximation $\widetilde{g}$ which takes an extra precision argument (a value in $\mathbb{F}$). So a **numerical pointer machine** N is defined by a sequence of these instructions; we assume a fixed convention for specifying a precision parameter p for such machines. N computes a partial function $\widetilde{f} : \mathcal{G}(\mathbb{F}) \times \mathbb{F} \dashrightarrow \mathcal{G}(\mathbb{F})$. Let $X = \mathcal{A}$ or $\mathcal{R}$. We say that $\widetilde{f}$ is an X-**approximation** of f if, for all $G \in \mathcal{G}(\mathbb{F})$ and $p \in \mathbb{F}$, the graph $\widetilde{f}(G, p)$ (if defined) is a p-bit X-approximation of $f(G)$ in this sense: their underlying reduced graphs are isomorphic, and each numerical value in $\widetilde{f}(G, p)$ is a p-bit X-approximation of the corresponding real value in $f(G)$. We say f is X-approximable if there is a halting numerical machine that computes an X-approximation of f. *This is the EGC notion of approximation.*

Transfer Theorems. Let $\mathcal{SN}_\Omega$ denote the class of semi-numerical problems that are Ω-solvable by real pointer machines. For instance, $\mathrm{Eval}_\Omega \in \mathcal{SN}_\Omega$. Similarly, $\widetilde{\mathcal{SN}}_\Omega$ is the class of semi-numerical problems that can be $\mathcal{R}$-approximated by numerical pointer machines. (Note that we use the relative approximation here.) What is the relationship between these two classes? We reformulate a basic result from [42, Theorem 23]:

Proposition 28. *Let Ω be any set of real operators. Then $\mathcal{SN}_\Omega \subseteq \widetilde{\mathcal{SN}}_\Omega$ iff $\mathrm{Eval}_\Omega \in \widetilde{\mathcal{SN}}_\Omega$.*

This can be viewed as a completeness result about Eval_Ω, or a transfer theorem that tells when the transition from STEP A to STEP B in our 2-stage scheme has guaranteed success. In numerical computation, we have a "transfer process" that is widely used: suppose M is a real pointer machine that Ω-solves some semi-numerical problem $f : \mathcal{G}(\mathbb{R}) \dashrightarrow \mathcal{G}(\mathbb{R})$. Then we can define a numerical pointer machine $\widetilde{M}$ that computes $\widetilde{f} : \mathcal{G}(\mathbb{F}) \times \mathbb{F} \dashrightarrow \mathcal{G}(\mathbb{F})$, where $\widetilde{M}$ simply replaces each algebraic operation $o(w_1, \ldots, w_m)$ by its approximate counterpart

---

[27] We might also say that recursive functions are an adequate basis for semi-numerical problems. But it is even less natural.

$\widetilde{o}(w_1, \ldots, w_m, p)$ where p specifies the precision argument for $\widetilde{f}(G, p)$. In fact, it is often assumed in numerical analysis that STEP B consists of applying this transformation to the ideal algorithm M from STEP A. We can now formulate a basic question: under what conditions does $\lim_p \widetilde{f}(G, p) = f(G)$ as $p \to \infty$?

The framework in this section makes it clear that our investigation of real approximation is predicated upon two choices: base reals $\mathbb{F}$, and computational basis Ω. Therefore, the set of "inaccessible reals" is not a fixed concept, but relative to these choices. When a real pointer machine uses primitive operations $g, h \in \Omega$, we face the problem of approximating $g(h(x))$ in the numerical pointer machine that simulates it. Thus, it is no longer sufficient to only know how to approximate g at base reals, since $h(x)$ may no longer be a base real even if $x \in \mathbb{F}$. Indeed, function composition becomes our central focus. In the analytic and algebraic approaches, the composition of computable functions is computable. But closure under composition is not longer automatic for approximable functions. This fact might be initially unsettling, but we believe it confirms the centrality of the Eval_Ω problem, which is about closure of composition in Ω.

7   Conclusion: Essential Duality

Our main objective was to construct a suitable foundation for the EGC approach to real computation. Eventually, we modified the analytic approach, and incorporated the algebraic approach into a larger synthesis. In this conclusion, we remark on a recurring theme involving the duality between the algebraic and analytic world views, and between the abstract and the concrete sets.

The first idea in our approach is that we must use explicit computations. This follows Weihrauch's [40] insistence that machines can only manipulate names, which must be interpreted. Our intrinsic approach to explicit sets formally justifies the *direct* discussion of abstract mathematical objects, without the encumbrance of representations. This has the same beneficial effect as our underbar-convention (Section 2). Now, interpreting names is just the flip-side of the coin that says mathematical objects must be represented. In Tarski's theory of truth, we have an analogous situation of syntax and semantics. These live in complementary worlds which must not be conflated if they are to each play their roles successfully. Thus, semantics in "explicit real computation" comes from the world of analysis where we can freely define and prove properties of $\mathbb{R}$ *without asking for their effectivity*. Syntax comes from the world of representation elements and their manipulation under strong constraints. Interpretation, which connects these two worlds, comes from notations.

A similar duality is reflected in our algebraic-numeric framework of Section 6: STEP A occurs in the ideal world of algebraic computation, STEP B takes place in the constructive world of numerical computation. A natural connection between them is the transfer process from ideal programs to implementable programs. The BCSS manifesto [5] argues cogently for having the ideal world. We fully agree, only adding that we must not forget the constructive complement to this ideal world.

The second idea concerns how to build the constructive world for real computation: it is that we must not take the "obvious first step" of incorporating *all* real numbers. Any computational model that incorporates this uncountable set $\mathbb{R}$ must suffer major negative consequences: it may lead to non-realizability (as in the algebraic approach) or a weak theory (as in the analytic approach that handles only continuous functions). The restriction to continuous functions is unacceptable in our applications to computational geometry, where all the interesting geometric phenomena occurs at discontinuities. In any case, the corresponding complexity theory is necessarily distorted. We must not even try to embrace so large a set as the computable reals: the Russian school did and paid the tremendous price of not being able to decide zero. Instead, we propose to only compute "approximations" in which all algorithmic input and output are restricted to well-behaved base reals. A natural and realistic complexity theory can now be developed. This complexity theory promises to be considerably more intricate than anything we have seen in discrete complexity. It is future work.

Consider the following natural reaction to our approximation approach: although we talk about real functions $f : \mathbb{R} \dashrightarrow \mathbb{R}$, our computational model only allows approximations, $\tilde{f} : \mathbb{F} \times \mathbb{F} \dashrightarrow \mathbb{F}$. Why not simply identify "real functions" with the partially explicit functions of the form $\tilde{f}$? This suggestion ("it would be more honest") is wrong for a simple reason: we *really* do wish to study the real functions f. All the properties we hold important are about f, not $\tilde{f}$. Indeed, the analytic properties of $\tilde{f}$ seems rather meager, and dependent on $\mathbb{F}$. If we discard f, and $\tilde{f}$ is all we have, then whenever $\mathbb{F}$ changes we would be studying new functions, which is not our intention. Or again, consider our transfer theorem concerning the inclusion $\mathcal{SN} \subseteq \widetilde{\mathcal{SN}}$. Such an inclusion can only be considered because $\widetilde{\mathcal{SN}}$ is defined to comprise semi-numerical problems $f : \mathcal{G}(\mathbb{R}) \dashrightarrow \mathcal{G}(\mathbb{R})$ that are approximable. The "honesty" suggestion would be to equate $\widetilde{\mathcal{SN}}$ with the set of approximations $\tilde{f} : \mathcal{G}(\mathbb{F}) \dashrightarrow \mathcal{G}(\mathbb{F})$.

We must avoid the intuitionistic (or formalists') impulse to discard the ideal world, and say that only the constructive world is meaningful. Nor must we believe that, with proper tweaking in the ideal world alone, we can recapture the properties of the world of computational machines and limited resources. No, we need both these complementary worlds of real computation, and fully embrace the essential gap between them. We can never exhaust the inaccessible reals, even though new advances in transcendental number theory continually make more reals accessible. This gap and tension is good and important: real mathematical progress is achieved at this interface.

Acknowledgments

I thank Andrej Brauer, Klaus Weihrauch and Martin Ziegler for discussions in Dagstuhl about real computation. Also, thanks to Vikram Sharma, Sung-il Pae and Sungwoo Choi for feedback on the matter of this paper. One referee pointed out the connection of Mal'cev's work on numbering of sets and algebraic systems. My deepest gratitude goes to Volker Bosserhoff and another anonymous referee, for their numerous insightful comments and catching many errors. They both caught a serious error in my original characterization of computable real functions.

References

1. Eigenwillig, A., Kettner, L., Krandick, W., Mehlhorn, K., Schmitt, S., Wolpert, N.: A Descartes algorithm for polynomials with bit stream coefficients. In: Ganzha, V.G., Mayr, E.W., Vorozhtsov, E.V. (eds.) CASC 2005. LNCS, vol. 3718, pp. 138–149. Springer, Heidelberg (2005)
2. Aho, A.V., Hopcroft, J.E., Ullman, J.D.: The Design and Analysis of Computer Algorithms. Addison-Wesley, Reading (1974)
3. Allender, E., Bürgisser, P., Kjeldgaard-Pedersen, J., Miltersen, P.B.: On the complexity of numerical analysis. In: Proc. 21st IEEE Conf. on Computational Complexity (to appear, 2006)
4. Beeson, M.J.: Foundations of Constructive Mathematics. Springer, Berlin (1985)
5. Blum, L., Cucker, F., Shub, M., Smale, S.: Complexity and real computation: A manifesto. Int. J. of Bifurcation and Chaos 6(1), 3–26 (1996)
6. Blum, L., Cucker, F., Shub, M., Smale, S.: Complexity and Real Computation. Springer, New York (1998)
7. Borodin, A., Munro, I.: The Computational Complexity of Algebraic and Numeric Problems. American Elsevier Publishing Company, Inc., New York (1975)
8. Brent, R.P.: Fast multiple-precision evaluation of elementary functions. J. of the ACM 23, 242–251 (1976)
9. Brent, R.P.: Multiple-precision zero-finding methods and the complexity of elementary function evaluation. In: Traub, J.F. (ed.) Proc. Symp. on Analytic Computational Complexity, pp. 151–176. Academic Press, London (1976)
10. Bürgisser, P., Clausen, M., Shokrollahi, M.A.: Algebraic Complexity Theory. Series of Comprehensive Studies in Mathematics, vol. 315. Springer, Berlin (1997)
11. Burnikel, C., Fleischer, R., Mehlhorn, K., Schirra, S.: Exact efficient geometric computation made easy. In: Proc. 15th ACM Symp. Comp. Geom., pp. 341–450. ACM Press, New York (1999)
12. Chang, E.-C., Choi, S.W., Kwon, D., Park, H., Yap, C.: Shortest paths for disc obstacles is computable. Int'l. J. Comput. Geometry and Appl. 16(5-6), 567–590 (2006); Special Issue of IJCGA on Geometric Constraints. (Gao, X.S., Michelucci, D (eds.))
13. Corman, T.H., Leiserson, C.E., Rivest, R.L., Stein, C.: Introduction to Algorithms, 2nd edn. The MIT Press and McGraw-Hill Book Company, Cambridge, Massachusetts and New York (2001)
14. Demmel, J.: The complexity of accurate floating point computation. In: Proc. of the ICM, Beijing, vol. 3, pp. 697–706 (2002)
15. Demmel, J., Dumitriu, I., Holtz, O.: Toward accurate polynomial evaluation in rounded arithmetic (2005) Paper ArXiv:math.NA/0508350, download from http://lanl.arxiv.org/
16. Du, Z., Sharma, V., Yap, C.: Amortized bounds for root isolation via Sturm sequences. In: Wang, D., Zhi, L. (eds.) Proc. Internat. Workshop on Symbolic-Numeric Computation. School of Science, Beihang University, Beijing, China, pp. 81–93 (2005); Int'l Workshop on Symbolic-Numeric Computation, Xi'an, China, July 19–21 (2005)

17. Fabri, A., Fogel, E., Gärtner, B., Hoffmann, M., Kettner, L., Pion, S., Teillaud, M., Veltkamp, R., Yvinec, M.: The CGAL manual, Release 3.0 (2003)
18. Fröhlich, A., Shepherdson, J.: Effective procedures in field theory. Philosophical Trans. Royal Soc. of London. Series A, Mathematical and Physical Sciences 248(950), 407–432 (1956)
19. Halmos, P.R.: Naive Set Theory. Van Nostrand Reinhold Company, New York (1960)
20. Tucker, J.V., Zucker, J.I.: Abstract computability and algebraic specification. ACM Trans. on Computational Logic 3(2), 279–333 (2002)
21. Karamcheti, V., Li, C., Pechtchanski, I., Yap, C.: A Core library for robust numerical and geometric computation. In: 15th ACM Symp. Computational Geometry, pp. 351–359 (1999)
22. Kettner, L., Mehlhorn, K., Pion, S., Schirra, S., Yap, C.: Classroom examples of robustness problems in geometric computation. In: Albers, S., Radzik, T. (eds.) ESA 2004. LNCS, vol. 3221, pp. 702–713. Springer, Heidelberg (2004)
23. Ko, K.-I.: Complexity Theory of Real Functions. Progress in Theoretical Computer Science. Birkhäuser, Boston (1991)
24. Kreitz, C., Weihrauch, K.: Theory of representations. Theoretical Computer Science 38, 35–53 (1985)
25. Lambov, B.: Topics in the Theory and Practice of Computable Analysis. Phd thesis, University of Aarhus, Denmark (2005)
26. Mal'cev, A.I.: The Metamethematics of Algebraic Systems. Collected papers: 1937–1967. North-Holland, Amsterdam (1971); Translated and edited by Wells, III, B.F
27. Mehlhorn, K., Schirra, S.: Exact computation with leda_real – theory and geometric applications. In: Alefeld, G., Rohn, J., Rump, S., Yamamoto, T. (eds.) Symbolic Algebraic Methods and Verification Methods, Vienna, pp. 163–172. Springer, Heidelberg (2001)
28. Mueller, N., Escardo, M., Zimmermann, P.: Guest editor's introduction: Practical development of exact real number computation. J. of Logic and Algebraic Programming 64(1) (2004) (special Issue)
29. Müler, N.T.: Subpolynomial complexity classes of real functions and real numbers. In: Kott, L. (ed.) Proc. 13th Int'l Colloq. on Automata, Languages and Programming. LNCS, vol. 226, pp. 284–293. Springer, Berlin (1986); I cite this paper for Weihrauch's broken arrow notation for partial functions... apparently, it is the older of the two notation from Weihrauch!
30. Müller, N.T.: The iRRAM: Exact arithmetic in C++. In: Blank, J., Brattka, V., Hertling, P. (eds.) CCA 2000. LNCS, vol. 2064, pp. 222–252. Springer, Heidelberg (2001)
31. Pour-El, M.B., Richards, J.I.: Computability in Analysis and Physics. Perspectives in Mathematical Logic. Springer, Berlin (1989)
32. Research Triangle Park (RTI). Planning Report 02-3: The economic impacts of inadequate infrastructure for software testing. Technical report, National Institute of Standards and Technology (NIST), U.S. Department of Commerce (May 2002)
33. Richardson, D.: How to recognize zero. J. of Symbolic Computation 24, 627–645 (1997)
34. Richardson, D., El-Sonbaty, A.: Counterexamples to the uniformity conjecture. Comput. Geometry: Theory and Appl. 33(1 & 2), 58–64 (2006); Special Issue on Robust Geometric Algorithms and its Implementations, Yap , C., Pion, S. (eds.) (to appear)
35. Rogers, H.: Theory of Recursive Functions and Effective Computability. McGraw-Hill, New York (1967)

36. Schönhage, A.: Storage modification machines. SIAM J. Computing 9, 490–508 (1980)
37. Spreen, D.: On some problems in computational topology. Schriften zur Theoretischen Informatik Bericht Nr.05-03, Fachberich Mathematik, Universitaet Siegen, Siegen, Germany (submitted, 2003)
38. van der Waerden, B.L.: Algebra, vol. 1. Frederick Ungar Publishing Co., New York (1970)
39. Stoltenberg-Hansen, V., Tucker, J.V.: Computable rings and fields. In: Griffor, E. (ed.) Handbook of Computability Theory. Elsevier, Amsterdam (1999)
40. Weihrauch, K.: Computable Analysis. Springer, Berlin (2000)
41. Yap, C.K.: Fundamental Problems of Algorithmic Algebra. Oxford University Press, Oxford (2000)
42. Yap, C.K.: On guaranteed accuracy computation. In: Chen, F., Wang, D. (eds.) Geometric Computation, ch. 12, pp. 322–373. World Scientific Publishing Co., Singapore (2004)
43. Yap, C.K.: Robust geometric computation. In: Goodman, J.E., O'Rourke, J. (eds.) Handbook of Discrete and Computational Geometry, ch. 41, 2nd edn., pp. 927–952. Chapman & Hall/CRC, Boca Raton (2004)

Author Index

Lecture Notes in Computer Science

Sublibrary 1: Theoretical Computer Science and General Issues

For information about Vols. 1– 4917
please contact your bookseller or Springer